CW00325489

THE
EARTH
AND
I

THE EARTH AND I

ARTHUR FIRSTENBERG

Skyhorse Publishing

Copyright © 2024 by Arthur Firstenberg

All Rights Reserved. No part of this book may be reproduced in any manner without the express written consent of the publisher, except in the case of brief excerpts in critical reviews or articles. All inquiries should be addressed to Skyhorse Publishing, 307 West 36th Street, 11th Floor, New York, NY 10018.

Skyhorse Publishing books may be purchased in bulk at special discounts for sales promotion, corporate gifts, fund-raising, or educational purposes. Special editions can also be created to specifications. For details, contact the Special Sales Department, Skyhorse Publishing, 307 West 36th Street, 11th Floor, New York, NY 10018 or info@skyhorsepublishing.com.

Skyhorse® and Skyhorse Publishing® are registered trademarks of Skyhorse Publishing, Inc.®, a Delaware corporation.

Visit our website at www.skyhorsepublishing.com.

Please follow our publisher Tony Lyons on Instagram @tonylyonsisuncertain.

10 9 8 7 6 5 4 3 2 1

Library of Congress Cataloging-in-Publication Data is available on file.

Hardcover ISBN: 978-1-5107-8183-2
eBook ISBN: 978-1-5107-8184-9

Cover design by David Ter-Avanesyan
Cover photograph by Getty Images

Printed in the United States of America

Contents

Illustrations   ix
Maps   x
Preface   xi

Which Way Is Forward?   1

PART ONE

THE BACKGROUND   7
1. Road to the Present (1990)   9
2. Self-Portrait   21
3. Thunder and Lightning   29
4. A Living Universe   32
5. Of Marvelous Form   37
6. Ecology (1991)   46
7. Gaia   57
8. Being and Becoming   61

THE STATE OF THE EARTH   71
9. We Are All Indians   73
10. Air   79
    In a Greenhouse   82
    Under a Blazing Sun   95
    The Gentle Rain   101
    Chemical Soup   104
11. Wood   109
12. Water   124
    Groundwater   125
    A Thirsty World   129

	Endangered Springs	130
	Toxic Wastes	133
	Hi-Tech Poisons	136
	Dams	138
	The Flush Toilet	149
	The Sea	150
	They Nuked Paradise . . .	159
	Strip-Mining the Seas	160
13.	Peace Walk (1991)	162
14.	Plants and Animals	165
	Oxygen	174
	Insects	175
	Birds	176
	Bats	179
	Amphibians	179
	Where Have All the Songbirds Gone?	183
	Forgotten Fliers	184
	The Last Refuge	187
	A World Park	189
15.	Space	195

THE FLAMES OF PROGRESS — 199
A Parable — 200

16.	Living on the Edge	201
17.	Chemistry	206
18.	Radiation	208
19.	Plastic	216
20.	Detergents	225
21.	Biocides	230
22.	Farewell to Silence	233
23.	Transportation	238
24.	Guns	240
25.	The Internet	242
26.	Community	245

PART TWO

DIGGING BELOW THE SURFACE — 251

27. Economics, and Human Diversity — 253
    Agricultural Societies — 255
    Foragers — 265
        Central Africa — 278
        South America — 279
        Europe — 286
    Industrial Societies — 289
        The Great Depression of the 1930s — 295
    Hippies — 297
28. Population, and a Tour of the World — 299
    Java — 306
    China — 306
    India — 309
    Sri Lanka — 311
    The Fertile Crescent — 313
    Japan — 315
    Mongolia and Turkestan — 319
    Siberia — 324
    Europe — 326
    Ireland — 328
    Africa — 330
    America — 341
    The Pacific Islands — 352
    Summary — 353
29. War — 358
30. Slavery — 368
31. Religion — 372
32. Sex — 382
33. Technological Trance — 388

TOWARD THE FUTURE   **391**
34. Slowing Down   393
35. The Earth and I   399

Notes   402
Bibliography   419
Index   461

Illustrations

The global electrical circuit	35
The layers of life	42
Graph of oxygen decline since 1990	81
Graph of cumulative CO_2 emissions since 1750	82
Graph of global temperature change since 1000	84
Wind farm	93
Number of active satellites from 1957-2022	98
The southern ozone hole	99
Size of the southern ozone hole since 1979	100
Deforestation in the Amazon	119
Hot springs	130
Canoeing	138
Outhouse	149
Dead birds	178
Monteverde Cloud Forest	182
The little curlew	187
Pine cone	198
Slaves carrying sacks of cobalt for the world's cell phones	339

Maps

Java, Borneo, and Southeast Asia   258
Major trade routes of Asia, Africa, and Europe   305
Atlantic and Pacific Ocean currents   340

Preface

No plums fruited in my garden last summer because no pollinators visited the flowers last spring. And the large flock of ravens who used to converse with me from the high branches of my neighbor's Siberian elms each winter and spring were nowhere to be seen this year. My ears are filled with an unaccustomed silence, and my heart with a growing hole of sadness.

Where did they all go, our birds and bees, and why? And what can we do to get them back? I asked related questions in 1989, after a journey by train, bus, ferry, and foot to the far Canadian north, a journey whose purpose was to find a refuge, an island of escape somewhere, from the technological insanity that was destroying all that was beautiful—to find some place on earth that was still pristine where it might yet be possible to live a simple, natural life. Instead, I found the destruction proceeding up there at a pace that was even faster than the chaos I had thought to leave behind, and I headed back south to the streets of Vancouver, which I wandered for five long days, crying my eyes out. Crying, and asking questions of the universe, the same questions I had been asking all my life, questions that in my fiftieth year of wandering I finally realized no one was going to answer for me.

For the next four years, I practically lived in the main branch of the New York Public Library, the library with the stone lions keeping watch over a treasure of new and ancient books. And I followed the winding trails of my questions into whatever realms they might lead me, however unfamiliar and strange. Who am I? I needed to know! What is the real nature of the relationship between me and my Earth? Why is it all coming to an end? How can we ensure ourselves and all of the other creatures we share this world with a future?

The result was this book, typed on my old Smith Corona manual typewriter. I submitted it to some publishing houses without success. I was just an ordinary, unknown man at that time, who had never published anything, and who had come up with answers that were very different from those everyone else was looking for. I was seeing a forest where the rest of the literary world, and the environmental authors they courted, were seeing only trees.

And then my personal world came crashing down, and I found myself in the middle of just one aspect of the collective fight for survival: electromagnetic pollution, known outside of America as electrosmog. The wireless revolution came to my home town with a vengeance in 1996, and I founded the Cellular Phone Task Force to defend my city against this new assault. We joined forces with over fifty other organizations around the country in taking the Federal Communications Commission to court in an attempt to address this emergency.

I have since become known for my expertise about electromagnetic fields (EMFs), and I have published a popular book on that subject titled *The Invisible Rainbow: A History of Electricity and Life*. But EMFs are only one element of the machine that has vanished bees from my garden and birds from my neighbor's trees. It is time to return to the earlier manuscript I had put aside, to paint the rest of the environmental canvas and to address not only the results, but the causes, of the catastrophe that is befalling our beautiful world—and to outline real solutions, if it is not too late.

Much of this book is timeless and essentially as I originally wrote it. Many chapters have been updated and a few have been added, to catch up with current reality. Where appropriate, I have indicated the year in which certain chapters were written. Although I have consulted my old manuscript frequently over the years and consider it my "bible" about most aspects of the environment, I was appalled when I sat down to review and revise it. Appalled because as bad as the world situation was in 1990 when I was first driven to do this research, I found upon updating the manuscript that the state of the world's ecology, in every aspect, is immensely worse today than it was then. Yet the world has not changed its course, or the lens through which environmental problems are seen, in any significant way, and has not even slowed its accelerating plunge toward oblivion. Worse, the environmental movement has become increasingly specialized and fragmented, and no organization that I am aware of is even tracking the catastrophe as a whole anymore. The World Watch Institute, which published the *State of the World* every year for forty-three years, ceased operations in 2017 and no organization has taken its place. What is needed today is *less*, not more, specialization and a refocusing of a nearsighted world's attention on the big picture. The painting of the environmental canvas from the earth's point of view, *considering Homo sapiens as part of, not separate from, nature*, is even more urgent today than it was when I put this book aside thirty years ago. Each chapter of this book weaves a portion of that canvas.

Arthur Firstenberg
Santa Fe, New Mexico 2024

Which Way Is Forward?

In the search for solutions to our environmental problems we face a great dilemma. If we continue down the path of industrial development, we risk worsening an already dire situation. Automobiles and airplanes have brought us tremendous mobility. Telephones, televisions, and computers have put us in easy touch with one another. We live long, and our lives are made easy by labor-saving devices. Our society continues to strive toward greater democracy, more universal education, increased political and religious freedom, and greater opportunity for individual achievement. But along with freedom, ease, and safety, we have also gotten dirty air, polluted water, disappearing wildlife, and huge piles of garbage everywhere.

There have been some calls to change direction—very few, and always roundly criticized. The objections raised against the Luddites 210 years ago still hold sway today. "There's no going back." "You can't fight progress." We hear it over and over again. And perhaps, after all, there is no going back. But the way we call backwards depends on which way we are facing. Mobility is not only a question of speed and power.

The deterioration of our environment is rapidly becoming an emergency. Pollution, deforestation, and species extinctions are accelerating, and the earth's life support systems are failing. The very survival of earthly life is in question. Yet this is not due to any lack of awareness or education or for want of individuals organizing together to protect our world. There is no lack of scientists studying the situation. There is no end to the steady flow of environmental books into our bookstores, and environmental courses into our schools and universities. We are all more or less aware of the problems and their apparent causes and no one wants them to continue, yet not only is our environment not improving, it is deteriorating more quickly than ever.

Concern for the environment did not begin in this century, nor in the last. Rachel Carson is often considered to be something of a pioneer. She warned us in 1962 that one day a silent spring might come, and that we must do something to prevent it. But she was by no means the first. A hundred years before Rachel Carson, Henry Brooks Adams, grandson of the sixth American president, wrote, "I firmly believe that before many centuries more, science will be the master of man. The engines he will have invented will be beyond his strength to control. Some day science may have the existence of mankind in its power, and the human race commit suicide by blowing up the world."[1] And Henry Brooks Adams was not the first. Over two thousand years before him, Plato complained of deforestation, soil erosion, and the drying up of rivers and springs in ancient Greece, and Han Fei Tzu worried about overpopulation in ancient China.

Nor do environmental problems know any national or cultural boundaries. It is easy to assign blame to one group or another, or to think that some other society possesses the wisdom that ours lacks. Dual tendencies—to think of modern civilization as a great improvement on the quality of human life, and the opposite tendency to idealize the noble savage—both prevent us from seeing one another for who we are. Be we Mohawk, French, Zulu, or Chinese; bankers, farmers, or hunter-gatherers, we are all fellow human beings struggling to get by, and we are all in this crisis together.

The impact of any group of people upon their environment depends much more on the technology at their disposal than on their culture or their religion or their traditions. This is not a moral problem, but a practical one. The important question that is not very often asked in environmental circles is how to choose our technology wisely. The question that has been beaten to death—how to *use* our technology wisely—has not brought us any closer to the solutions we seek.

The immediate causes of environmental destruction are clear: burning fossil fuels causes global warming; chlorinated chemicals destroy ozone; toxic wastes pollute air and water; habitat destruction decimates wildlife. In some cases, what needs to be done would appear to be so obvious and simple that one wonders why we are not doing it. I asked thirty years ago: "If our refrigerators are eating a hole in the ozone layer, why aren't we turning them off? For the convenience they provide, we are trading away our atmosphere!" We have since found alternative chemicals to use in our refrigerators, but ozone depletion is getting worse. Rachel Carson asked sixty years ago: "Can anyone believe it is possible to lay down such a barrage of poisons on the surface of the earth without making it unfit for all life?" She later added, "We allow the chemical death rain to fall as though there were no alternative, whereas in fact there are many, and our ingenuity could soon discover many more if given opportunity."[2] But the question of finding alternatives is becoming increasingly irrelevant. The question we really have got to decide is this: Shall we destroy our world while we wait for alternatives to become available? Or shall we stop?

The difficulty, I think, is not so much in finding alternatives, but in making decisions. And while the immediate causes of our troubles are known, the deeper roots of the problem are clouded with confusion. There is wide disagreement, for example, about the role of economics, and little understanding of the causes of overpopulation. Most importantly, are we in control of our technology, or is it, as Henry Brooks Adams implied, in control of us?

This book will explore both the obvious and the mysterious in order to achieve a more unified understanding. Most of its subjects have been examined at great length elsewhere, but from quite another perspective. My purpose here is to leave the modern industrial viewpoint and its mythologies behind. Instead of change and separation, I look for commonality and continuity. They are just as easy to find, and have much more to teach us about peace and survival.

PART ONE

THE BACKGROUND

CHAPTER 1

Road to the Present (1990)

A pair of small dark eyes looked down at me curiously, and their owner returned to his inspection of my dinner bowl and the remains of my evening meal. He was making his nightly rounds and, I thought, keeping up on the activity of his large temporary guest. Before the night was over, he would know what I had eaten and drunk during the day, what new objects I had brought in from outside, and how I had rearranged his furniture since the night before.

I was as curious about him as he seemed to be about me and my possessions. What separates me from a mouse? I wondered as he climbed down into the box of firewood. What is it about me that my kind destroys our environment, and his kind does not?

It is questions such as these that had brought me to a remote corner of eastern Ontario, to spend a few weeks alone in an old cabin with no electricity or plumbing, just a stream in the backyard to drink from and wash in, and the sun and a few candles for light. Alone, that is, except for a resident field mouse and a bat.

I had thought to gain a fresh perspective on my civilization by removing myself from it for a time and doing without most of its technology. Environmental degradation plainly had something to do with human technology, I was convinced of it, for without technology none of this trouble would exist. And yet, what was technology but the application of human intelligence to human dreams?

We are a species of dreamers, but we are not unique in that. When that mouse slept, he might dream as well. I imagined he dreamt of being

Mighty Mouse, of flying through the air like a bat and vanquishing his hawk enemies, of enslaving them and forcing them to swoop down on unsuspecting cats and peck their eyes out. Probably he dreamt of nothing of the sort, but most likely he did dream, of something unknown to me, when, asleep in his nest, his eyes raced back and forth under their lids and his little legs twitched.

My race of dreamers, I thought, we have built a world that is changing at a dizzying, frightening pace within a world that seems eternal. I thought back over the previous twelve months. Human politics had changed tremendously. Communism had fallen in Eastern Europe, the Berlin Wall had been torn down, and the Soviet Union was disintegrating. The Cold War was over.

Technological change seemed just as great. Records and record players were suddenly gone, made obsolete by compact discs. Almost overnight, an invasion of computers and fax machines had ousted typewriters and messengers from offices all across North America. A century ago, I thought, automobiles did not even exist, and electricity was only a novelty. Today we are sending rockets into outer space on a regular basis.

Even the eternal, I thought with sadness, is now changing on a grand, unimaginable scale. Rainforests are tumbling down all over the world, the ozone layer has a hole in it, and even the great oceans are badly polluted.

As I lay there in my cabin, pondering these and other changes, the stars were shining in on me through an open window, shining from the sky as they had always done. And I tried to imagine life as it may have been long ago when my race was young, and the world as it may have looked through their eyes. They were dreaming of different events and different changes as they lay in front of their caves, or their shelters, with the stars shining down on them from the sky. But they were human beings just the same.

The world I pictured is long and irretrievably vanished, but some of its remains have survived until the present day. Stones and bones dug up from millennia past provide us with clues to that long-lost world, and archaeologists and anthropologists have come far in piecing together the outline of the human puzzle. The decay of time has left room for the imagination to fill in the details, and to include the dreams that accompanied the objects. The human story, which is where any search for understanding of environmental problems must begin, may have gone something like this:

A million and a half years ago—about fifty thousand generations—our ancestors in Africa were very much like us, sharing with us not the details of our lives, but our humanity. The tools they used to shape their environment were not even so unlike our own as to be unrecognizable, and they took considerable skill to make, more skill, certainly, than I myself have as a tool-maker. I am skilled mainly as a tool-user and would be relatively helpless if stranded in the woods with nothing but my hands and my wits about me. They fashioned their axes and their knives from local stone, and their spears, clubs, and digging sticks were hewn from the forest. Perhaps large leaves, or gourds, or the eggshells of big birds, or the stomachs or bladders of animals served as vessels for carrying food and water, and one may imagine skins for carrying babies. It was the latest in modern technology.

What they looked like—the color of their skin, hair or eyes—any clothing they may have worn, the sound of their speech—as a scientist it would be fruitless to speculate about such matters, but as a man I am free to picture what I like. Perhaps jet-black hair, dark eyes, and olive-brown skin would be suitable for a forest-dweller. Probably naked, in a tropical climate. One may imagine a society of nomadic hunters in a thickly forested world abundant with a diversity of large and small animals, and numberless varieties of fruits, nuts, and roots to be had for the picking. And in the hearts and minds of the inhabitants dwelled human souls with the same yearnings that you and I have. They would be concerned with making friends, their version of getting married, having children, and educating the kids. They would think about making a living, protecting and providing for their family, and being respected by the community. And they would worry about staying healthy and being taken care of in their old age.

The world was a dangerous place. Venomous snakes, spiders, and scorpions lurked in the forests. Leopards and lions roamed the hills. Thunder and lightning, the four winds, sun, and rain were powers to be respected, and sometimes feared. It was to be expected that half of one's children would never live to have children of their own.

Compared to the pace of "progress" today, technological change was immensely slow. But over centuries and millennia, people who dreamed of making their lives safer and more comfortable built shelters that were secure from wild beasts and from the elements. They invented many new tools and improved upon old ones, and they found new varieties of plants and animals that they used to satisfy their desires. Their food supply expanded, and their children were safer. Human populations began to grow.

As populations grew larger, some groups emigrated to new lands to seek their fortunes, in Africa, southern and southeastern Asia, and eventually to lands where the climate wasn't so gentle. The earth was changing as well. Glaciers were forming in the mountains, and the northern ice fields were advancing southward. The planet was becoming cooler. People in the higher elevations, and in northern Europe and Asia, were forced to wear clothing, and so became much more dependent on killing animals for their valuable hides as well as for food.

At least 500,000 years ago human beings everywhere began to make fire, and it became an intimate friend. Fire cooked their food, kept them warm, and cheered their hearts at night. It frightened away wild animals and lit up their caves and shelters. It enabled them to live anywhere that was not covered by ice year-round, and our ancestors spread over a large portion of what is now called the Old World. By at least 300,000 years ago some of them had colonized North and South America as well, and by 130,000 years ago some had made it to Australia.

In dense forests and in open woodland, in the warm tropics and in the colder north, in inland valleys and on coastal plains, people made efficient use of the resources at hand. Skilled stoneworkers gradually crafted smaller, thinner, and more specialized tools for the sawing, scraping, boring, and shaping of wood, bone, antler, and leather. They attached stone tips to their wooden spears, and wooden handles to some of their stone knives and axes. Archaeologists have discovered as many as sixty different types of tools that were used by some cultures that lived between 100,000 and 40,000 years ago.

As people continued to diversify their food supply and protect themselves more and more from natural dangers, human populations continued to grow. Our ancestors migrated further into northern Europe and Asia, blazed trails into Siberia, and learned to navigate the seas. The frontiers were filling up, and territorial conflicts became more serious and less avoidable. Some people, perhaps for the first time, began to turn their hunting weapons against one another.

Until this time we haven't any clues to the religious beliefs early human beings may have held. But sometime before 40,000 years ago some people, in some places, began to bury their dead. These were people who were beginning to live more settled lives. Perhaps they remained in each place for a few years at a time, or they returned year after year to the same seasonal camps, and they had to bury their dead as a matter of sanitation and so as not to attract wild beasts into their villages. Their more nomadic neighbors simply abandoned camp when someone died, leaving the body, and perhaps the soul, to nature's housekeepers. Here the dead dwelled in graveyards near the living, and our ancestors began to give more thought to an afterlife. In the graves of their loved ones they also buried some possessions and provisions for the hereafter. One day their descendants would think of embalming the body to preserve it forever. And so was planted the germ of the recycling problem from which we suffer so extremely today.

Innovations in technology continued. People who dreamed of hunting animals with more power and accuracy invented the sling and the boomerang, and later the blowgun, the spear-thrower, and the bow. They made finely carved spearheads and arrowheads of bone as well as flint. They discovered poisons that could be applied to the tips of their weapons to paralyze their prey. They also produced finer knife blades, chisels, saws, and drills, and shaped antler, horn, bone, and teeth into harpoons, fishhooks, needles, awls, buttons, beads, and other objects of hearth and hunt.

Our ancestors were now well equipped to hunt birds and catch sea creatures. They were able to hunt and trap larger, more dangerous land animals, and instead of going after the weak and the young, they took pride in felling the most vigorous and healthy animals. They learned to drive large herds of their prey to their deaths, over cliffs, or into marshes or human ambushes. Sometimes they set fire to the forest to drive a herd of horses, or buffalo, or elephants to their demise.

Human beings were beginning to upset the balance of nature in a serious way. Over the centuries repeated burning of forests created more open woodlands, and great expanses of prairie where trees once stood. Populations of grazing animals grew, and hunting became easier. But the hunting of healthy, mature animals, and the slaughter of whole herds at a time could not be withstood forever. By 10,000 years ago, all across northern Eurasia, and throughout North and South America, many of the very large herbivores, among them mammoths, mastodons, giant buffalo, giant deer, woolly rhinoceros, ground sloths, giant beavers, several kinds of camels, horses, and asses, and many others, had been driven to extinction along with their predators—dire wolves, saber-toothed tigers, giant varieties of bears, hyenas, and lions, and others.

Some people turned their herding skills to the domestication of their prey. In different parts of Europe, Asia, Africa, and America, people domesticated reindeer, or horses, or cattle, or goats, or camels. They bred dogs as pets, protectors, and hunting companions. They also planted edible roots, grasses, and other food and fiber plants that previously they had only gathered. The wilderness was being tamed. People began to alter the biological community in which they lived by removing the plants and animals they did not want and replacing them with crops and cattle. Those who kept flocks of animals and planted grain ensured themselves a large annual food supply and kept valuable grassland from being lost to the encroaching forests.

By about 10,000 years ago, as the ice retreated from the lower latitudes, human beings occupied every ice-free piece of land on earth except for Polynesia and a few other oceanic islands like Mauritius, Bermuda, and the Azores. The rapid advances in technology that were occurring have been named the Agricultural Revolution. Not all peoples decided to go this route. But some did, all over the world at more or less the same time. In the Southeast Asian tropics, people planted taro, yams, bamboo, and bananas, and raised pigs, chickens, and geese. In the drier regions of southern Asia, from the Indus Valley westward to the Mediterranean Sea, people cultivated wheat, barley, peas, and lentils, and grazed sheep and goats. Early Mexicans planted maize, beans, squashes, and chile peppers, and bred turkeys. South Americans grew manioc and sweet potatoes in the lowland forests, and those in the Andean highlands grew potatoes, tomatoes, and lima beans, and bred guinea pigs, llamas, and alpacas. The ancestors of the Chinese grew millet and soybeans and kept pigs and ducks. Early North Africans raised donkeys, and planted sorghum, African millet, sesame, and cowpeas.

Those new farmers took over a lot of the work of nature. Before, nature did all of the work to produce food, and people only gathered it when it was ready to eat. Now, people had to clear land, prepare the soil, fertilize the soil, plant the seeds, care for the seedlings, irrigate the crops, pull up weeds, harvest the crops, and save seed for next year. They had to live in more permanent settlements in order to tend their fields. Herders with their flocks were still nomadic.

The first farmers were semi-nomadic, and many still are today. Those who became settled had to build sturdy houses that would last many years. They made hoes, and later plows, to cultivate the soil. Digging sticks would no longer do.

Life was not easier for people who started to farm. In fact their workload increased tremendously. Their diet became much more limited. Their crops were much more vulnerable to droughts, floods, and disease than the fruits of nature they used to gather. Like their modern-day counterparts, food gatherers probably spent relatively little time, perhaps two or three hours a day, getting food. The total workday for both men and women among foraging peoples today in Africa, Australia, and Asia may average eight hours, including tool-making and housework. 1964 was the third year of a severe drought in southern Africa's Kalahari Desert. While many of their agricultural and pastoral neighbors were starving, !Kung San hunters and gatherers fed themselves well, though they spent just one day out of every three looking for food. Theirs was a society where children did not work, old people were cared for, food was so abundant that it was never stored for the future, and there was plenty of leisure for relaxing, socializing, and dancing.[3]

The change from gathering food to producing it had repercussions throughout human society. Since our ancestors could now grow seemingly limitless amounts of food, their populations soared. And with their soaring numbers they were in turn able to plant larger crops and tend more cattle. Their children became a source of wealth, and large families increasingly desirable.

Farming villages became bustling towns, and specialized crafts developed. People became potters, weavers, woodworkers, basket makers. Trade routes flourished, and trading centers grew in size. Jericho, one of the world's first cities, was established some 10,500 years ago at the crossroads of three continents.

Unlike their nomadic ancestors, and their nomadic neighbors, farmers accumulated possessions. They were no longer limited by the amount of weight they could carry with them from place to place.

Dense populations, the accumulation of wealth, and the acquisition of permanent territories increased the jealousies between neighboring clans, and the tensions between neighboring peoples. Painters and sculptors of this period have left us clear evidence that for the first time warfare was a regular part of human life.

The need to organize ever more complex societies led to the establishment of more formalized and more powerful governments, together with bodies of codified laws, and courts, judges, and one day police forces to enforce them. Human beings had considerably less freedom.

We have come a long way further on the road of progress in the past 10,000 years. In many parts of the world people learned to smelt copper, and to make bronze. In some places they learned how to smelt iron, and then to make steel. With every new advance in technology, they had to move increasing amounts of raw materials and finished products over longer distances. More large cities grew up as centers of production and trade. Transportation and food storage were improved, and populations grew again as people filled the cities.

In the countryside, people continued to improve the technology of agriculture. They built large irrigation systems, and they invented the plow and the wheel, two very powerful instruments for subduing the earth and asserting human control over nature.

Our ancestors were working harder than ever, and they dreamed of finding ways to ease their growing burdens. Gradually they came to the idea of using other than human power to do their work for them. Their ability to transform their environment began to increase exponentially, until today the average human being in an industrial society has at his or her command machinery that can do the work of several hundred people. The entire earth has something over 8 billion people living on it, but our combined impact on the earth is as if we were 160 billion strong.

People first used the strength of large animals to replace human brawn on and off the farm. Animals pulled their plows, ground their grain, carried their loads, and provided transportation. Later people harnessed the power of the wind and the force of flowing water. For smelting metals they relied on the burning of charcoal, derived from wood, but eventually the supply of trees could not keep up with the demands of industry, and in modern times we have come to rely more on coal, oil, and natural gas, as well as, recently, the power released in nuclear reactions.

The Industrial Revolution is usually dated from the invention of the steam engine in England in the early 1700s. In substituting machine power for human power for providing more of the necessities of life, even more people could now be fed, clothed, and housed. The large families that people continued to have supplied the factories with abundant labor and brought money back from swelling cities to expanding farms.

Warfare was still a fact of life, and it remains so today, when we wage our battles with warships, submarines, tanks, aircraft, missiles, bombs, and chemical and biological weapons. Our most sophisticated technology is no longer used for hunting animals at all, but instead for hunting each other.

The improvement of technology has been going on for so long, it appears to have a life of its own. But behind it all in most people's minds is the continuing desire to make life safer, longer, easier, and more comfortable, and to free us from the limitations of the human form.

In those parts of the world where the wilderness was tamed, where the deep dark forests were gone, and with them the bears and the wolves, and where human beings were living in well-defended towns and cities, many of our most ancient fears and dangers had been largely conquered. Wild animals in particular were no longer a menace. We were well-sheltered from the elements in heated homes. We had warm clothing on our bodies.

Food was growing plentifully in the backyard or on the farm, we no longer had to hunt for it.

Yet there were still dangers to conquer, and dreams to fulfill. We had made little progress, for example, against disease, and three out of every ten children still died before reaching adulthood.

In the eighteenth and nineteenth centuries we took great strides in improving sanitation and providing pure water to more people. There followed the discovery of bacteria, and the beginning of vaccination, pasteurization of milk, and other public health measures. In the early 1900s we began to chlorinate our drinking water. Widespread use of antibiotics followed the discovery of sulfa drugs and penicillin in the 1930s.

Fewer people were dying of smallpox, typhoid, diphtheria, and pneumonia, and we turned our efforts to eradicating insect enemies. DDT was synthesized in 1943, and thousands of other potent pesticides since.

Impossible dreams have become reality. Human beings are now able to fly through the air, and swim deep beneath the seas. We can transmit our words, even our voices, through empty space across thousands of miles and have them arrive instantly at any desired location.

The modern world is truly a miracle of technology designed to fulfill our fantasies, make our lives easy, and keep us safe from any possible dangers. We have built a worldwide transportation and communication network of ever-increasing scope, speed, and efficiency. Fire engines will come within minutes to put out a fire. Ambulances or helicopters will rush to any emergency with medical aid, and they can be summoned instantly from almost anywhere on earth by stationary or mobile phone. Food distribution is so efficient, and food preservation techniques so highly developed that in most parts of the world no one need starve. Today we have almost every type of labor-saving device imaginable in our homes. For the first time in history most children who are born on the earth live to be adults. Our life expectancy is longer than it's ever been.

The world, seemingly, has been transformed. But the concerns, goals, fears, and longings of human beings are basically the same, and our drive to improve our technology continues without letup—not just those recent aspects of our technology that have put airplanes in our skies, televisions in our homes, and cell phones in our hands, but even those functions of our technology that are the very oldest, that provide us with comfort, ease, and safety.

We are still working hard to increase our food supply, and our population is still rising out of control.

We are still trying to make life easier for ourselves, because with all of our labor-saving devices we work more and have less leisure than ever before.

Our world is still dangerous. Perhaps we don't have to watch out for leopards and snakes. But automobiles are deadlier and more common than any wild animals ever were. And human beings with knives and guns stalk the streets of our towns and cities. Perhaps we are well sheltered from the heat and cold, from thunder and lightning, wind and rain. But inside our homes are pipes full of explosive gas and wires that can electrocute us. And under our sinks are deadlier poisons in greater quantities than were ever found out in the wild.

Although our homes are warmer, drier, and more comfortable than ever before, we are still inventing bigger and better central heating systems, air conditioners, and dehumidifiers, and getting rid of the old ones.

We spend billions of dollars fighting heart disease and cancer and on longevity research, even though we already live a very long time. Infant mortality is down around five in every thousand births, which is extraordinary, but the rate of caesarean sections is still going up in an effort to do better.

Our use of insecticides continues to increase exponentially.

We continue to hunt the few remaining mountain lions in the American southwest. The timber wolf is practically extinct in the lower forty-eight United States, but we continue to hunt it in Alaska and Canada. We have exterminated the North American white bear in its entire range except for its last sanctuaries in the extreme Arctic north, and we continue to hunt it there.

Even when the last wild animal is gone, the last insect destroyed, the last germ wiped out, the whole earth is climate controlled, all danger is eliminated, and people live forever, I suspect we will not be satisfied, because we ourselves have not changed, not even after a million years and more. We are, after all, still human, and no amount of technology is going to still the yearnings of the human soul.

Perhaps I am no more to blame for my kind's impact on the earth than the mouse in my cabin was for the destruction his kind has caused to human homes throughout history. He was just being a mouse. How can a mouse be anything else than who he is?

I think back, sometimes, to my far-distant forebears in their caves, resting with full bellies after a successful day's hunt. The same stars shine in my sky that shone in theirs; only their arrangement is somewhat different. The tools that I use are so much more powerful than theirs were, but I have neither made nor designed them. I have never even learned how to make a fire. My ancestors would be appalled. What separates me from them is mainly human history. And the power that is mine for harm or harmony with respect to the earth lies, I have come to believe, in my ability to choose from among the many tools and resources that my society, with its history, has given me.

CHAPTER 2

Self-Portrait

Sometimes, in learning something about myself, an animal has been my greatest teacher. I was walking one early autumn day in the Canadian north, and as dusk was approaching, I made myself a pine needle bed underneath a tall tree. Before I settled down in my sleeping bag for the night, I took off my "Two Hawks" tee-shirt and set it down beside me. In the morning when I woke up it was full of large holes, and a squirrel was chattering at me loudly from above in the tree beneath which I lay. I was startled—and furious—to realize that it had spent part of the night within inches of my sleeping body, chewing holes in my favorite shirt. Now normally I am a peaceable guy, inclined to treat all animals as my brothers and sisters. Once, while picking blackberries during hunting season, I surprised a very large buck who was hiding in the blackberry bushes, and who nearly knocked me down in his panic to flee. I have on occasion, when sleeping out, been woken up by bears rummaging through my camp. My curiosity usually gets the better of my fear. But this time, I felt some primal stirrings inside of me, that as a human being I wanted to kill this non-human creature who had destroyed a prized possession and who was mocking me from above, quite out of reach. Did it want the nice soft cotton for its nest, I wondered? Or was it merely showing its annoyance at me for trespassing under its tree? I did not know, and for a moment I did not care. I felt sudden kinship with hunters, with those who think of nature as a threat rather than a delight, and animals as nuisances to be exploited, rather than as neighbors.

It is a personal question, not just one for philosophers, theologians, and scientists: What distinguishes human beings from the beasts? What divides us from the rest of nature? One answer, which I have already noted, is human history, and since I believe that is an important answer, a portion of this book will be devoted to a reexamination of our history. But one may still ask, what is it about the human animal that has led us to this kind of history? There are, after all, other differences between me and other beasts besides the tools my society has given me. Even without those tools, I could have attacked that squirrel, if I had liked, from a distance, by throwing rocks at it. I would not have had to chase it down and bite it to death, like other animals do. Not if my aim were sharp and the strength of my throw were sufficient.

A portrait of the human animal might begin with a description of some of our most unique traits. "Fire-maker" and "tool-maker" are terms that apply to us, and to us only. "Two-legged" applies also to birds, but even among birds, only the aquatic penguin stands bolt upright like we do. No animal besides us puts on clothing, wears jewelry, or decorates its body in any way.

A physical description rapidly reveals our weaknesses. Having neither fangs nor claws, tough hide nor great speed, our bodies are relatively vulnerable and defenseless—unusual in a world in which only the fit survive. We make up for our vulnerability with some other highly developed traits, which are related to each other: our intelligence, our language, and our culture. We are also blessed with opposable thumbs, with which we manipulate our tools and weapons to defend ourselves and shape our environment. Monkeys and certain other animals have hands too—beavers' hands are almost as flexible as ours—but none have the fine motor coordination that allows human beings to aim rocks at distant squirrels.

In one classification of animals, we would find ourselves grouped with the land dwellers; in another, with the social animals. Taxonomists classify us as mammals—warm-blooded vertebrates with hair on our bodies who suckle our young. We are, more specifically, primates, most closely related to the great apes—the chimpanzees, gorillas, and orangutans. Like skunks, raccoons, bears, and pigs, we are omnivorous, able to eat both plants and animals. And like turtles, elephants, and parrots, we have long life spans.

The quality we are most proud of is our intelligence. But although we are among the most intelligent of animals, comparisons are often difficult to make. Animal languages are not well understood by us, and scientists are by no means in agreement about the meaning of brain size. The horse, the elephant, and the whale all have bigger brains than we do. Dolphins and many whales have brains which are as big for their body size as ours, and there is some controversy about whether they may be as smart, or even smarter than we are. Many birds are turning out not to be "bird-brained" at all. Parrots, for example, are often credited only with being excellent mimics, but it now appears they can be taught to understand the words they say. Among the invertebrates—animals with soft bodies—the octopus has both a large brain and a large learning capacity, but its perception of the world must be so different from ours that comparing its intelligence with ours is difficult if not meaningless.

We are not the only intelligent animals, but no other animals depend on their wits to the extent we do. Elephants, being big, strong and thick-skinned, fear no enemies, at least they didn't until we humans came along. Dolphins and whales, in addition to having size and strength, are sleek and fast and equipped with sonar. Octopi are good swimmers, have eight arms with suckers, black ink to squirt at predators, and a poison that paralyzes prey. They are all better equipped physically than we are for getting food and avoiding danger. Intelligence for them is a luxury. For human beings it is a necessity.

Although all animals communicate with their own kind, language is much more developed among animals who are social by nature than among those who live solitary lives. Dolphins and whales who, like us, are both social and large-brained, may well have complex languages. So may elephants, whose great bass voices are so low that human beings often can't hear them as they call to one another across the vast savannah.

What makes humans unique is that because we are less secure physically, we use our intelligence, our language, and our hands to make weapons and tools with which to manipulate our environment. We have come to dominate the earth not only because we are so smart, but also because we are so vulnerable.

We could never have stood up and walked erect, exposing our soft bellies to the world, if we did not already have weapons in our hands with which to defend ourselves. An important survival rule for all vertebrates, from fishes to reptiles to birds to mammals, is "Protect your belly." An animal normally presents its belly to a stranger of its own or another species only when it is attacking or defending itself, or as a signal of defeat or submission. Standing up is in effect saying to everybody, "I am not afraid. I am your master."

Desmond Morris, in *The Naked Ape*, has this to add: "Because of his vertical posture it is impossible for a naked ape to approach another member of his species without performing a genital display . . . The covering of the genital region with some simple kind of garment must have been an early cultural development."[4]

The portrait of ourselves is being fleshed out. One might now imagine an encyclopedia entry that begins this way: "Human beings are thin-skinned, almost hairless primates who no longer dwell in trees and who don't have the strength or speed of other animals their size that live on the ground. They stand completely erect with their bellies and genitals exposed in front of them. They live in organized, complex social groups. Most aspects of their social behavior and their complex language are not inherited, but are learned during a prolonged, dependent childhood. They have large brains, and opposable thumbs with fine motor coordination, which they use to manipulate their environment and make weapons, tools, and fire."

Two other human characteristics will be of particular interest in a book about the environmental crisis: we are the only species on the planet whose population seems not to be regulated; and we are the only species that regularly kills each other in warfare and other fighting.

Anthropologists, social planners, economists, historians, and others who study overpopulation are in confusion and disagreement over the causes, cures, and even the significance of the problem. Unfortunately, the human population problem is often discussed as if we did not belong to the animal kingdom and obey the same laws of biology as everybody else.

In the natural world, neither extinction nor overpopulation is a very common event. Predation, starvation, and disease are among the factors that control animal numbers, but most predator and prey species remain in relative balance even in times of plenty, when their populations are healthy and well-fed. The mechanisms that maintain the balance have not been well understood. Zoologist Marston Bates, for example, wondered how it was that the lions in the African savannahs were always sleek and fat. "The lions keep the antelope from becoming numerous enough to ruin the grass, but what keeps the lions from becoming so numerous that they kill off their own antelope food supply? I keep asking the question of friends who might know, but the answers are not very clear."[5]

It is now becoming clear that various methods of what might be called birth control are the rule among many species. Animals do not usually wait until they are starving to limit their numbers. Rather, they have instinctual behaviors that link their population to the carrying capacity of their environment, that is, the number of individuals that can be indefinitely supported by that environment. There are some exceptions, like the Arctic lemmings, whose populations grow in boom-and-bust cycles of four or five years—they multiply like crazy for five years, and then countless numbers migrate in search of greener pastures, most fall victim to predators or starve to death, and the cycle begins again. But other animals keep their numbers more stable by a variety of means.[6]

Many birds have a fixed nesting area. Breeding pairs defend their territories, and when the nesting area is filled up, no more birds get to breed until the next season; even though there would seem to be additional breeding habitat nearby, it goes unoccupied. Grey seals are among the animals known to limit their numbers in a similar way.

Predation may combine with territoriality in population control. Red grouse, for example, establish territories on the English heather moors each autumn; any birds without territories become easy targets for predators.

Female mice produce an actual birth control scent; when too many females are together they become infertile. The scent of a strange male will also put a halt to the development of mouse embryos.

Birth control scents seem to be widespread in the animal world. Meal beetles reproduce rapidly in mills and granaries, but their excrement contains a chemical that renders them infertile as soon as they become crowded enough. Tadpoles will simply stop eating when there are too many of them in a pond. In thirty gallons of water one large tadpole will compel six smaller ones to stop eating even though there is plenty of food for all. If one merely pours the water, in which several large tadpoles have been swimming, into a tank containing small tadpoles, the small tadpoles will cease to eat. Many freshwater fish regulate their populations in a like manner.

When elephants become crowded, they respond by spacing out their children. Under the crowded conditions in Murchison Falls Park in Uganda during the 1960s, female elephants averaged six years and ten months between the birth of a calf and the conception of another one. Whether they became infertile or just abstained from sex is not clear.

Even in the case of lemmings the stimulus for emigration is overcrowding and not starvation. The local population remains healthy and well-fed and does not destroy its environment.

Even some vegetables practice a kind of birth control. Cacti space themselves out; so do roses, brome grass, and citrus, apple, and peach trees. The roots of these plants secrete inhibitory substances; even the presence of dead roots is often enough to prevent the germination of nearby seeds of the same kind of plant.[7]

Whatever the means of population control, it is always linked to the carrying capacity. We humans have been able to continually increase our numbers because with our tools we have been able to continually increase the carrying capacity of our environment for our species—by our farming and industrial practices we have been able to supply an increasing number of people with food, clothing, shelter, and the other necessities of life. Our species has many physiological and cultural methods for regulating our numbers, which will be explored in chapter 28, but none of them is effective so long as we can increase the ability of our environment to support us at will.

In *On Aggression* Konrad Lorenz suggested that it is because we make tools and weapons that we kill so many of each other. Without a weapon it is simply not easy for one person to kill another. As we have developed more and more powerful and varied weapons, and more dangerous tools, killing has gotten easier, and we are required to exercise more and more social restraint in order not to harm one another. In other words, aggression, which is a normal part of being alive, has no safe outlet today among human beings. In addition, he said, the natural inhibitions that we all have against killing one another are bypassed by our long-distance weapons:

> The distance at which all shooting weapons take effect screens the killer against the stimulus situation which would otherwise activate his killing inhibitions. . . . No sane man would even go rabbit hunting for pleasure if the necessity of killing his prey with his natural weapons brought home to him the full, emotional realization of what he is actually doing.[8]

Perhaps. But all kinds of animals—squirrels, bears, and monkeys alike—have their bullies. Even chickens have a pecking order. Peace is generally kept by staying out of the bullies' way when they are on the rampage. Maybe we are not an especially warlike species—we have just put ridiculously powerful weapons into the hands of our bullies.

Desmond Morris, too, says that with the remoteness of our weapons of war, "rivals, instead of being defeated, are indiscriminately destroyed . . . Defeat is what an animal wants, not murder; domination is the goal of aggression, not destruction, and basically we do not seem to differ from other species in this respect." Our cooperativeness, he adds, which served us so well in hunting, "has now recoiled upon us . . . Loyalty on the hunt has become loyalty in fighting, and war is born. Ironically, it is the evolution of a deep-seated urge to help our fellows that has been the main cause of all the major horrors of war. It is this that has driven us on and given us our lethal gangs, mobs, hordes and armies. . . . They attack now more to support their comrades than to dominate their enemies, and their inherent susceptibility to direct appeasement is given little or no chance to express itself."[9]

I think there is some truth to what these authors say, but the need to explain human warfare results in part from a failure to look beyond our own culture and our own time. Historically, human cooperation has not led to violence and warfare until we were densely populated and had many possessions and territories to defend. Even in today's world, if we look beyond the narrow confines of our own culture we see many peoples with stable populations who have never known warfare: the eastern Inuit of North America; the Penan of Borneo; the Kubu of Sumatra; the Semang and the Sakai of Malaya; the Arapesh of New Guinea; the Veddahs of Sri Lanka; the Mbuti of the forests of the Democratic Republic of Congo. The facts about human warfare and overpopulation, and their causes, will be explored in greater depth in later chapters.

We turn now from the study of ourselves to a look at nature as a whole, of which we are only a tiny part.

CHAPTER 3

Thunder and Lightning

In the far-horizoned Arizona desert I have sat and watched mid-summer storms of great power, and without microscope, telescope, or any machinery at all, I felt that important secrets of existence lay revealed all around me, in the green sahuaro cacti reaching skyward, in the lizards darting under rocks and the owls nesting in cactus holes, in the black pillow clouds above me, in the flooding downpour, in the tremendous echoing booms and the flashing fingers reaching earthward from the clouds, in the huge winds and the hailstones, and especially in the electricity that seemed to bristle the air and alert the cells of my body.

They are in every man's experience, and every woman's, and they are important symbols in every culture, in every land. In my culture, we speak the name of Thunder once a week, on Thursday, and we give the name of Jupiter, hurler of Lightning, to the largest planet in our night sky. These oldest and mightiest symbols of all-powerful nature are deeply embedded in our psyches, representing to us our own wild male and female selves that civilization cannot eradicate. This tension between wildness and civilization is the ecological dilemma of our time.

Civilization will have no part of wildness, and so Thunder and Lightning remain largely ignored and little understood by the builders of our cities. Physicists, biologists, and chemists acknowledge their presence surprisingly little. Electrical engineers have harnessed but not tamed them. Modern science and technology have given us general relativity, quantum mechanics, and nuclear fusion; the theory of evolution and genetic engineering; cybernetics and computers.

But the spark and the thunder of the living world are missing from all of them.

I am not speaking only metaphorically. The dilemma that prevents a civilized person from seeing what he or she is doing to the wild world is the same dilemma that prevents scientists from acknowledging the wild nature of the forces they study. We have built a world that is pervaded by artificial electric wires, but natural electricity—the life force that quickens our bodies, freshens the air, and orders the stars—is practically a taboo subject in mainstream science.

Gravity holds the universe together. Electricity makes it alive. Electricity gives us negative and positive, male and female, all the choices and movement that animate life. The basic units of positive and negative charge in the universe, and the building blocks of all matter, are the proton and the electron. The electric repulsion between two protons is fantastically stronger than their gravitational attraction—stronger by a ratio of 10^{38} to 1. The cosmos is almost entirely a vast ocean of such charged particles, weaving on their journeys an ever-changing magnetic web that molds and shapes and colors our living, pulsating world.

Astronomers tell us that we live in an expanding universe. The entire universe, they say, was created out of nothing between ten and twenty billion years ago in a cosmic orgasm they call the "Big Bang," and they say it is still expanding from the force of the explosion. That is one model of the world, but it is a model that may well tell us more about the turmoil of the civilized mind than about the living reality around us. It is a model that leaves out electricity. The actual galaxies in the real universe have shapes and sizes that are not easily explained by the laws of gravity alone, and their motions apparently obey quite other forces. And in all parts of the heavens one finds galaxies clustered together in groups so enormous that twenty, or even forty billion years is not nearly enough time for them to have come into existence.

Hannes Alfvén and a small group of other plasma physicists have developed a model of the universe in which electricity and magnetism are restored to a prominent role as cosmological forces. Their theories successfully predict many of the galactic shapes and other common structures we see in the heavens. All of the forms we observe in the sky have had plenty of time to evolve, they say, because the universe had no beginning, and furthermore it has no reason to end. The universe of twenty billion years ago is completely unknowable to us, they say, because not only gravity, whose effects are easily calculated, but also electricity and magnetism have been operating on a grand scale, and their effects over such a long period of time are beyond analysis.

Our ecological dilemma cannot be properly understood without an appreciation of just what it is that we are destroying. We must find out not only who we are, but where we fit in the scheme of nature. The tendency of modern society is to fragment nature in an attempt to control it. We must learn once more to see the connections of all the parts. The story in the next two chapters has not been told before, but all of the facts it is based on are prominent in one or another scientific specialty or subspecialty. It begins with the stars and galaxies that shine in our night skies.

CHAPTER 4

A Living Universe

All things by immortal power,
Near or far,
Hiddenly
To each other linked are,
That thou canst not stir a flower
Without troubling of a star.

— Francis Thompson, in
The Mistress of Vision

When I look at a flower, what I see is not the same as what a honey bee sees, who comes to drink its nectar. She sees beautiful patterns of ultraviolet that are invisible to me, and she is blind to the color red. A red poppy is ultraviolet to her. A cinquefoil flower, which looks pure yellow to me, is to her purple, with a yellow center luring her to its nectar. Most white flowers are blue-green to her eye.

When I look upon the night sky, the stars appear as points of color twinkling through earth's atmosphere. Everywhere else, except for the moon and a few planets, is blackness. But it is the blackness of illusion.

If you could see all the colors in the world, including the ultraviolets that honey bees can see, the infrareds that snakes can see, the low electric frequencies that catfish and salamanders can see, the radio waves, the X-rays, the gamma rays, the slow galactic pulsations, if you could see everything that is really there in its myriad shapes and hues, in all of its blinding glory, instead of blackness you'd see form and motion everywhere, day and night.

Almost all of the matter in the universe is electrically charged, an endless sea of ionized particles called plasma, named after the contents of living cells because of the unpredictable, lifelike behavior of electrified matter. The stars we see are made of electrons, protons, bare atomic nuclei, and other charged particles in constant motion. The space between the stars and galaxies, far from being empty, teems with electrically charged subatomic particles, swimming in vast swirling electromagnetic fields, accelerated by those fields to near-light speeds. Plasma is such a good conductor of electricity, far better than any metals, that filaments of plasma—invisible wires billions of light-years long—transport electromagnetic energy in gigantic circuits from one part of the universe to another, shaping the heavens. Under the influence of electromagnetic forces, over billions of years, cosmic whirlpools of matter collect along these filaments, like beads on a string, evolving into the galaxies that decorate our night sky. In addition, thin sheaths of electric current called double layers, like the membranes of biological cells, divide intergalactic space into immense compartments, each of which can have different physical, chemical, electrical, and magnetic properties. There may even, some speculate, be matter on one side of a double layer and antimatter on the other. Enormous electric fields prevent the different regions of space from mixing, just as the integrity of our own cells is preserved by the electric fields of the membranes surrounding them.

Our own Milky Way, in which we live, a medium-sized spiral galaxy one hundred thousand light-years across, rotates around its center once every two hundred and fifty million earth years, generating around itself an enormous magnetic field. Filaments of plasma five hundred light years long, generating additional magnetic fields, have been photographed looping out of our galactic center.

Our sun, also made of plasma, sends out an ocean of electrons, protons, and helium ions in a steady current called the solar wind. Blowing at three hundred miles per second, it bathes the earth and all of the planets before diffusing out into the plasma between the stars.

The earth, with its core of iron, rotates on its axis in the electric fields of the solar system and the galaxy, and as it rotates it generates its own magnetic field that traps and deflects the charged particles of the solar wind. They wrap the earth in an envelope of plasma called the magnetosphere, which stretches out on the night side of the planet into a comet-like tail hundreds of millions of miles long. Some of the particles from the solar wind collect in layers we call the Van Allen belts, where they circulate six hundred to thirty-five thousand miles above our heads. Driven along magnetic lines of force toward the poles, the electrons collide with oxygen and nitrogen atoms in the upper atmosphere. These fluoresce to produce the northern and southern lights, the aurorae borealis and australis, that dance in the long winter nights of the high latitudes.

The sun also bombards our planet with ultraviolet light and X-rays. These strike the air fifty to two hundred and fifty miles above us, ionizing it, freeing the electrons that carry electric currents in the upper atmosphere. This, the earth's own layer of plasma, is called the ionosphere.

The earth is also showered with charged particles from all directions called cosmic rays. These are atomic nuclei and subatomic particles that travel at velocities approaching the speed of light. From within the earth comes radiation emitted by uranium and other radioactive elements. Cosmic rays from space and radiation from the rocks and soil provide the small ions that carry the electric currents that surround us in the lower atmosphere.

In this electromagnetic environment we evolved.

We all live in a fairly constant vertical electric field averaging 130 volts per meter. In fair weather, the ground beneath us has a negative charge, the ionosphere above us has a positive charge, and the potential difference between ground and sky is about 300,000 volts. The most spectacular reminder that electricity is always playing around and through us, bringing messages from the sun and stars, is, of course, lightning. Electricity courses through the sky far above us, explodes downward in thunderstorms, rushes through the ground beneath us, and flows gently back up through the air in fair weather, carried by small ions. All of this happens continuously, as electricity animates the entire earth; about one hundred bolts of lightning, each delivering a trillion watts of energy, strike the earth every second. During thunderstorms the electric tension in the air around us can reach 4,000 volts per meter and more.

When I first learned about the global electrical circuit, thirty years ago, I drew the following sketch to help me think about it.

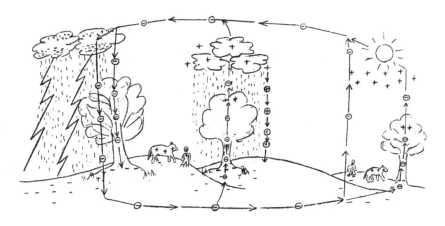

Living organisms, as the drawing indicates, are part of the global circuit. Each of us generates our own electric fields, which keep us vertically polarized like the atmosphere, with our feet and hands negative with respect to our spine and head. Our negative feet walk on the negative ground, as our positive heads point to the positive sky. The complex electric circuits that course gently through our bodies are completed by ground and sky, and in this very real way the earth and sun, the Great Yin and the Great Yang of the *Yellow Emperor's Classic of Internal Medicine* of ancient China, are energy sources for life.

It is not widely appreciated that the reverse is also true: not only does life need the earth, but the earth needs life. The atmosphere, for example, exists only because green things have been growing for billions of years. Plants created the oxygen, all of it, and very likely the nitrogen too. Yet we fail to treat our fragile cushion of air as the irreplaceable treasure that it is, more precious than the rarest diamond. Because for every atom of coal or oil that we burn, for every molecule of carbon dioxide that we produce from them, we destroy forever one molecule of oxygen. The burning of fossil fuels, of ancient plants that once breathed life into the future, is really the undoing of creation.

Electrically, too, life is essential. Living trees rise hundreds of feet into the air from the negatively charged ground. And because most raindrops, except in thunderstorms, carry positive charge down to earth, trees attract rain out of the clouds, and the felling of trees contributes electrically towards a loss of rainfall where forests used to stand.

"As for men," said Loren Eiseley, "those myriad little detached ponds with their own swarming corpuscular life, what were they but a way that water has of going about beyond the reach of rivers?"[10] Not only we, but especially trees, are the earth's way of watering the desert. Trees increase evaporation and lower temperatures, and the currents of life speeding through their sap are continuous with the sky and the rain.

We are all part of a living earth, as the earth is a member of a living solar system and a living universe. The play of electricity across the galaxy, the magnetic rhythms of the planets, the eleven-year cycle of sunspots, the fluctuations in the solar wind, thunder and lightning upon this earth, biological currents within our bodies—the one depends upon all the others. We are like tiny cells in the body of the universe. Events on the other side of the galaxy affect all life here on earth. And it is perhaps not too far-fetched to say that any dramatic change in life on earth will have a small but noticeable effect on the sun and stars.

CHAPTER 5

Of Marvelous Form

Let us continue a bit longer looking at life in terms of electricity. Consciousness itself, it has been discovered, is linked to the electromagnetic state of the body. The electric currents that our bodies generate when we are awake keep our hands and feet negatively charged compared to our brain and spine, and the front of our brain negative with respect to the back. The areas of greatest positive electrical potential within our bodies appear to be the same as the wheels of energy, or "chakras," that yogis and mystics tell us about. The base of the spine, the chest between the shoulder blades, and the brain have the highest potentials of all.

These electrical patterns are essential to waking life. Consciousness cannot exist without them. During sleep or under anesthesia, the body's electrical fields are reversed in polarity. And a strong enough external field that directly opposes the body's own fields will also take away sensation and awareness. Acupuncture is an ancient way of conducting current from the atmosphere into the body using fine needles, and it too can produce anesthesia.

The polarization of our bodies' tissues also guides all growth and repair processes. Orthopedic surgeon Robert O. Becker's pioneering work with broken bones, for example, showed that the area around a fracture becomes strongly positive, and that the electric field does not return to its usual form until healing has occurred. This research led to the design of medical equipment, now available in most hospitals, that helps bones to heal by electric stimulation. Tiny electric currents of the correct polarity, as small as 500 picoamperes, can accelerate not only fracture repair, but also the healing of other wounds and the regeneration of cartilage and peripheral nerves.

The acupuncturist's art helps the body to heal in a similar way, balancing the energy flows within the body by making use of natural connections with the currents of the earth itself. Practitioners use the Chinese term "chi" or "qi" instead of "electricity," and they call the polarities "yin" and "yang" instead of "positive" and "negative," but they are referring to essentially the same thing.

To look at life as an electromagnetic phenomenon is only one way of seeing. If we had eyes to see it, we would observe that the telephone, radio, and television are modeled after living systems, and that the science of electronics is really the science of life. But life is much more beautiful. The glowing halo of colors I once saw around a friend, layers of shimmering yellow and purple surrounding her body, is a sight I will never forget. Mystics call it the aura, and they say we are all capable of seeing it.

There are still other ways of seeing. If we had the compound eyes of a honey bee, we would perceive that daylight is polarized in a direction that depends on the position of the sun. Unlike us, an insect can tell direction simply by looking at the sky, even on a cloudy day. If we had such eyes, we might become very aware of the crystalline nature of life, because one of the properties of crystals is that they polarize and rotate light.

Crystals share many of the properties we usually associate with living organisms. Anyone who has ever grown a crystal garden has watched orderly structures grow from seeds. Crystals are very responsive to their surroundings. Their form and their growth are influenced by the ambient temperature, and by light, pressure, electric and magnetic fields, and chemicals in their environment. Crystals store energy in their structure, and they also transform heat and mechanical stress into electricity and vice versa. It is these properties that make quartz crystals so important in electronics.

Crystals also have memory. In their structure they record everything that has ever happened to them, and they reproduce this record faithfully as they grow.

It is not surprising that many forms of living tissue—bone, collagen, cellulose, horn, wool, wood, tendon, blood vessel walls, muscle, nerve, fibrin, and DNA—possess crystalline properties, in some cases equaling quartz in their abilities. These properties have very much to do with how living organisms grow, function, and evolve in an electromagnetic environment.

Only a few kinds of living tissue, like bone and horn, are solid like a true crystal. But there is a state of matter intermediate between solid and liquid called the liquid crystal. Liquid crystals are structurally ordered like solid crystals, but they form droplets and pour like liquids. They respond to external stimuli like light, sound, temperature, electric and magnetic fields, and changes in the chemical environment—all properties associated with living cells. The chemicals that life is made of—proteins, nucleic acids, lipids, polysaccharides, pyrroles, carotenoids—all form liquid crystal systems in water.

Now let's change our vision again and look at life in terms of solution chemistry, the model that is most familiar to molecular biologists, geneticists, and biochemists. How are the different types of molecules we are made of related to the phenomenon we call life?

"Organic chemistry" usually means carbon-based chemistry. The properties of carbon atoms, particularly their ability to bond together in long chains, forming very large and complex molecules, are what give life the form that we know.

However, life is not all carbon. We are made of all the elements found in ocean water and the atmosphere. The world ocean in which life originated can still be found ebbing and flowing, inside of us. Our blood and the fluid that bathes all our cells is predominantly salt water, with dissolved sodium, chloride, calcium, magnesium, bicarbonate, and phosphate ions. The ocean is not very different, containing in addition potassium, sulfate, bromide, and strontium in significant quantities. The potassium has found its way to the interior of our cells, where it replaces sodium.

The large molecules that are necessary for growth, metabolism, reproduction, and the structure and function of all living things, are mostly proteins, carbohydrates, lipids, and nucleic acids, and are made largely out of the elements carbon, hydrogen, oxygen, and nitrogen, with a smaller component of sulfur. The carbon, oxygen, and nitrogen come from the atmosphere; the hydrogen and sulfur from the sea—all courtesy of global cycles that circulate the constituents of life throughout the land, sea, and air.

The sun shines on the oceans and evaporates the water, which rises into the air as water vapor. Air currents blow the moist air over the land, where it condenses into clouds. It rains and snows, and the water seeps into the ground, and collects into springs, and lakes, and streams, and rivers, and eventually the rivers flow back into the sea. Plants absorb water from the soil into their bodies and they evaporate water from their leaves back into the air. Fishes and aquatic animals and land animals and you and I drink the water that collects in the springs and wells and streams and rivers and lakes, and it circulates through our bodies and we urinate it back into the soil.

The sun shines on the oceans and on the land, and algae and plants the world over use the energy of sunlight to split carbon dioxide from the air, and water from the sea and the soil, and manufacture sugars and other carbohydrates and put pure oxygen into the atmosphere. You and I, and all the other animals in the world, eat those plants, or we eat animals who eat those plants. We and all living things breathe oxygen from the air to burn those carbohydrates for energy, and we breathe carbon dioxide back into the air and urinate water back into the soil, for use again by the plants.

Thunderclouds gather over the oceans and over the continents, lightning oxidizes nitrogen in the atmosphere, and rain washes the resulting nitrates into the oceans and into the soils. Algae in the sea and bacteria in the soil split nitrogen from the air to make ammonia, nitrites, and more nitrates. Plants take up nitrates from the water and the soil through their roots and use them, together with carbon, hydrogen, and oxygen, to make amino acids to build proteins, and nucleotides to build DNA and RNA. Animals get the nitrogen we need to build our proteins and nucleic acids from the plants and other animals we eat. Some nitrogen cycles through our bodies rapidly and is excreted in our urine and feces and returned to the soil for more plants to use. Some is borrowed for our lifetime, and passed on to organisms who eat us after we die, or returned to the atmosphere by the bacteria that decay our bodies.

Proteins are the largest and most complex molecules in living organisms. They regulate metabolic processes, serve as enzymes for chemical reactions, and are involved in nearly every biological function. In general, proteins are what regulate the growth, metabolism, repair, and other functions of living things.

Nucleic acids—DNA and RNA—are the stuff of chromosomes and genes. They are responsible for heredity, reproduction, and directing the synthesis of proteins. The "blueprints" for making a living being are contained in its DNA; the mechanisms for carrying out the instructions are built of proteins; the energy to do all this is provided mostly by carbohydrates; and the structure of the organism is composed of lipids, and more carbohydrates and proteins.

Let us now come back to ourselves and look through our own human eyes, with the aid of a microscope if necessary, at the physical structure of living things. We can start by looking at ourselves. A human being has a head, two arms, two legs, and a torso. He or she has many different organs—a heart, two lungs, a liver, a spleen, two kidneys, a bladder, a pancreas, a stomach, and so forth. We can also see that we are composed of many different types of tissue—skin, bone, cartilage, muscle, nerve, blood, fat, and so on. When we look under the microscope, we can see in addition that all of our tissues are made of cells.

Every cell of your body contains a complete and identical set of your chromosomes; no matter what type of cell it is, it contains within it the full set of blueprints for the making of the whole human being that is you. Every plant and every animal on earth is similarly built of cells that have chromosomes made of DNA. A human being is different from a lily, basically because of the difference in the DNA content of our chromosomes.

The world is full of large multicellular organisms, like us, and tiny one-celled organisms like paramecia, amoebae, and blue-green algae that you can see only with a microscope. A drop of pond water or sea water has lots of little critters swimming or floating around in it. One-celled organisms are further divided into eukaryotes, like paramecia, whose chromosomes are packaged in a nucleus, and the even smaller prokaryotes like bacteria and blue-green algae, that have no nucleus.

Looking under a still more powerful microscope, one can see the viruses, which are nothing more than crystals of naked DNA or RNA with a protein coat. Viruses are incapable of reproduction or metabolism, in fact they show no signs of life at all unless they are inside a living cell.

In a eukaryotic cell, not only the nucleus, but certain other organelles, like mitochondria, which supply energy to the cell, and chloroplasts, which carry on photosynthesis in plants, contain their own chromosomes with their own DNA. Nuclei, mitochondria, and chloroplasts are just like bacteria or blue-green algae, except that they live inside a larger cell.

We begin to see the layers of life:

Virus – bare DNA crystal

Bacterium, Blue-Green Alga – One Cell without a Nucleus

 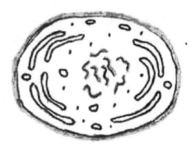

Bacterium   *Blue-green alga*

Eukaryotic cell – contains a nucleus and other organelles

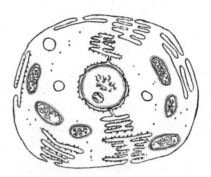

Multicellular organism (you or me)

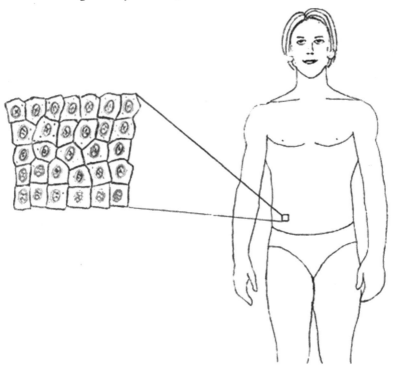

It is as if each form of life is nested inside of a more complex one. A surprising question must then be asked: are viruses and bacteria our enemies, or are they part of our very core?

The world of bacteria and viruses is remarkably fluid. When a bacterium finds itself in a new environment, it can alter its form completely, it can even lose its cell wall and change into tiny particles that pass through filters like viruses, and it can change back again. When I was in medical school in the late 1970s, this changeability or "pleomorphism" of bacteria was mentioned once in a lecture, briefly in passing, almost as an embarrassing fact, because different bacterial forms have traditionally been regarded as different species entirely.

Genes are commonly thought to be fixed and unchanging, subject only to random mutation. But the world of the gene is also surprisingly fluid, revealing an essential unity of all life on earth. It is amazing that there are just four DNA nucleotides in the whole world: adenine (A), guanine (G), thymine (T) and cytosine (C). RNA has uracil (U) in lieu of thymine. There are also just twenty amino acids from which all proteins in the whole plant and animal world are made. The genetic code has been worked out, and it is the same for all species:

UUU }	Phenyl-	UCU }		UAU }	Tyrosine	UGU }	Cysteine
UUC }	alanine	UCC }	Serine	UAC }		UGC }	
UUA }	Leucine	UCA }		UAA }	terminator	UGA	terminator
UUG }		UCG }		UAG }		UGG	Tryptamine
CUU }		CCU }		CAU }	Histidine	CGU }	
CUC }	Leucine	CCC }	Proline	CAC }		CGC }	Arginine
CUA }		CCA }		CAA }	Glutamine	CGA }	
CUG }		CCG }		CAG }		CGG }	
AUU }		ACU }		AAU }	Asparagine	AGU }	Serine
AUC }	Isoleucine	ACC }	Threonine	AAC }		AGC }	
AUA }	Methionine,	ACA }		AAA }	Lysine	AGA }	Arginine
AUG {	initiator	ACG }		AAG }		AGG }	
GUU }		GCU }		GAU }	Aspartic	GGU }	
GUC }	Valine	GCC }	Alanine	GAC }	acid	GGC }	Glycine
GUA }		GCA }		GAA }	Glutamic	GGA }	
GUG }		GCG }		GAG }	acid	GGG }	

It is a universal language, written small in the molecules of our cells. So a strand of RNA that looks like this:

A U G A C C U U G U G C C A A

(reading from left to right in groups of three) is an instruction to put together the amino acids methionine, threonine, leucine, cysteine, and glutamine, in that order. This instruction would be translated exactly the same way by a bacterium, a rhododendron, a whale, a human being, or a mushroom.

Genes can move around between organisms and be traded back and forth, even between unrelated individuals of different species, in a process biologists call genetic recombination. So-called "genetic engineering" is but the human adaptation of processes that occur everywhere in nature.

- Bacteria are known to pass on genes to other bacteria—even to different kinds of bacteria—in a kind of sexual relation. In this way, the genes for resistance to antibiotics often spread rapidly through many different kinds of bacteria in a hospital.
- Cells that are simply grown in a medium containing foreign DNA are sometimes "transformed" when they incorporate pieces of foreign DNA into their own chromosomes.

- Genes may be transferred from one organism to another by a virus. We usually think of viruses as foreign invaders that destroy our cells and make us ill. But before they destroy our cells they may copy some of our genes to take with them. Or they may remain latent within us for so long that they cease to be viruses at all, becoming instead a permanent part of our own chromosomes.
- There is recent evidence that genes have been transferred from one species of fruit fly to another by a tiny parasitic mite.
- Nucleotide sequences similar to the human hemoglobin gene have been found in leguminous plants. This and many other such examples are providing researchers with abundant evidence of gene transfers that have occurred between unrelated species throughout the living world.

Evolution has been going on here on earth for perhaps three billion years. Yet every cell of every single living organism on the whole planet, every bacterium, every mushroom, every snail, every sponge, every crocodile, every insect, every flower, every tree, every horse, every man, and every woman has the exact same genetic code, coding for the exact same twenty amino acids. Apparently not one molecule has changed, even in species that diverged from each other three billion years ago. I can think of only one explanation for this: the millions of species that inhabit this planet are so interconnected that we really form one single organism with a single gene pool. The genetic variability that is so crucial to species adaptation and evolution does not rely just on random mutation; vitally important is genetic recombination through the action of viruses and bacteria, which circulate DNA among all life on this planet.

The extinction of species, the loss of genetic diversity on the scale that is occurring today, particularly in the tropics, is a tragedy not just because it will make the world less beautiful; not just because we are losing plants and animals that might someday be useful to us for medical purposes, or for food, or other human purposes; not just because other species have as much of a right to exist as we do; but because it threatens the global *ecology*, the interrelationships and interdependence of all life on the earth, and because the gene pool that is being impoverished is our own.

CHAPTER 6

Ecology (1991)

Ecology is the study of the mutual relationships between living things and their environment. The word comes from the Greek "oikos" meaning "house" and "logos," "study." The earth, whatever part of it we inhabit, is our house, its soil and rocks our floor, its blue sky our roof, its streams and rivers our running water. The birds that sing and fly, the buzzing insects, the four-legged animals large and small, the snakes that slither through the brush, the fragrant flowers and the towering trees—all are our sisters and brothers with whom we share our home. Each one occupies its own room in the house, its own biological niche.

Recently I paid an autumn visit to Meares Island, off the west coast of Vancouver Island in British Columbia. Meares is still-undisturbed old growth forest—giant cedar, Sitka spruce, and all the other plant and animal species that have always been there. My Danish friend Michael Andersen and I paddled across in a rented canoe from Tofino one morning and spent all day exploring a small part of this ancient, beautiful place. Except for an occasional bird song or the chatter of a squirrel, it was quiet in there, as the mass of the trees kept out the jarring sounds of motorboats ferrying human residents to and from the more outlying islands. I felt like an invited guest in a well-kept home. Shaded by the forest canopy far above, the ground was covered with a thick layer of decaying humus. My view of the forest around me was dominated by cedar and spruce trees up to sixty feet around, two hundred feet high and perhaps two or three thousand years old, some straight and majestic, others twisted, battered, and with burned-out hollows big enough to stand inside of. Lightning had struck some of these giants. Fires had charred and hollowed them out, perhaps hundreds of years ago, and still they stood guard over an almost unchanging forest. Here and there seedlings sprouted from the decaying logs of ancient fallen trees, but mostly the open undergrowth allowed for easy walking over ground broken by roots of living trees and crossed by tiny rivulets. This is a forest where the cycle of life goes on as it has always done, where change still occurs very very slowly. By stark contrast, nearby mountains on Vancouver Island were being rapidly stripped bare of all vegetation by chain saws working seven days a week, twenty-four hours a day,

During my visit to the Queen Charlotte Islands, off northwestern British Columbia, I walked in another expanse of old growth forest on the north shore of the Tlell River. This area of old growth was not so large as the one on Meares, and the cedar, spruce, and hemlock that grew here were not such giants. But here, too, the ground was open and easy to walk on. I soon approached an area that had once been logged, and it was like running into a brick wall! This second growth forest had not been cut in probably sixty or seventy years, and the trees were quite large, but here the undergrowth was an impenetrable tangle of brush and debris. It looked to me just as if a bunch of logs and branches had been thrown down every which way, one on top of the other, and in between the cracks a jumble of bushes and weeds had grown up. This was not the well-kept home it once had been. Clear-cut logging had very seriously disturbed this ecosystem and this community of plants and animals. If human beings left it alone, it would be centuries before the equilibrium of a mature forest was regained.

But human beings are not leaving forests alone, not anywhere on the earth. They continue to replace them with houses, cities, farms, roads, and airports all over the planet, at a fast-accelerating rate, not just in the tropics, but everywhere. People are still clearing the remaining old growth in North America's Pacific Northwest for housing and construction; flooding vast areas of forest land for giant hydroelectric projects in Canada's James Bay area; clearing the entire Canadian North of its ancient boreal forests to feed into pulp mills to make our newspapers, books, and toilet paper.

Foresters who "manage" our remaining forests seem to believe that a tree farm is as good as a forest, that they can clear a huge mountainside of all its trees and vegetation, replant with a monoculture of the desired species of tree, and in eighty years the forest will be as good as new. In the Pacific Northwest they believe they can "harvest" a new crop of trees every sixty to eighty years indefinitely, with no detrimental effects on soil, wildlife, or the health of their trees and the quality of their wood. They seem sincere enough, but my experience tells me otherwise.

Meares Island has a little patch of forest that was clear-cut some thirty years ago. It is again green—sort of—a sickly pale color compared to the rich green of its magnificent surroundings. It's a permanent scar on the mountainside, much like any severe injury to your or my body that heals over with a visible scar.

On some of the land around Clayoquat Sound near Meares Island I saw extensive five-year-old clear-cuts that showed no signs of life at all.

The enormous mountains that dominate the view around Ucluelet, some twenty-five miles down the coast from Tofino, are now mountains of bare rock, showing no signs of regeneration whatsoever. It is one of the ugliest sights I ever want to see. Most of the residents of Ucluelet make their living in the logging industry, but how can they stand to wake up every morning and look out their windows at that view?

The Queen Charlottes receive enough rain that their clear-cut forests do "green up" in a few years and good-sized trees do grow to harvestable size in sixty to eighty years. So do the Douglas firs of Washington and Oregon, and the redwoods of coastal northern California. But a tree farm is not a forest, any more than a field of broccoli is a meadow, or a field of corn is a tall-grass prairie—or a storeroom full of furniture is a home.

The plants and animals of a region, together with the soil, rocks, land formations, rivers and lakes, and air, are an integrated ecosystem. Within the ecosystem is the biological community of living things, and within the community are populations of related individuals. No individual exists by itself; each interacts with others of its kind. Similarly, every population has many different sorts of relationships with other populations in the biological community. And no ecosystem on planet earth exists by itself; the cycling of air, water, soil, and nutrients is a planet-wide system. Rivers flow down mountainsides, across deserts, and into the sea. Many birds, fishes, mammals, and insects migrate seasonally from one ecosystem to another.

The interconnectedness of all living things begins to be apparent from the difficulty of even defining what an individual is.

It is fairly clear what an individual human being is—but just how clear is it really? Am I one individual human being or am I a collection of individual hands, feet, and other parts? Or am I a collection of trillions of individual cells? Some of my cells, such as my white blood cells, even seem to have a life of their own, swimming long distances to forage for a meal whenever bacteria invade my body. My white cells, someone might say, have the same genes as all my other cells, and therefore all are part of one large organism: me. How about the bacteria that live in my intestines, that are so much a part of me that I could not digest my food without them? Are they "really" part of me or are they separate individuals? Is someone who receives an organ transplant two individuals or one? What about a tree graft? Is a nectarine tree with a branch of peaches grafted onto it one individual or two?

Is an ant a separate being? Or a honey bee? All of the insects in one hive or one colony are offspring of the same mother, they have different physical forms according to their functions, and they quickly die if separated from their fellows. All are sterile except for the members whose sole purpose in life is to reproduce. Are they much different from cells in an organism?

When a redwood tree is cut down, a "fairy ring" of new trees surrounding the stump grows up from its undying roots. Are these one organism or many? Many trees, including aspen, buckthorn, cherry, oak, and chestnut often propagate by sprouting new trees from the same root system rather than from seed; new trees may grow up a hundred feet away from the "parent" trunk.

Even separate trees often join up underground by self-grafting their roots where they are pressed together, so that a forest containing one kind of tree may have a single interlocking web of roots. This is one reason cut trees often sprout again so rapidly—their roots receive nutrients from their neighbors. Again, where does one individual stop and another begin?

The plant world is full of similar examples—bamboo, bracken, conch-grass, ground elder, bindweed, horsetail—where what you see above ground looks like many separate plants, but below ground all are connected as one. A colony of blackberries, strawberries, or box huckleberries spreads its runners indefinitely across the countryside, sending down roots at intervals.

Some animals, too, live in colonies or clones that maintain a connection. In the sea a clump of coral is a colony of individual "animals" building a common skeleton. So is a sponge. That biologists consider the coral a colony and the sponge an individual is largely a matter of convention.

The Portuguese man-of-war is a sea creature resembling a very large jellyfish that upon close examination seems to be a colony of animals having different forms and functions. One serves as the float; hanging beneath are three types of polyps, one with tentacles for capturing prey, a second carrying out the digestive functions, and a third having the reproductive function. Are these "really" many individuals or just one? The parts do not survive if they are separated from one another.

Back on land, a lichen is an association of fungus and alga whose lives are so closely connected that neither exists on its own in nature. Their collaboration resembles leaves in both structure and function, and the fungus and alga reproduce together as one organism.

The gourmet's black truffle grows in association with the roots of oak trees, to the mutual benefit of both. The fungus helps the tree to absorb water and nutrients from the soil, and in return the tree supplies it with carbohydrates. But these truffles never grow except on the roots of oak trees; they might be considered to be part of the tree.

In Mendocino where I lived, we used to go out into the forest after the first rains in the fall to pick edible Boletus mushrooms. No matter how young these reddish-brown mushrooms were, it seemed they always had a particular kind of worm eating them; the trick was to pick the mushroom young enough that the worm hadn't eaten too much of it. Friends of mine usually considered the worm "part" of the mushroom. Whether this is a one-sided parasitic relationship, or whether the mushroom benefits too, I do not know.

Every species of fig tree is associated with its own species of fig wasp, and neither can live and reproduce without the other. The wasps spend most of their life cycle inside of their figs (called "gall-figs"). Their eggs are laid there, the larvae develop there and are nourished by the fig, and when the fig is ripe, the adult male and female wasps mate inside their fig. The females then push out through the fig's orifice, where they get dusted with pollen from male flowers. They then fly on, some to pollinate female figs and others to deposit their eggs in other gall-figs and repeat the cycle.

The huge organ-pipe cactus of Mexico, in the manner of many plants of tropical America, has purple flowers that open up at night, providing nectar for the leaf-nosed bats which are their pollinators. These small nectar-drinking bats have long snouts and longer tongues, with which they reach down into the flowers much as hummingbirds do with their daytime flowers, Neither plant nor bat can survive one without the other.

Ants have many sorts of mutual relationships with plants and other animals. Many trees and tree-climbing vines in tropical rainforests provide ants with nesting sites in hollow stems or other parts. In return, the fierce ants protect the plants from attack, keeping off harmful insects, caterpillars, snails, grazing mammals, and leaf-cutting ants. "In effect an ant-inhabited tree maintains a standing army," says Anthony Huxley.[11]

Some trees also provide pastures for aphids and other sap-sucking insects that the ants herd and "milk." There are leaf-cutting ants here in New York as well as in the tropics that grow carefully weeded fungus gardens in large underground chambers. This species of fungus grows only where cultivated by the ants, and it forms the sole food of their colonies. Other ants grow flower gardens in the tops of tropical trees. "The Bucket Orchid is very difficult to cultivate successfully in the absence of its ants . . . If such an orchid is placed in a greenhouse any ants around will colonize its roots."[12] These plants are well-protected; the ants emerge and attack the instant one of their plants is touched.

An ecosystem is the totality of all of the complex relationships among all of its living residents. Everyone plays his or her part, everyone's story is just as unique and interesting as the brief examples I have given. The tropical forest, the coral reef, the continental shelf, the desert, the grassland, the boreal forest, the river, the lake—each neighborhood is an integrated unit of life.

Every organism is part of a "food chain" or "food web" that supports all living things in the area. In the forest, green plants get their food directly from the air and the soil, using the energy provided by sunlight. Herbivores—mice, rabbits, squirrels, deer—eat the green plants. Carnivores—snakes, wolves, mountain lions, hawks—eat the herbivores, and large carnivores eat smaller ones. Scavengers—vultures—eat the remains of dead animals. Decomposers—fungi, bacteria, termites, worms, centipedes, mites—take care of all the leftovers, animal wastes, and dead plants, and turn them back into rich soil.

The food web is not simple, even here In North America where the biodiversity is so much less than in the tropics. One of the ancient forests I have known in California and Oregon is the Douglas fir forest, home of the famous northern spotted owl that has innocently found itself the object of a long and bitter war pitting loggers against environmentalists. Here, perhaps more than in any other ecosystem North America, the complex relationships among hundreds of species of plants and animals have been studied in detail.

A few of the smaller vegetarians of the forest, for example, include the deer mouse, who lives in burrows beneath the soil, foraging by night for Douglas fir seeds, truffles, and berries, and who is up and about throughout the winter; the Townsend chipmunk, who also eats truffles and berries, but is active during the day, and hibernates in underground nests during the winter; the Oregon red tree vole, who spends its life in the tops of older Douglas fir trees—it is nocturnal and feeds primarily on Douglas fir needles and twigs; the red squirrel, or chickaree, who is active from dawn until dusk, and makes its nest on tree limbs or, in winter, in natural hollows or abandoned woodpecker holes—it feasts preferentially on Douglas fir seeds and truffles, but its diet also includes a wide variety of fruits, berries, greenery, and the needles, inner bark, twigs, and sap of the Douglas fir, all depending on the time of year. The chattering chickaree is one of the sentinels of the forest, loudly warning everyone who will listen of approaching predators and intruders.

Every animal has its own place—its own niche—in the forest, and it is a little different from everyone else's. The hours it keeps, the place it resides, its diet—who it eats and who eats it—its mating habits, and all the other peculiarities of its life determine how every species makes a living that is different from everyone else's.

Different plants may need different amounts of light, or water, or certain nutrients. Their roots may penetrate to different depths in the same soil, or they may form root associations with different species of fungus. Their flowers may be pollinated by different insects.

Normally there is not too much competition between different species, because everybody has a different job to do. It is misleading to speak, as ecologists often do, of "dominant" species. There are species which are more or less common in any ecosystem, but this is due mostly to the size of the different niches. To carry on with my metaphor, some job categories have more vacancies than others. When a species disappears, it is usually because its niche disappeared, not because it was competing with a different type of animal or plant. If the environment is changing, and it is going to survive, an organism may have to change the way it makes its living. This occurs through competition *within* the species, and is the stuff of evolution. Doctors compete with other doctors to become better at what they do, but doctors do not compete with shoemakers. Of course, if changing its way of making a living brings one species into competition with another, this will have repercussions throughout the ecosystem; but over the long term, two separate species do not occupy the same niche.

Nature accommodates incredible densities of life. Innumerable different species can live side by side in the same small space so long as they are not in direct competition with one another. In Oregon most of the diversity is below the forest floor, where eight thousand different kinds of insects, spiders, mites, centipedes, and other arthropods may live amongst one another in the soil. In the tropics the diversity is above ground. A rainforest may have thousands of kinds of insects inhabiting hundreds of kinds of trees.

If the traveller notices a particular species and wishes to find more like it, he may often turn his eyes in vain in every direction. Trees of varied form, dimensions and colors are around him, but he rarely sees any of them repeated. Time after time he goes towards a tree which looks like the one he seeks, but a closer examination proves it to be distinct. He may at length, perhaps, meet a second specimen a half a mile off, or he may fail altogether, till on another occasion he stumbles on one by accident. (Alfred Russel Wallace, in *Tropical Nature*)[13]

These are the very forests we are razing to plant grass for our cattle. Thousands of species of plants and animals that have lived together for untold millennia will be gone forever. The poor rainforest soil will support just three species—grass, cattle, and human beings—for a few scant years, and then it will be sun-baked desert for untold millennia to come.

In the original forest, every species plays its part in supporting the whole ecosystem, indefinitely; the neighborhood evolves with the slowly changing relationships among all its plants and animals, and with changing climate and terrain, over hundreds and thousands of years.

The plants of a forest are just as dependent on the animals for their long-term survival as the animals are on the plants for their sources of food. The truffles that grow underneath Douglas firs can't even reproduce unless mice and squirrels sniff them out, dig them up and eat them, later spreading their spores when they defecate elsewhere in the forest. This, in turn, inoculates the soil with the fungus that is essential for the growth of new Douglas fir trees in areas that have been burned by fire. Chickarees stash away many more Douglas fir cones than they will eat, in hoards all over the forest, and so help spread the seeds of future trees.

Predators and prey are kept in dynamic balance. The owls that swoop down on mice and voles, and the mountain lions that stalk deer and elk— not only do they help to control the numbers of their prey, but they keep prey populations strong and healthy by weeding out the weak and the sick. In this respect human beings differ from all other predators: we prefer to kill the strongest and healthiest animals; we weaken the populations of our prey and drive them toward extinction.

Even dead trees, or "snags," are vitally important parts of the ecosystem, providing food and shelter for many living things. The decaying wood is home to a large variety of insects, and therefore a rich source of food for birds. Woodpeckers come to eat wood-boring beetles and to excavate nesting cavities in the wood. Brown creepers and bats find homes under the loosening bark. When woodpeckers abandon their cavities, other creatures use them for nests and winter refuges. High hollows are chosen by tree swallows, lower ones by western bluebirds. Vaux's swifts nest on the inside of hollow trees. Mammals such as flying squirrels, chickarees, woodrats, and martens may either use natural cavities or those made by woodpeckers.

After a tree falls, it continues to provide food and shelter for a wide variety of animals. Wood-boring beetles and other insects continue its decomposition. Lichens and mosses grow on it. Carpenter ants and termites help create inner spaces that are inhabited by mites, springtails, salamanders, and later squirrels and other small mammals. Finally, bacteria and fungi convert the tree back into soil.

In its decay a tree may also function as a nursery for young trees of both its own and other species. The rotting wood provides a good environment for seedlings that might otherwise have a hard time among the debris on the forest floor.

An ecosystem exists in time as well as in space. When trees fall over, forming gaps in the forest canopy, or when an area has been burned by fire, a new community of plants and animals moves in to take advantage of the new conditions, After a Douglas fir forest is burned, it is initially recolonized by grasses, and herbs such as dandelion, thistle, lupine, and wild strawberry. The roots of lupines harbor nitrogen-fixing bacteria, increasing the nitrogen content of the soil. Gophers burrow in the soil, turning it over and making it porous and able to hold more moisture.

The grasses, herbs, and gophers alter the soil until grasses and herbs no longer thrive in it. Shrubs such as snowbush and vine maple now appear, and in a similar way the shrubs prepare the soil for Douglas fir trees. In the absence of fire, Douglas firs eventually give way to western hemlock, western red cedar, grand fir, and other trees, which grow in the shade of Douglas firs and replace them when the ancient trees fall. The Douglas fir species cannot grow under its own shade and therefore requires fire for renewal.

And so, there are always patches of grasses, herbs, and shrubs scattered throughout an old growth forest. Douglas fir may be the most common kind of tree, but here and there are stands of hemlock and red cedar, and along rivers and streams are populations of other plants and other animals, forming living corridors into pine woods, and redwood forest, and still other kinds of communities in other neighborhoods farther away.

> To the wood products man an old-growth forest, with its many dead and dying trees, is an over-mature forest, a decadent forest, a forest in decline doing no human being any good . . .
>
> He doesn't see the slow exuberant dance the forest does through time. He doesn't see the intricate webwork of fungi that strands through the ground, drawing its food from the roots of trees and helping the roots draw food from the soil. He doesn't see the red-backed vole that eats the fungi's fruiting bodies and disperses their spores, sheltering itself in downed rotting wood. He doesn't see the spotted owl that eats the red-backed vole, hunting in the dark through thousands of acres of trees, nesting high in a standing snag and feeding her owlets, this brood and all her broods, as the Douglas firs keep growing and growing, each in its turn going down, melting into the ground, sheltering the vole and feeding the fungi and holding the cold meltwater in its fragrant sponge. (John Daniel, "The Long Dance of the Trees")[14]

CHAPTER 7

Gaia

The Earth travels around the solar system in a society of nine planets, thousands of asteroids and comets, and its own moon. As it rotates, its daytime face receives nourishing light and heat from the sun and is bathed in the solar wind. From its night side it gives off waste heat, which it radiates out into space. It feels the motions of the moon and sun as tides. Communications from the stars come in the form of electromagnetic waves and cosmic rays.

Our planet's skin is protected by a deep ocean over more than two-thirds of its surface, and a blue enveloping blanket of air. A great circulatory system mixes the world's waters so that the sea is about 3.4 percent salt everywhere on earth. Global wind and weather patterns distribute air and moisture east and west, and from tropics to poles.

Normal skin temperature of the earth is about 57°F. The polar ice caps reflect a lot of the sun's warmth back out into space and, together with ocean currents and weather systems, regulate the planetary temperature. The Gulf Stream brings tropical heat northward to moderate the arctic climate, while the Labrador and East Greenland currents, and the water flowing in the ocean deeps, bring arctic chill southward towards the equator. In like manner does the Southern Ocean wash the southern tropics and the cold Antarctic. The exchange of heat allows the polar ice caps to grow or melt in response to excess cooling or heating of the earth. And water vapor that rises from the world ocean under a hot sun blows poleward to fall out as snow, building up the ice over Antarctica and Greenland and increasing their capacity to cool the planet.

The coming and going of ice over large portions of the earth—the great Ice Ages—is caused by the interaction of astronomical, geological, and biological cycles whose details we can only guess. The shifting continents; the changing coastlines; the building up and erosion of mountain ranges; the shifting tides; and the sun's energy—all help to shape global wind and ocean currents. The precession, or slow wobble of the earth's axis; the slow changes in the earth's elliptical orbit around the sun; the atmospheric concentration of greenhouse gases; the albedo, or reflective power of the earth's surface; the amount of cloud cover; and the fluctuating fires of the sun itself—all modify the intensity of solar radiation reaching the different parts of the world. The respiration of living organisms determines the concentration of greenhouse gases in the earth's atmosphere. The humidifying effect of trees and other living things increases cloud cover.

Earth's atmosphere is 78% nitrogen, 21% oxygen, 1% argon, and .03% carbon dioxide; it has this, and not some other composition, because living things are breathing here. From its volcanoes, the earth exhales mostly water vapor, carbon dioxide, sulfur dioxide, and nitrogen. Volcanic water has become the world ocean. Volcanic sulfur has become dissolved in the ocean and buried under the ocean floor by bacteria. All of our oxygen has been manufactured by living plants from carbon dioxide; the coal, oil, and other fossilized carbon that is buried all over the planet represents oxygen that has been liberated by photosynthesizing plants during their two billion years or so of residence on earth.

The salt content of the sea, which bathes the earth and runs through our veins, cannot have changed very much since life began. Rain and rivers constantly wash more minerals from the continents into the seas, and living cycles constantly remove them and bring them back to the land, or bury them under the sea bed. Excess sulfur is removed by bacterial production of iron pyrites. Marine algae return sulfur and chlorine to the land in the form of dimethyl sulfide and methyl chloride gases. The ceaseless rain of the shells of foraminifera and other small creatures onto the ocean floor removes large amounts of calcium from the seas, and the animals that build the great coral reefs remove still more. Silicon is removed from the ocean by the continuous rain of the siliceous skeletons of tiny plants known as diatoms, and tiny animals called radiolaria. The regulation of sodium, potassium, and magnesium in sea water may also be linked to this constant fall of organic debris onto the ocean floor, and probably is also connected to chemical equilibria involving clay minerals.

Some minerals are also cycled back to the land in the bodies of fishes and aquatic birds that migrate and re-enter terrestrial food chains; at the seashores, very large amounts of organic material cross the borders in this way between ocean and land.

The air and sea would be very different without organic life; the earth's soil would not even exist. This porous nourishing layer that clothes the rocks and hills of our planet like a soft carpet was created by the worms, bugs, plants, and the billions of microbes that inhabit every inch of it, munching on compost and minerals and slowly changing old rock into new dirt, replacing that which is blown and washed away by the winds and rains of centuries.

That Mother Earth is alive is a very ancient and widespread notion. James Hutton, the founder of modern geology, gave it scientific form more than two centuries ago:

> Here is a compound system of things, forming together one whole living world, a world maintaining an almost endless diversity of plants and animals, by the disposition of its various parts, and by the circulation of its different kinds of matter... This earth, like the body of an animal, is wasted at the same time that it is repaired. It has a state of growth and augmentation; it has another state, which is that of diminution and decay. This world is thus destroyed in one part, but it is renewed in another; and the operations by which this world is thus constantly renewed, are as evident to the scientific eye, as are those in which it is necessarily destroyed.[15]

In our own times, the late James Lovelock again popularized the idea of the earth as a living being. He adopted the name the ancient Greeks gave to our world—Gaia:

> The entire range of living matter on earth, from whales to viruses, and from oaks to algae, could be regarded as a single living entity,

capable of manipulating the Earth's atmosphere to suit its needs and endowed with faculties and powers far beyond those of its constituent parts.[16]

In my own view, the oceans, atmosphere, and soils, and all of the plants and animals, are integral parts of a dynamic, changing, living, evolving organism. The atmosphere and oceans have exactly the compositions that are needed right now by life on earth, not because of any manipulation on the part of organic life, but because the atmosphere and oceans are *part* of organic life. We are all one body, and we all evolved together.

CHAPTER 8

Being and Becoming

I

In the beginning, I was taught in school, was a prebiotic soup—simple chemicals floating in the warm seas under a fiery sky. The sun's powerful rays, and lightning flickering out of a caustic sky, ignited the chemicals in the soup, and created the simple molecules of life. Hundreds of millions of years of random mixing in the early oceans, so the story went, by chance brought together the right molecules and the first primitive cell was born. After that, evolution took its course, starting with the simplest bacteria and algae and producing, with the passage of eons, more and more complex plants and animals . . . invertebrates . . . insects . . . fishes . . . reptiles . . . mammals . . . primates . . . culminating in nature's greatest achievement, *Homo sapiens sapiens*.

Charles Darwin's gift to us was to show that life is constantly changing its form. But the belief that complex forms must always spring from simpler ones stems from the quite different notion that ours is a Created universe.

Evolution is still often confused with Creation. Creation is a human act, and human beings create complex objects out of simple ones. We create paintings out of paints and canvas; chairs out of pieces of wood; automobiles out of steel parts that are made out of iron ore that is made out of rock.

In nature, however, we see the opposite: living cells may turn into dead cells, wood into sawdust, paper into ashes; the complex may become simple, but never the other way around. Physicists will recognize this as the second law of thermodynamics.

The truth seems to be quite otherwise than I was taught. Even our own acts of creation always simplify nature. We get our pieces of wood from trees that are infinitely more complex than the houses and furniture we make out of them. In excavating rock for iron ore, we destroy ecosystems that are far more complicated than the automobiles we manufacture, or ever will manufacture.

Human beings are indeed the end result of three billion years of evolution, but every other organism alive today—every worm, every cockroach, every mushroom, every bacterium, every tree, every dog, every fish—is also the end result of evolution. We are all brothers and sisters on the earth. And we are all equally complex. The complex could not have evolved from the simple because *there was no simple.*

We have seen that to truly understand life we must look at living things in many different ways:

1. As an electromagnetic phenomenon in a world full of electric and magnetic fields.
2. As a crystalline phenomenon in a sea of liquid crystals.
3. As organic chemistry in a chemical universe.
4. Structurally, as physical organisms in a world of air, water, earth, and fire.
5. Functionally, as eating, breathing, growing, reproducing beings in a world of movement.
6. Ecologically, as integrated parts of a living earth.

With a lot of imagination I might weave a tale of the birth of life some three billion years ago as a semiconducting, photosynthesizing, liquid crystalline phenomenon that organized itself out of amino acids, nucleotides, carbohydrates, and lipids synthesized on the surface of clay minerals in the warm oceans of a newly formed, electrically active, radioactive, volcanically active, spinning earth with a strong magnetic field under a sea of intense solar and cosmic radiation. As soon as volcanic gases had formed an appreciable atmosphere, and the newly formed earth had cooled to a temperature where water was a liquid, the earth was already a complex, living planet and evolution was on its way—long before the appearance of the first cells. Ocean currents and weather patterns were circulating the globe, rocks were eroding under the action of wind and water, the sun was shining, and the oceans were accumulating dissolved minerals, organic chemicals, and liquid crystals. The drama that followed is called life.

Life cannot in principle be synthesized in a laboratory. Neither you nor I nor any other part of the earth is truly alive in isolation from the earth. No space colony will ever be able to maintain itself indefinitely away from our own planet, any more than a hand or a foot can function, or indeed have any meaning, away from the rest of its own body.

Nor can evolution be truly understood only in terms of competition between individuals. Competition is only half of the story; the other half is cooperation. Competition makes it possible to distinguish between individuals; cooperation, as we have seen, makes the distinction somewhat hazy. Again, I would make the analogy to the cells of an organism. Cells that are weak or diseased do not long survive. But even healthy cells do not live unless they all work together.

Except in times of scarcity, cooperation between individuals in a species is as much in evidence as competition—and, as we discovered in chapter 2, populations of animals and plants tend to keep their numbers below the carrying capacity of their environment, so that scarcity is avoided.

It is likely that most of the incredible diversity of life on the planet today came about under relatively peaceful conditions in times of plenty, and that natural disasters, which bring on intense competition in the struggle for survival, are times when species go extinct rather than evolving to new forms. It is during times of plenty that nature can afford to experiment with a great many new varieties. Thus, the lush environments of the earth's tropical rainforests have produced by far the greatest diversity of life, and the harsh competitive environments of the far north and south have the fewest species. Evolution does not seem to be primarily about competition.

It should be emphasized that predation is not the same thing as competition, that is, "dog eat dog" is not at all the same thing as "dog eat deer" or "cat eat mouse." An animal does not compete with its prey, it only eats it. When prey become *scarce*, animals then compete with their fellows.

Mutation and genetic recombination occur at a fairly constant rate, regardless of environmental conditions. Species are thus changing all the time, and those that cooperate sufficiently survive. Competition is also at work, and inherited traits are passed on to future generations because they are advantageous to individuals in a particular environment. But it is often the case that individuals with similar traits simply hang out together and mate with one another, and they survive because of mutual support and association. Weak and vulnerable individuals nevertheless thrive as a species because of *social* evolution—indeed humankind is the best example of this.

It is because human beings have so thoroughly disrupted natural systems all over the world that we see only scarcity and competition everywhere we look. But this has not been the normal state of affairs on the planet. We are now in a time of great extinction, not evolution. In wilderness that human beings have not disturbed—and there is precious little left—the animals show no fear of humans and live mostly in peace with each other as well.

Charles Darwin himself commented on the extreme tameness of the wildlife in the Galapagos Islands:

> A gun is here almost superfluous, for with the muzzle I pushed a hawk off the branch of a tree. One day, whilst lying down, a mocking-thrush alighted on the edge of a pitcher, made of the shell of a tortoise, which I held in my hand, and began very quietly to sip the water. Formerly the birds appear to have been even tamer than at present. Cowley (in the year 1684) says that the "Turtle-doves were so tame, that they would often alight upon our hats and arms, so as that we could take them alive."[17]

Darwin remarked also on the "extraordinary tameness" of the birds of the Falkland Islands, Bourbon, and Tristan d'Acunha in the south Atlantic.

I myself, in remote areas of the United States, have had wild birds perch in the palm of my hand, and I have also fed oats from the palm of my hand to wild deer.

Darwin wrote about African monkeys of distinct species that lived in community together, and European rooks, jackdaws, and starlings that flew together in united flocks. He cited numerous examples throughout animalkind of members of the same or different species living with one another, warning one another of danger, and otherwise coming to one another's aid.[18]

Russian geographer Piotr Kropotkin observed cooperation in nature even under the harshest conditions:

> Two aspects of animal life impressed me most during my journeys which I made in my youth in Eastern Siberia and Northern Manchuria. One of them, was the extreme severity of the struggle for existence which most species of animals have to carry on against an inclement Nature; the enormous destruction of life which periodically results from natural agencies, and the consequent paucity of life over the vast territory which fell under my observation. And the other was, that even in those few spots where animal life teemed in abundance, I failed to find—although I was eagerly looking for it—that bitter struggle for the means of existence, among animals belonging to the same species, which was considered by most Darwinists (though not always by Darwin himself) as the dominant characteristic of the struggle for life, and the main factor of evolution. . . .
>
> Wherever I saw animal life in abundance, as, for instance on the lakes where scores of species and millions of individuals came together to rear their progeny; in the colonies of rodents; in the migrations of birds which took place at that time on a truly American scale along the Usuri; and especially in a migration of fallow-deer which I witnessed on the Amur, and during which scores of thousands of these intelligent animals came together from an immense territory, flying before the coming deep snow, in order to cross the Amur where it is narrowest—in all these scenes of animal life which

passed before my eyes, I saw mutual aid and mutual support carried on to an extent which made me suspect in it a feature of the greatest importance for the maintenance of life, the preservation of each species, and its further evolution.

And finally, I saw among the semi-wild cattle and horses in Transbaikalia, among the wild ruminants everywhere, the squirrels, and so on, that when animals have to struggle against scarcity of food . . . the whole of that portion of the species which is affected by the calamity, comes out of the ordeal so much impoverished in vigour and health, that no progressive evolution of the species can be based upon such periods of keen competition.[19]

In ant communities, cooperation is so much the rule that the individual ant is more like a cell in a larger organism—an organism that may grow its own food crops, or herd aphids like cattle, that builds large nests and long highways. Every ant is obligated to share its food, by regurgitation, with any hungry member of its own community who begs for it. Ants are relatively defenseless as individuals, but in their large numbers they can overcome much bigger insects—crickets, grasshoppers, spiders, beetles, butterflies, even wasps; and army ants, organized by the thousands, can kill most any animal of any size that isn't fast enough to escape. The popular conception is that different species of ants are constantly at war with one another, but in the tropics as many as forty-three species of ants have been found living in peace on a single tree. Separate species of ants may even live together in the same nest, share a communal ant garden, and defend one another from common enemies.

On my fire escape in Brooklyn, I have observed sparrows, house finches, mourning doves, and even an occasional squirrel eating birdseed together from a communal dish. The adult sparrows regurgitate food for any fledgling sparrow who begs for it.

A few blocks away I have been privileged to observe a flock of noisy green parrots that have built huge communal stick nests high in the lights above a college football field. I suppose they think they have found the tallest trees in the urban jungle. On sunny days I have seen groups of them hanging out with starlings and mourning doves on windowsills and fire escapes. All parrots are gregarious, and in the wild they usually die of old age—not because they are so strong as individuals but because they are so well organized to defend their group and to warn one another of approaching danger.

I have read that even little sparrows will gang up in numbers to attack and chase away a large hawk. Lapwings have been observed to protect not only their own species, but also other aquatic birds from the attacks of predators.

In the springtime, the shores of almost any lake or pond in North America, Europe, or Asia once teemed with thousands of migrating birds of dozens of species, all watching out for one another.

A century ago, pelicans used to gather in South America in flocks of forty to fifty thousand birds, some of which would keep watch while others slept, and still others were fishing. Howard Stansbury told of an old blind pelican in Utah which was fed, and fed well, by other pelicans with fishes that had to be brought to it a distance of over thirty miles.[20] Darwin mentioned also a case of crows that fed their blind companions.[21]

Cooperation and sociable living are evident, in all degrees and variations throughout the animal world. Many fish travel together in great schools, not always of just one species. Schools of yellowfin tuna swim with schools of dolphins. Grazing mammals congregate in enormous herds of thousands of animals. The great migrating herds of North American buffalo used to number in the millions; so did the caribou of the far north. Prairie dogs live in enormous underground villages, posting sentries above ground to keep watch; three hundred million prairie dogs once lived in one 25,000-square-mile area of Texas. Kropotkin observed that "when the Russians took possession of Siberia they found it so densely peopled with deer, antelopes, squirrels, and other sociable animals that the very conquest of Siberia was nothing but a hunting expedition which lasted for two hundred years."[22]

Carnivores are far fewer, but they too are mostly social animals, many of the cats being exceptions in preferring solitude. On today's impoverished earth, foxes, weasels, and bears live and hunt mostly alone, but when they were more numerous and less harassed by human beings, they too used to live and hunt in groups.

Most primates, including humans, are also intensely social beings. Some, such as gorillas and orangutans, live in small family groups. But most form large clans, and some, as Darwin observed, even join with species other than their own.

Even plants cooperate. Sugar maples in eastern forests, for example, draw water up from deep tap roots at night, releasing it from shallow roots to water the surrounding area and make it hospitable for thirsty trilliums and goldenrods.

> In the great struggle for life—for the greatest possible fullness and intensity of life with the least waste of energy—natural selection continually seeks out the ways precisely for avoiding competition as much as possible. The ants combine in nests and nations; they pile up their stores, they rear their cattle—and thus avoid competition; and natural selection picks out of the ants' family the species which know best how to avoid competition, with its unavoidably deleterious consequences. Most of our birds slowly move southwards as the winter comes, or gather in numberless societies and undertake long journeys—and thus avoid competition. Many rodents fall asleep when the time comes that competition should set in; while other rodents store food for the winter, and gather in large villages for obtaining the necessary protection when at work. The reindeer, when the lichens are dry in the interior of the continent, migrate towards the sea. Buffaloes cross an immense continent in order to find plenty of food. And the beavers, when they grow numerous on a river, divide into two parties, and go, the old ones down the river, and the young ones up the river—and avoid competition. And when animals can neither fall asleep, nor migrate, nor lay in stores, nor themselves grow their food like the ants, they do what the titmouse does . . . they resort to new kinds of food—and thus, again, avoid competition. (Piotr Kropotkin, in *Mutual Aid*)[23]

Competition is surely one of the forces of evolution. But a greater number of species and a greater number of individuals can thrive side by side in the same territory at the same time when there is also a large measure of cooperation.

II

I have heard a great many people talk about becoming more "evolved" human beings, as if evolution were a goal to be pursued, and as if it were a process we had control over. It is neither. Evolution has no goals. It is only a gradual change of scenery in the continuing drama of life, in which we are all actors but not directors. Nor can we ever know or control all of the effects of our actions.

This last statement is known in physics as the Uncertainty Principle, a law that has been misunderstood and misused by those who would "create their own reality." What the Uncertainty Principle actually says is that the observer, by his or her actions, changes reality *in an unpredictable way*. It is really a mathematical statement of the extent to which we do *not* create our own reality.

Ecological systems are normally in states of slowly changing equilibrium which results from the behavior of all concerned; when that behavior changes, the system is altered, and that alteration is called evolution. It is meaningless to speak of the evolution of a species without reference to its environment.

An elephant, being a vegetarian, eats a lot of vegetation, and when it is hungry enough, it even pulls down trees. That's the way elephants are, and have been for a very long time. They don't suddenly decide to be different. So their neighborhood adjusts to having elephants in it. Trees become less abundant, and grass grows up in the open spaces. The African savannah is a savannah, and not a forest, partially because elephants and other large grazing herbivores live in it. And elephants are elephants partially because they live in the savannah. The whole ecosystem, and everyone who lives in it, evolved together.

Evolution is a long-term affair, and the changes that happen are much more gradual and gentle than human-caused changes have recently been. Organisms change their form and behavior little by little. The first fish that took a breath of air and walked on land set in motion quite a chain of events, leading to the birth of all the amphibians, reptiles, birds, and mammals on the earth today, including you and me, and to the evolution of most of the plants that we live with, especially the flowering plants that make up most of the food supply for us mammals. But it took about two hundred million years from the time the first fish walked to the time its progeny first flew. The earth had plenty of time to adjust and evolve.

Human beings are changing our behavior so rapidly that nothing has time to adjust. In just a hundred years we have changed from being a species that moves more slowly than most other animals to one that moves many times faster than any other living thing on land, under sea, and in the air, and as predators our hunting methods have become invincible. Evolution is at an end, the earth cannot possibly cope.

There are those who believe that there is a technological fix for every problem; that we will survive and prosper in a human-controlled and human-regulated world. But human beings are fallible, and we will continue to fail at our efforts to regulate the whole world. We fail because our attempts at regulation are only crude imitations of the natural systems we study; we fail because natural systems are not regulated according to any plan at all, but are just what happens when life, in all its myriad forms, is left to itself. So if we attempt to control and plan our whole environment, we will succeed only in keeping it in a state of perpetual unbalance, which is always further out of balance than it used to be, requiring more and more energy to maintain by more and more effort on the part of more and more people.

In our arrogance, we often think that if only we know *enough* and plan carefully *enough*, we can control everything in the world—as if we were gods and goddesses. But we are only a small part of a larger organism, the earth, and we cannot control the whole earth any more than a finger can control a whole body. We can vastly simplify the earth, as we are doing, and we will have more of a chance of exerting our will over fewer parts, but if we continue on this path we will wake up one day and discover that the earth's heart and lungs no longer function, and that we cannot survive without them.

THE STATE OF THE EARTH

CHAPTER 9

We Are All Indians

Many look to the Indian nations and other "native peoples" of this world for wisdom and solutions to modern problems. They are looking in the right direction but rarely in the right places. Even native people, by and large, have not understood the real causes of their past success, and our present failure, at sustainable living with nature.

The answers we seek are both more obvious and more disturbing than the ones we would like to find. Native societies have truly lived more harmoniously with their environment, but they are not different from other societies in any important respect except one: their technology. And that one difference is vanishing.

If we look deeply, we must find that the very term "native" has its roots in racism, and in a power inequality: when the technology of warfare changed in the fifteenth century, they are the ones who did not have the guns. We mean nothing more than this when we refer to "native," "aboriginal," "indigenous," or "Indian" people. Their ancestors had no guns.

A powerful mythology has come into being among white people about the Indians they have conquered, and it is reflected in the environmental movement, nowhere more clearly than in the myth of Chief Seattle.

He was a leader of the Suquamish people who lived in what is now Washington State from 1786 to 1866. And in 1854 he is supposed to have delivered a speech in response to an offer from the United States to buy his people's land. One version of this speech that has circulated is such a clear and compelling ecological statement that it became a sort of manifesto for the environmental movement. It is also a bitter rebuke of the white race for polluting and plundering the Earth. "The end of living and the beginning of survival," so is the white man's world portrayed.

I hold similar feelings within my heart. When we have to wear shoes so as not to notice that the earth we walk on is all slimy with tar and grease; when we have to filter and chlorinate the waters in our rivers and lakes so they will support life and health, and we drink water not because it is delicious but because we are thirsty; when air that smelled so good it was a pleasure to breathe it is only a memory from the past—this to me is "the end of living and the beginning of survival."

But these are modern sentiments born of the present environmental conditions, and they belong not to any one race of people, but to us all. The speech in question was written not by Chief Seattle, but by Ted Perry for a Southern Baptist film on pollution that was shown on television in 1972. The speech became famous, however, not because Ted Perry wrote it, but because everyone thought an Indian Chief wrote it. Chief Seattle was not such a prophet. The speech he actually delivered in 1854, at a reception for Governor Isaac Stevens, the new commissioner of Indian affairs for Washington Territory, expresses mostly the love for homeland that every human begin must feel, and the regrets of a conquered, once-powerful people. It is an eloquent voice from a turbulent past, and although it contains none of the ecological sentiment of the television version, it deserves some attention.

It is important for us to hear the voices of the indigenous people, not because they have been superior in their feelings toward nature, but because they have been so ordinary, and they had no guns. We have forgotten that ordinary people can live that way—not only without guns, but without many other terrible things.

> Yonder sky that has wept tears of compassion upon our fathers for centuries untold, and which to us looks eternal, may change. Today it is fair, tomorrow it may be overcast with clouds . . .
>
> The Great—and I presume—good White Chief, sends us word that he wants to buy our lands but is willing to allow us to reserve enough to live on comfortably. This indeed appears generous, for

the Red Man no longer has rights that he need respect, and the offer may be wise, also, for we are no longer in need of a great country.

There was a time when our people covered the whole land as the waves of a wind-ruffled sea covers its shell-paved floor, but that time has long since passed away with the greatness of tribes now almost forgotten. I will not dwell on nor mourn our untimely decay, nor reproach my paleface brothers with hastening it, for we, too, may have been somewhat to blame.

Youth is impulsive. When our young men grow angry at some real or imagined wrong, and disfigure their faces with black paint, their hearts also are disfigured and turn black, and then they are often cruel and relentless and know no bounds, and our old men are unable to restrain them . . .

It is true that revenge by young braves is considered gain, even at the cost of their own lives, but old men who stay at home in times of war, and mothers who have sons to lose, know better . . .

Your God seems to us to be partial. He came to the white man. We never saw him, never heard his voice. He gave the white man laws, but had no word for his red children whose teeming millions once filled this vast continent as the stars fill the firmament . . .

Our dead never forget this beautiful world that gave them being. They still love its winding rivers, its great mountains and its sequestered vales, and they ever yearn in tenderest affection over the lonely-hearted living, and often return to visit, guide and comfort them . . .

A few more moons, a few more winters—and not one of all the mighty hosts that once filled this broad land and that now roam in fragmentary bands through these vast solitudes or lived in happy homes, protected by the Great Spirit, will remain to weep over the graves of a people once as powerful and as hopeful as your own!

But why should I repent? Why should I murmur at the fate of my people? Tribes are made up of individuals and are no better than they. Men come and go like the waves of the sea. A tear, a tamanamus, a dirge and they are gone from our longing eyes forever. It is the order of Nature . . .

> Every part of this country is sacred to my people. Every hillside, every valley, every plain and grove, has been hallowed by some fond memory or some sad experience of my tribe. Even the rocks, which seem to lie dumb as they swelter in the sun along the silent sea shore in solemn grandeur thrill with memories of past events connected with the lives of my people . . .[24]

When I was growing up, the teeming millions of the red race had become very few indeed, but the green land that was their home still dominated my world. The world as I experienced it was mostly nature, with a few islands of civilization here and there in the cities. When you left the city, it was on narrow, winding, often dirt roads through beautiful countryside, and it took a long time to get where you were going. The pathless woods were silent, the air fresh and delicious, and you could drink wonderful-tasting water from just about any lake or swift-running stream. Airplanes were uncommon, motorboats were a novelty, and chain saws were almost unknown.

Even in the city the air was clear and good to breathe. "Visibility 20 miles" was a regular part of the weather forecast in the 1950s, and when my aunt Rose took me up to the observation deck of the Empire State Building, I could see forever. On the streets where I lived in Brooklyn, the gutters and parts of the sidewalks were not paved, and a block away stood a farmhouse shaded by a centuries-old purple beech tree. In front of our apartment grew fruit trees and pussy willows. The summer breeze was enlivened by the chirps of crickets, the songs of birds, and many colors of butterflies that came to visit the neighborhood flowers.

I vividly recall the electric power blackout of 1965 in New York City. It was nighttime, the streetlights were dark, and so were all the houses and buildings. A bright full yellow moon illuminated the city, and as I walked through the streets, I thought I had never seen New York more beautiful. Nature was taking back its own, if only for a few hours. The arrogant, frenetic city was being humbled by the living earth surrounding it, and the moon smiled down on a suddenly dark and peaceful place.

In 1967 I left home to go to college, upstate in Ithaca, and one of the first things I did was join Cornell's Outing Club. The Club owned an old secluded farmhouse thirteen miles away in Caroline, and many were the nights I spent there, alone or with friends—sometimes with the house dog Schlep—sleeping under the pine trees on a soft pine needle bed. On moonless nights, a sky as black as the deepest mystery was pierced by the light of uncountable millions of stars.

Fall Lake George was a gathering of tribes. Every autumn all the Outing Clubs from all the colleges on the east coast of the United States and Canada used to gather for a weekend of camping, folk singing, square dancing, and general merriment on Turtle Island in the middle of Lake George. We would all arrive on the lake shore on Friday evening and paddle ourselves and our supplies to the island by canoe. In 1967 the lake was our drinking water, as it had always been the drinking water for a thousand generations of human beings before us. The next fall, in 1968, we were told for the first time to boil the lake water before we drank it. In 1970 we were warned not to drink Lake George water at all. We had to bring city water to the wilderness!

I watched a lot of changes happen in a few short years. Fall Creek and Cascadilla Creek, which run through the Cornell campus at the bottom of deep rugged gorges, were clean whitewater rivers that were delightful to swim in when I was a freshman, and they became yellow and full of detergent foam by the time I was a senior.

I used to go down into the gorge and lie in the sun for hours while I studied, but by the time I was a senior, clear blue skies were rare over Ithaca as air traffic increased spectacularly. I would stretch out on a rock early in the morning while the sky was still blue and watch jet after jet leave vapor trails behind them, blending one into the other to form a think chaotic ugly cloud cover by early afternoon.

Cross-country skiing was my favorite winter activity in the white snows around Caroline, but after about 1970 it was not so peaceful anymore—you had to watch out for snowmobiles that not only wrecked the silence of the woods but might run you over!

They straightened Route 17, the pretty, meandering road we used to travel to get to Ithaca, and made a sterile, boring freeway out of it, wide, fast, and deadly.

I couldn't stand it! The earth was being ruined right before my eyes, and nobody else even seemed to notice the changes. Or if they did, it didn't seem to bother them. I could not understand.

CHAPTER 10

Air

The earth has been breathing for at least three billion years, but these days it is breathing extra deeply, and it is running a fever. The news reports about air pollution, the greenhouse effect, and the ozone hole have not, I think, managed to impress most people with just how sick the earth really is. Air is invisible and, after all, we are still breathing and birds are still flying. How bad could it really be?

I will begin with the most obvious, and least noticed aspect of the problem: we are burning up our air supply.

The atmosphere is a gift of life, and it is also a gift *from* life, courtesy of photosynthesizing plants. Wherever plants are born and growing, there oxygen is being created. It is consumed again in the breath of the animals who eat them, and by the bacteria that decay them, and in the flames of forest fires. The coming together and taking apart of carbon and oxygen—this is part of the endless dance of life and death.

If all life on earth died and completed its cycle, there would be no more oxygen left in the air. And so the amount of oxygen now in the atmosphere corresponds to the total amount of carbon that has ever lived and breathed on the earth that has not yet completed its cycle. This includes everything that is now alive plus everything once living that has escaped the cycle of birth and decay, whose carbon has become fossilized in the soil and rocks underground.

In other words, the oxygen now in our atmosphere was put there by the same biological processes that buried peat, coal, oil, natural gas, and all the other sedimentary carbon, beneath the ground and under the sea, all over the earth, slowly, for hundreds of millions of years. If we human beings succeed in burning up all the "fossil fuel" on this planet, we will have no oxygen left to breathe, and there is absolutely no possibility of replenishing it.

Few scientists are taking this seriously. Their estimates vary, but the consensus in 1990 was that the eventual burning of "all recoverable reserves" of fossil fuels would use up between one and two percent of our oxygen supply, which apparently was considered an acceptable sacrifice to maintain our standard of living. Three decades later the number is closer to three percent.

The key word is "recoverable." In 1990, oil was being extracted from sand and shale in several parts of the world with fossil carbon contents as low as five percent, which was then considered the lower limit of what was economically recoverable. Today, that limit has been lowered tenfold; oil is being extracted from sand and shale in a great many more parts of the world with carbon contents as low as 0.5 percent. As human beings burn up more of the earth's peat, coal, oil, and natural gas, they are developing more and more efficient techniques for extracting carbon out of rocks, and the definition of "recoverable reserves" is continuing to expand. If *all* of the sedimentary carbon buried underground and beneath the ocean floor were burned as fuel, it would use up all of the oxygen in our atmosphere fifteen to thirty times over.[25]

How fast are we in fact burning fossilized hydrocarbons on the earth today?

Annual world petroleum consumption is currently 37 billion barrels, coal consumption is about 8 billion metric tons, and natural gas consumption is 4.2 trillion cubic meters. This amounts to burning 13 billion metric tons of fossilized carbon every year, which uses up 49 billion metric tons of oxygen (up from 24 billion tons in 1990). At this rate, it would take us approximately twenty-five years to deplete our supply of oxygen by 0.1 percent, and twenty-five thousand years to burn our oxygen all up. Actual monitoring of oxygen levels at research stations around the world confirms that these figures are not far off. Numerous stations in the tropics, temperate latitudes, and polar regions—from Barrow, Alaska to American Samoa to the South Pole—have all measured a drop in atmospheric oxygen since 2020 of between 4.8 and 6.4 parts per million per year. The average decrease is 5.4 parts per million per year, a rate which would deplete 0.1 percent of our oxygen in thirty-eight years.

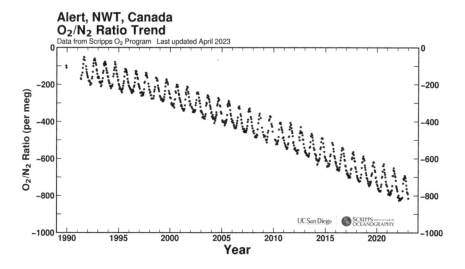

Part of the difference between my calculations and their measurements may be due to the fact that a warming ocean (see next chapter) cannot hold as much oxygen as it used to, and it is releasing some of it into the atmosphere. And some petroleum is not burned right away, but is made into asphalt, plastics, and thousands of other industrial and consumer products. Regardless, both ocean life and land-based life are gradually being deprived of oxygen. Researchers in London have even predicted that as the number of oxygen-producing plants on earth decreases and the number of oxygen-consuming animals (including humans) increases, oxygen will be depleted at an exponentially steeper rate. If current trends do not change, they project a 50 percent decline in atmospheric oxygen in just 3,600 years, which would be incompatible with life.[26]

Our oxygen supply has already diminished some during the past few centuries. Since 1750, the use of coal, petroleum and natural gas has produced 1.737 trillion metric tons of carbon dioxide which has consumed about 1.77 trillion metric tons of oxygen, or 0.15 percent of our oxygen supply. This has affected the quality of life throughout land and oceans. An additional 0.15 percent depletion in just one-fifth the time, as is happening today, will impair our vitality further and more quickly, and even a healthy earth would require 7,500 years to regenerate what we burn up in twenty-five. "A breath of fresh air" will not be quite the same for a very long time to come, unless we stop—and stop soon—burning fossil fuels. We must cease digging this stuff up, because it is down there for a very good reason.

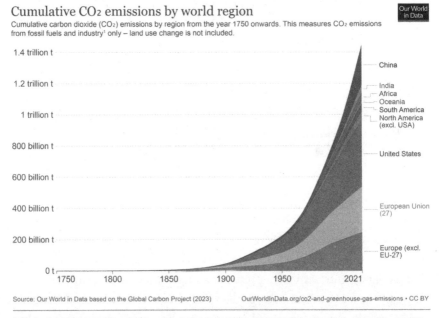

1. **Fossil emissions**: Fossil emissions measure the quantity of carbon dioxide (CO_2) emitted from the burning of fossil fuels, and directly from industrial processes such as cement and steel production. Fossil CO_2 includes emissions from coal, oil, gas, flaring, cement, steel, and other industrial processes. Fossil emissions do not include land use change, deforestation, soils, or vegetation.

In a Greenhouse

There is another reason to leave coal and oil in the ground, a reason that scientists and the media are much more excited about: the burning of these substances is raising the surface temperature of the earth.

Photosynthesis, it will be remembered, is the process by which green plants, using the energy in sunlight, convert carbon dioxide and water into carbohydrates and oxygen. Respiration is the opposite: oxygen is used up to burn carbohydrates and regenerate carbon dioxide.

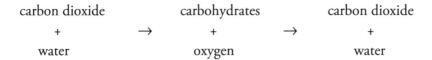

In nature these two processes are in balance. The amounts of oxygen and carbon dioxide in the atmosphere remain almost constant for enormous periods of time; relatively small changes have always had dramatic effects on global life.

Carbon dioxide is called a "greenhouse gas" because like glass on a greenhouse it lets the light from the sun pass through to the earth's surface but prevents heat from escaping back out into space. This was pointed out as a property of the atmosphere by Joseph Fourier in 1824, and as a specific property of carbon dioxide by John Tyndall in 1862. In small quantities, therefore, carbon dioxide in our air helps make the earth a comfortable, livable place for its residents. For most of human existence, carbon dioxide has occupied less than 0.03 percent of the atmosphere, an amount that sounds insignificant. But without that tiny amount of carbon dioxide in the air, the earth's surface would average about zero degrees Fahrenheit instead of the 59 degrees of today and the entire earth's surface would be frozen year-round instead of the paradise we have taken for granted.

Water vapor is also a greenhouse gas. So are methane, nitrous oxide, and chlorofluorocarbons (CFCs). But it is the rapid buildup of carbon dioxide in the past century and a half which is most responsible for global warming.

The large-scale burning of coal, oil, and natural gas, as we have seen, is depleting the earth's enormous oxygen supply slowly by a tiny amount every year. It is also increasing the earth's carbon dioxide supply by the same tiny amount, but since the amount of carbon dioxide in the air is very small to begin with, the consequences are rapid and catastrophic. For all of our talk about the greenhouse effect, we as a society are engaged in very massive denial. Because of our inability anymore to imagine life without cars, airplanes, electricity, and all the rest of industrial technology, most of us cannot see the magnitude of the disaster befalling us.

The evidence is clear. By examining the layers of ice deep within Antarctica, scientists have determined that for the past eight hundred thousand years, the carbon dioxide level in our atmosphere has fluctuated between 190 and 295 parts per million. Lower carbon dioxide levels always occurred during cold ice ages, and higher levels during periods of warmth. Since 1850 carbon dioxide has increased so dramatically that it is now up to about 420 parts per million and rising, much higher than it has been on the earth in eight hundred thousand years for sure, and most likely much much longer. On a healthy earth the oceans can absorb any excess, but we are pumping out carbon dioxide so fast that saturated oceans cannot keep up. Earth's forests and marine algae also inhale carbon dioxide and exhale oxygen for us, but we are presently burning carbon at least three hundred times as fast as natural cycles can keep up with. And we are simultaneously cutting down the life-giving forests as fast as we possibly can. The balance between photosynthesis and respiration has been completely overwhelmed.

1970s earth was about 1°F warmer on average than 1870s earth. Temperatures have gone up about another degree and a half since then.

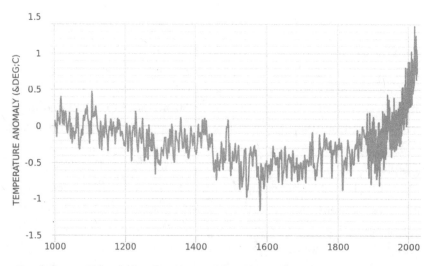

Graph from 2 Degrees Institute, www.2degreesinstitute.org.

We would all be getting warmer a lot faster were it not for the destruction of tropical forests, the near-daily rocket launches, and the moderating effect of the oceans. Two to three billion tons of plant matter annually is still going up in smoke; it travels around the world creating global shade. So does the black soot from all the rockets that are launching satellites into space. These particulates remain in the upper atmosphere for years, partially canceling the greenhouse effect—but only partially.

The consequences of increasing carbon dioxide, and increasing temperatures, are all around us. Melting ice is flowing swiftly from the Greenland ice cap into the northern ocean, and from the Antarctic ice cap into the Southern Ocean. The melting has created lakes of water beneath the ice, forming a system of waterways eventually draining into the sea. The Arctic tundra has warmed as much as 12°F since 1900. Glaciers are vanishing from the Himalayas, the Alps, and elsewhere. Glacier National Park in Montana had 150 massive glaciers in 1850; just 25 remain today, and those are all expected to disappear by 2030. All this melting is causing the oceans to rise by more than half an inch per year. Sea levels have risen about 10 inches since 1880, and all of the world's coastlines, coastal cities, and low-lying islands are slowly being drowned.

We have become so dependent on the burning of coal, oil, and natural gas that one might be tempted to throw up one's hands and prepare for the inevitable new conditions. It would be easier, one might think, to get ready for warmer weather and move our cities further inland as it becomes necessary, than to stop burning fossilized carbon. Unfortunately, the greenhouse effect is not just a matter of a rise in temperature and some coastal flooding. Earth is a living being, and its entire structure and functioning are being affected. Earth's trees, for example, breathe in carbon dioxide to make carbohydrates; there is so much carbon dioxide in the air now that tree leaves the world over have on average 40 percent fewer breathing pores than in pre-industrial times. Many plants and wild animals that must live within certain ranges of temperature and moisture have already had to move northwards, displacing other species. Many are not surviving the transition; unlike us, they cannot build walls against their changing environment and install air conditioners and humidifiers.

The greenhouse effect warms the lower atmosphere more than the upper, and it warms the oceans more than the land. Earth is experiencing larger temperature differences and therefore an increase in the forces that drive earth's weather systems. Westerly winds have expanded and are looping far to the north and south. Hurricanes and typhoons have become more powerful. It is no longer unusual for a storm to pack winds of 180 miles per hour. Tornadoes have become more frequent and are appearing in places they rarely visited before, including my home state of New York. New York averaged one tornado per year when I was a child. Now it averages ten per year, and some years has had as many as twenty-six.

The heating of the planet causes more water to evaporate from the surface of the tropical oceans, contributing to increased cloud cover the world over and greater rainfalls and snowfalls during storms.

Earth's weather is therefore not just getting warmer, it is getting more extreme. Intense heat waves, cold waves, droughts, floods, and fierce storms have become commonplace, and forest fires have become hotter and more destructive. The greenhouse effect is only part of a story that includes global deforestation, the dying off of plankton on the world ocean, and the extinction of living species all over the earth. Earth's living systems buffer our climate, make our weather gentle, and in general make the planet a comfortable place to live. With the great diminution of life that is taking place, impoverished ecosystems are less able to absorb heat and moisture; dwindling forests no longer serve as barriers to the winds; soils are less able to absorb rainfall and hold moisture; vegetation is less fire resistant.

The world has known famines, floods, and fires throughout human history, but rarely has the weather been as erratic as it has become in recent years.

We might well think the extreme weather is confined to where we are, because the extreme weather elsewhere goes unreported in the local news. But it is extreme all over.

GLOBAL WEATHER REPORT, 2020–2023

2020

A six-month heat wave set in, in the middle of winter in January 2020, in one of the coldest inhabited places on earth: northern Siberia. On June 20 it reached 38 degrees Celsius (100.4 degrees Fahrenheit) in Verkhoyansk, the highest temperature ever recorded north of the Arctic Circle.

The winter heat was not confined to Siberia. Moscow's average temperature for December, January and February 2019–2020 was above freezing for the first time in 200 years of record-keeping. In Helsinki, no snow fell in January or February for the first time on record. The whole of France had its warmest winter on record. Temperatures in Sweden were so high ski resorts could not even make artificial snow. Cherry trees bloomed in January in New York City's Central Park.

Records were likewise broken in the southern hemisphere, where it was summer. On January 4, the temperature reached 49 degrees Celsius in Sydney, Australia, breaking the previous record set in 2018, and a record 43.6 degrees in Canberra. On February 6, 18.4 degrees Celsius, the highest temperature ever recorded on the continent of Antarctica was registered at Argentina's Esperanza research station, and on February 9, an even higher temperature, 20.75 degrees, was registered on Seymour Island, just off the Antarctic coast.

On January 1, 2020, 377 millimeters (14.8 inches) of rain fell on Jakarta, Indonesia, the highest daily rainfall ever recorded in that city. It caused rivers to spill over and flooded more than 30,000 people out of their homes. During January, 100 cities in three states of the U.S. were under a state of emergency due to a month of extreme rainfall, flooding, and landslides. On March 2, 320 millimeters (more than a foot) of rain fell in a single day—a record—in the Baixada Santista metropolitan region of São Paulo state, Brazil.

During June and July, the middle and lower reaches of China's Yangzi River received up to 1.72 *meters* of rain, causing severe flooding. From early July to mid-August, South Korea had the largest number of heavy rain events on record, causing deadly floods and landslides. In just six days, from July 3–8, parts of Kyushu District in Japan had more than one meter of rain.

The Atlantic hurricane season of 2020, lasting from mid-May through November, had 30 named hurricanes and tropical storms, a record. Super Typhoon Goni struck the Philippines on November 1, 2020, with sustained winds of 195 miles per hour, the strongest tropical cyclone to make landfall in the recorded history of the world.

2021

In February 2021, Europe experienced record cold followed by record heat. In the week beginning February 8, temperatures fell below -20°C at many locations in Germany. In Göttingen the temperature swung from -23°C to +18.1°C in less than a week. On February 22, it hit 21.1°C in Hamburg, the highest ever February temperature in that country.

Sweden saw its highest ever February temperature on the 25th day of that month at 16.8°C in the city of Kalmar. So did Poland, with a temperature of 21.7°C in Maków Podhalanski, and Slovakia, with a temperature of 20.8°C in Hurbanovo. Records also fell in Austria and the Czech Republic.

In June 2021, a record heat wave seared what had always been a mild, cloudy, rainy climate in the heavily forested Pacific Northwest of the United States and Canada. It reached 116 degrees Fahrenheit in Portland, Oregon; 108 degrees in Seattle, Washington; and an incredible 121.3 degrees (49.6 degrees Celsius) in Lytton, British Columbia. More than a thousand people died of the heat.

In July 2021, one-third of Bangladesh was underwater after prolonged monsoon rains. 1.1 million people were flooded out of their homes. While on July 20, 650 millimeters (25.3 inches) of rain fell on Zhengzhou, China in twenty-four hours, with up to 200 millimeters (8 inches) of rain falling per hour during the storm's peak, flooding the world's biggest manufacturing center for iPhones.

On October 4, 2021, almost three feet of rain fell on parts of northern Italy in twenty-four hours. Up to 29.2 inches (740.6 mm) fell on Genoa province in twelve hours, the greatest twelve-hour rainfall ever recorded in Europe.

2022

Eastern Australia was flooded by record rainfall from February to April 2022. Greater Brisbane received 677 millimeters of rain (26.7 inches) in three days, from February 28 to Mar 2. Brisbane, Sydney, and many other cities suffered major flooding. 415 millimeters of rain (16.3 inches) fell in ten hours in the metropolitan area of Petrópolis, Brazil on March 20, flooding many neighborhoods. More than 300 millimeters of rain (12 inches) fell in a twenty-four-hour period from April 11–12 on parts of KwaZulu-Natal province in South Africa, causing flooding that destroyed roads, bridges, and homes.

In the summer of 2022, record rainfall left one-third of Pakistan underwater for months. While in China's Guangdong province, two feet of rain flooded 177,600 people out of their homes.

The United States had 233 tornadoes in the month of March 2022, triple the average of years past. On July 25–26, 7.68 inches of rain fell in just six hours in St. Louis, Missouri, and on July 27–78, six to nine inches of rain fell in few hours in eastern Kentucky. In August, parts of Texas received up to fifteen inches of rain in one day. In September, Hurricane Fiona dropped over twenty-seven inches of rain on Puerto Rico in one day. Also in September it reached 116 degrees in Sacramento; 117 degrees in Ukiah, California; 110 degrees in Redwood City; and 107 degrees in Salt Lake City, Utah. While Death Valley, the hottest place on earth and the driest place in North America, received 1.7 inches of rain, the most it has ever received in a single day and close to what it normally receives in a year. In October and November, it stopped raining altogether in much of the U.S., and the Mississippi River dropped to the lowest levels ever recorded and became unable to carry normal ship and barge traffic. In November, various cities in New York State received between five and seven feet of snow in a single snowstorm. In December, Buffalo, New York received two feet of snow in a single day.

By the end of 2022, the Upper Nile region of Africa had experienced four consecutive years of record rainfall, leaving up to two-thirds of South Sudan underwater.

2023

On February 19, 2023, in the midst of Carnival season, up to two feet of rain fell in a single day in parts of São Paulo state, Brazil, a record for the entire country.

From February 6 to March 15, 2023, Cyclone Freddy, the longest-duration cyclone in recorded history, displaced half a million people in Malawi, Mozambique, Madagascar, and Zimbabwe.

In the middle of April 2023, a heat wave swept over China, India, Thailand, Laos, Bangladesh, and other countries across the Asian continent, with temperatures up to 45 degrees Celsius, breaking temperature records in over a dozen countries.

In May 2023, Typhoon Mawar, the most powerful northern hemisphere cyclone ever recorded in the month of May, flooded Guam with more than two feet of rain.

In February 2023, in Chile, after a decade of extreme drought, record temperatures topping 40 degrees Celsius sparked one of the deadliest forest fires in the country's history, consuming 270,000 hectares (667,000 acres). While, three months later, fires began to consume both eastern and western Canada on a previously unknown scale. By the end of June, almost 18 million acres of Canadian forests and prairies had burned. June 2023 was the warmest June ever recorded in the 174-year global climate record. The average temperature in the world on July 4, and again on July 5, 2023, at 62.9°F (17.2°C), was the warmest ever recorded. The June-July-August season for 2023 was the warmest on record globally by a large margin. On July 15 in Bucks County, Pennsylvania, seven inches of rain fell in forty-five minutes. On July 16 in Sanbao township in China on July 16 it was 52.2°C (126°F), the highest temperature ever recorded in China—after a winter in which the city of Mohe recorded -53°C (-63.4°F), the coldest temperature ever recorded in China. On July 21-22, 2023, more than ten inches of rain fell on parts of Nova Scotia in just 24 hours. On July 28, 2023, the average surface temperature of the North Atlantic Ocean reached a new record high of 24.9°C (76.8°F). The average global sea surface temperature exceeded 21°C (69.8°F), the highest ever recorded, in early April and throughout the month of August. On October 25, Hurricane Otis, with sustained winds of 165 miles per hour, hit Acapulco, Mexico. It was the most powerful hurricane ever to make landfall in the western hemisphere. On December 16–17, more than nine inches of rain fell on coastal South Carolina in one day. On December 21, parts of Southern California received more rain in one hour than they normally receive in one month. Parts of Victoria, Queensland, and New South Wales, Australia recorded their wettest Christmas ever. 2023 globally was the warmest year ever recorded.

That is only a report from the one-third of the earth that is covered by land. The two-thirds that are covered by water are also heating up. Average sea surface temperatures of the world have risen about 1.5°C since 1950.

The above weather report is meant to give the reader a gestalt, a general impression of the world's weather that is impossible to get from reading a newspaper or watching TV. The impression I get from looking at the weather reports from around the world since the 1980s, is that something very terrible is happening to my earth. But I don't think it is so very difficult to find out why.

It is sometimes difficult to separate out the greenhouse effect from all the other causes of ecological chaos: deforestation, soil erosion, air and water pollution, and the general impoverishment of earth's protective blanket of life. What is very clear, however, is that the temperature *is* rising, and that not only is this in itself a catastrophe, but it is also making all of the other problems much worse, and the prospects for ecological recovery much poorer. Rather than give up in hopelessness, it is necessary to focus on the problems at hand, one at a time. They are not as complicated as they might appear.

Let us begin with the subject of this chapter. The only way to stop global warming—and to stop burning up our oxygen supply—and to give the earth an opportunity to recover, will be to stop burning our precious fossil fuels, and leave the rest of them in the ground where they belong. There is no longer time to wait for alternatives, indeed it is not reasonable to expect people to look for alternatives when coal and oil work perfectly well! What is needed first is the *decision* to stop burning these substances; when we no longer consider coal and oil options, the alternatives will then be forced upon us.

No one is going to force us to make the decision. It must be made by each individual, against societal pressure to do otherwise, until the idea begins to catch on. The question then will confront us, like it or not: If I can't put gasoline in my automobile, what am I going to feed it with? What will heat my house, if not oil or gas? Where will my electricity come from?

The answers are obvious: I must either learn to do without these conveniences, as fifty thousand generations of my ancestors did, or I must substitute renewable sources of energy. In some cases I will have no options but to do without, because alternatives are not available. And in some cases I will do so cheerfully, because I will discover, for instance, that life without an automobile is so much more peaceful and less expensive.

We must keep in mind that even renewable sources of energy are not necessarily a good answer, if they are used on too large a scale. For example, factories on both sides of the North Atlantic are discharging enough heat into the Gulf Stream to hasten the melting of polar ice. Even solar or wind energy would do this; there are simply too many factories.

And they are not a good answer, even though they protect the atmosphere, if they do irreversible harm to our planet in other ways. Running all cars on alcohol made from corn, for example, would require stealing even more of the world's prairies and forests from the birds, insects, and animals we are trying to save. Offshore wind farms are beginning to devastate seabird populations the world over. Windmills themselves, whether offshore or on land, generate low-frequency sound, called infrasound, as well as widespread earth currents of low-frequency electricity—both of which devastate birds and wildlife and severely damage human health.

Air 93

Photo by Travis Heying, *The Wichita Eagle*.

Solar farms shade the earth beneath them so plants cannot grow and wildlife cannot exist. The manufacture of solar cells requires hundreds of poisonous chemicals, metals and gases and is massively and permanently polluting the groundwater wherever they are made.

Some Chemicals Used in Manufacturing Solar Cells

1,1,1-trichloroethane	Copper
Acetone	Copper indium gallium diselenide
Aluminum	Copper sulfide
Ammonia	Cuprous chloride
Ammonium chloride	Diborane
Ammonium fluoride	Diethyl silane
Ammonium fluoroborate	Dimethyl zinc
Arsenic	Ethyl acetate
Arsine	Ethyl vinyl acetate
Boron trichloride	Gallium
Cadmium	Germanium
Cadmium chloride	Germanium tetrafluoride
Cadmium sulfide	Gold
Chlorosilanes	Hydrochloric acid
Chromate	Hydrofluoric acid

Hydrogen fluoride	Silicon nitride
Hydrogen selenide	Silicon tetrachloride
Hydrogen sulfide	Silicon tetrafluoride
Indium	Silicon trioxide
Ion amine catalyst	Silver
Isopropanol	Sodium hydroxide
Isopropyl alcohol	Stannic chloride
Lead	Sulfur
Methane	Sulfur hexafluoride
Methanol	Sulfuric acid
Molybdenum	Tantalum pentoxide
Molybdenum hexafluoride	Tellurium
Nickel	Tertiarybutyl arsine
Nitric acid	Tertiarybutyl phosphine
Nitrogen trifluoride	Thiourea
Phosphine	Tin
Phosphoric acid	Titanium
Phosphorus	Titanium dioxide
Phosphorus oxychloride	Trichloroethylene
Phosphorus trichloride	Trichlorosilane
Polyethylene terephthalate	Triethyl gallium
Polyvinyl butyral	Trimethyl aluminum
Selenium	Trimethyl gallium
Silane	Tungsten hexafluoride
Silicon	Zinc
Silicon dioxide	Zinc fluoroborate
Silicon monoxide	

Large amounts of fossil fuels are also used in their manufacture, and some of the chemicals released are greenhouse gases themselves. Sulfur hexafluoride, for example, is 22,800 times more potent a greenhouse gas than carbon dioxide, and nitrogen trifluoride is 17,000 times more potent. All of this defeats the purpose for which photovoltaics are being advocated.

And the electrification of all of society—our cars, homes, factories, and everything else—is simply trading one form of pollution for another. It is enveloping the whole earth in more and more layers of electrosmog, a potent pollutant in itself. It is electrocuting all life on the planet even more rapidly than the heat is cooking it.

We should also keep in mind, someone is sure to say, that economics, and the necessity of feeding eight billion people, now, and ten billion or more in the near future, are important elements of the problem. We are constantly reminded that our environmental problems are complex and many-sided.

But they are also simple. Economics is fundamentally a fiction, a mathematical construct of human activities that are built on the products of nature. And the driving force behind both economics and population growth, as I will show in later chapters, is the technology we have chosen.

Our cars are among the biggest consumers of fossil fuels in today's world, and they are certainly the biggest consumers that are under personal control. If, then, we wish our children to survive—and we wish them not only to survive, but to live full, healthy lives in a beautiful world—we must either learn once more to do without cars or run them on renewable fuels. With regard only to the greenhouse effect, replacing gasoline with either alcohol or solar or wind electricity will solve this portion of the problem. None of these alternatives, however, will do less damage to the earth than burning coal and oil. Renewable does not mean sustainable. Sustainable alternatives are not yet available, and we have not got the time to wait.

Under a Blazing Sun

Sunbathing, in my youth, was one of life's most pleasurable activities. Each spring I, like millions of young men and women, gradually increased my daily time in the sun, until with a dark tan I could enjoy whole days outdoors without burning. The sun is still an important source of nourishment for each one of us. Our skin soaks up its ultraviolet rays, needed to maintain healthy bones and teeth. Our blood is bathed in its white light, red arteries and blue veins absorbing the life-giving colors we need. Our eyes are light-portals to our brain, which regulates the rhythms of our daily lives by the coming and going of the sun.

Nowadays, however, most of us nourish ourselves poorly with artificial light. We get our bone-maintaining vitamin from laboratories instead of from the sun. And the fires of the sun have become menacing, burning, no longer comfortable in early spring, no longer tolerable in mid-summer. Comforter has turned enemy, the source of all life has become a danger to all life.

Up in the stratosphere, beginning about eight miles above our heads, is a layer of air rich in ozone that shields us animals and plants from a lethal overdose of the sun's UV-B, or short wavelength ultraviolet rays.

About ninety years ago, human beings began to manufacture large amounts of some gases and liquids never before seen in nature called chlorofluorocarbons (CFCs). They are nearly indestructible and were extremely useful technologically, particularly as refrigerants, as propellants in aerosol cans, as fire extinguishing agents, and in the making of plastic foam. As solvents, they were used to clean computer chips, to dry clean clothes, and to clean metals and plastics in many industries. Unfortunately all the CFC solvents, all the CFCs in our refrigerators and air conditioners, all the CFCs in all the aerosol cans and fire extinguishers, and all the CFCs in the air cells of all that plastic foam, eventually all of it leaked or evaporated into the atmosphere where it mixed with the rest of the air, and when it reached the stratosphere in six to eight years, it destroyed ozone. And since they stay in the air for decades to centuries, a large amount of the CFCs ever produced are still up there.

It appears that CFCs destroy ozone by carrying chlorine atoms into the stratosphere. The chlorine released by the CFCs reacts with ozone to form chlorine monoxide which, in turn, reacts with atomic oxygen to regenerate chlorine. The process is catalytic—the chlorine is not used up in the reactions, and so every chlorine atom that reaches the stratosphere will destroy as many as a hundred thousand molecules of ozone before it leaves.

During the 1970s, humanity woke up to at least part of the ozone layer problem, and in 1987 the world's governments gathered in Montreal, Canada, and signed a treaty phasing out CFCs for most purposes. During the next three decades the treaty was amended until all of the world's governments finally agreed to end the production and use of not only CFCs but other chlorine- and bromine-containing chemicals including hydrochlorofluorocarbons (HCFCs), hydrobromofluorocarbons (HBFCs), halons, methyl bromide, bromochloromethane, carbon tetrachloride, and methyl chloroform. Most production was ended by 2010. This has been celebrated by environmentalists and has been considered such a huge victory that few scientists are paying attention to ozone depletion anymore, and the media no longer reports on it at all. But ozone is actually being depleted more than ever. Part of the reason is that chlorinated and brominated compounds that remain in the atmosphere for a hundred years cannot be called back and are still doing damage. And older refrigerators, air conditioners and plastic foam are still slowly outgassing these chemicals and will continue to do so for a while. But CFCs and related compounds are only part of the picture.

Nitric oxide is another gas that destroys ozone catalytically. It exists in nature in tiny concentrations; as new ozone is continuously made by the action of solar radiation on stratospheric oxygen, so it is continuously destroyed by nitric oxide and other naturally occurring gases. But this natural balance has been overwhelmed by human-made chemicals. Including nitrogen fertilizers.

Nitrous oxide is released from agricultural fields wherever nitrogen fertilizer is used. And like some of the chlorinated compounds, nitrous oxide persists in the atmosphere for about 150 years. In the stratosphere it reacts with ozone to create nitric oxide, which catalytically destroys the ozone. The 200 million metric tons of fertilizer used by the world every year have therefore become a large contributor to ozone depletion. But there are others.

From 1976 to 2003, British Airways and Air France operated a supersonic commercial airplane called Concorde that injected large amounts of nitric oxide *directly* into the stratosphere. This was unavoidable since nitric oxide is created whenever air is heated to more than 2500 degrees Fahrenheit, including in the engines of all automobiles and airplanes. Only 14 Concorde aircraft were ever in service, and the stratosphere has been spared this assault for the past two decades. But that is about to change in a big way. American Airlines, United Airlines, Virgin Atlantic, Japan Airlines and other airlines have already placed orders with a company called Boom for delivery of—so far—130 new supersonic jets. Test flights are planned for 2026, with commercial service to begin in 2029. Numerous other companies are getting into the act. Spike Aerospace is taking orders for supersonic business jets. Hermeus is partnering with Raytheon and the U.S. Air

Force. Exosonic is designing supersonic military transports. NASA has built the X-59, a single-engine supersonic research plane that is supposed to begin flying in 2024. Lockheed Martin is designing a forty-seat supersonic airplane. A Russian supersonic passenger aircraft was due to begin construction in 2023. Chinese aerospace firm Space Transportation is developing a supersonic business jet that will fly from Beijing to New York in an hour. Japanese companies IHI, Mitsubishi, Kawasaki Heavy Industries, Subaru, and Japan Aircraft Development Corporation have joined together as Japan Supersonic Research to build a large fleet of supersonic passenger jets. They project a demand for 1,000 to 2,000 supersonic aircraft to serve corporate executives, high-level government officials, and wealthy travelers.

And then there is the satellite industry. Rocket exhaust, too, depletes ozone, and this is an industry that is *completely* unregulated.

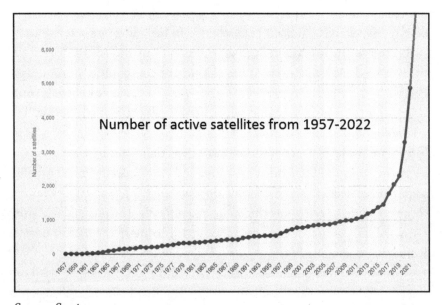

Source: *Statista*

There are already about 10,000 operational satellites in orbit around the earth, with plans by governments and corporations worldwide for 1,000,000 more,[27] and rockets of all kinds, using all types of fuel, are being launched somewhere on earth about every other day. All rockets, like supersonic jets, put large amounts of nitrogen oxides directly into the stratosphere, which by itself destroys ozone. Solid rocket fuel, in addition, puts both chlorine and aluminum into the stratosphere. Kerosone and hypergolic fuels put black soot into the stratosphere. All types of rocket fuels put water vapor into the stratosphere. Every one of these pollutants destroys ozone.[28]

So while the environmental community is celebrating the phase-out of CFCs and pretending the ozone problem has been solved, in real life it is getting worse.

One reason most people are content to ignore this problem is they think there is a "hole" in the ozone layer that is confined to Antarctica, where no one lives, and that the protective layer of ozone is intact elsewhere.

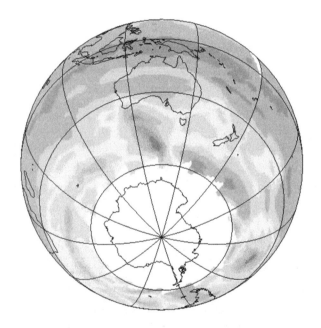

And in fact Antarctica is where the most dramatic declines in stratospheric ozone—declines of up to about 60 percent—have been observed. But ozone levels today are by convention compared to levels in 1979, which is when the world was just waking up to the problem. The level of stratospheric ozone over the Antarctic as observed by NASA's TOMS (Total Ozone Mapping Spectrometer) satellite in 1979 has been decreed to be "normal" even though ozone levels had obviously been declining steeply for many years previously.

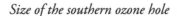

Size of the southern ozone hole

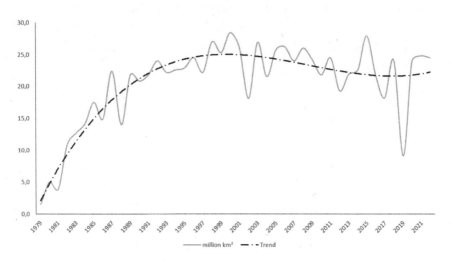

If the annual southern ozone hole were calculated based on 1960 ozone levels instead of 1979, it would be closer to reality. In fact, southern hemisphere ozone levels have been decreasing, annually, for longer and longer periods, not just since 1979 but since at least 1960; not just during the spring but to a smaller extent during the rest of the year; and over not just Antarctica but New Zealand, Australia, and portions of South America as well. And for purposes of yearly comparisons its size is always calculated between September 7 and October 16. But this too is deceptive and does not measure its real impact on the earth, because the hole is lasting longer and longer. In each of the last three years—2020, 2021 and 2022—the hole has lasted well into December, with the longest-lasting ozone hole on record in 2020 "closing"—i.e., returning to 1979 levels—on December 28.

There has also been up to about 25 percent ozone depletion in the northern hemisphere, the so-called Arctic ozone hole, first noticed in 2011, which is not confined to the Arctic, and which has occurred most years since then during winter and spring. The lowest ozone levels over the Arctic, so far, occurred in 2020. From March through May, there was nearly 43 percent ozone depletion over the Arctic and up to 15 percent ozone depletion over all of Europe. All with respect to 1979 levels, therefore understating the problem.

And, almost unnoticed by the scientific community, there is significant ozone depletion over the tropics as well, and not just during one season but throughout the year. Tropical ozone in the lower stratosphere is up to 60 percent less, all year, than it was in the 1980s, and up to 86 percent less than it was in the 1960s.[29] This has not been considered important by the scientific and environmental communities because historical ozone levels in the tropics were much higher than those over Antarctica, and even with such a large reduction, tropical ozone levels are still not below the level that has been established as "normal" over Antarctica, and so it is ignored. This disregards the fact that the tropics contain a vastly richer living community of plants and animals which require much greater protection from ultraviolet radiation than does the South Pole, and which evolved over billions of years with such protection. Qing-Bin Lu warns that "[t]he tropics (30°N–30°S) constitutes 50 percent of the Earth's surface area, which is home to about 50 percent of the world's population." He writes that measured UV radiation in the tropics is far greater than expected, with the UV index reaching as high as 24 in Quito, Ecuador. A UV index greater than 10 is considered extreme. "The annual mean ozone depletion in the lower stratosphere over the tropics is strikingly large," he says, "which is 77 percent vs 47 percent over the Antarctic."

Humans can take refuge from the sun in houses and put on sunscreen when outside. Animals and plants can't.

The Gentle Rain

In pre-industrial times most rain fell gently, neutrally on the world's land masses, giving life, and returning it, by way of streams and rivers, to its oceanic source. At most the rain was slightly acidic, because carbon dioxide in the atmosphere combines with rainwater to form weak carbonic acid.

Today, however, our rain contains nitric and sulfuric acids, courtesy of the industrial burning of coal and oil, the smelting of metals, and the polluting engines of our automobiles. The rain, as well as the air we breathe, also contains microplastics (chapter 19), and radioactive fallout (chapter 18), emitted by each of the 436 nuclear power plants in the world. It contains varying amounts of the fifty million different chemicals produced by the world's manufacturing plants (chapter 17).

The only two of these millions of contaminants the world's governments have ever paid much attention to are the acids—nitric and sulfuric—that lowered the pH of the rain, acidified our soil and our lakes, ate away at our forests, and—though few seemed to notice—slowly burned our lungs. Acid rain is the reason for more and more efficient scrubbers in our smokestacks and catalytic converters on all of our cars.

In the decades following the Second World War, the industrializing nations of the world, the United States foremost among them, built hundreds of gigantic smokestacks up to 1200 feet tall, whose purpose was to spew their polluting gases into the prevailing winds far above the ground. Their enormous height prevented the massive pollution of local landscapes that would have otherwise resulted, instead transforming hundreds of local problems into one global one. The pollution now remained in the atmosphere for much longer periods of time, where it was carried by winds to distant parts of the globe, to fall out as acid rain over places like Greenland and Hawaii. Eventually, acid rain became both a local *and* a global problem.

The smelting operations of Inco, Ltd. at Sudbury, Ontario, producing 5,000 tons of nickel, 1,200 tons of copper and 2,500 tons of iron every day for the world's markets, did so at a tremendous cost to its own and the world's skies. Sulfuric acid from the world's tallest smokestack killed off all the trees within several miles of Sudbury, leaving only scattered, bleached white stumps standing on a black, barren plain. Sudbury looked like hell when I first saw it in 1970.

In Scandinavia, as many as twenty thousand lakes became too acidic for fish to survive in. In Adirondack State Park in New York State, about half of all high mountain lakes supported no fish. High in the Great Smoky Mountains of North Carolina and Tennessee, once-healthy forests of majestic 400-year-old red spruce and Fraser fir were transformed into diseased stands of dying 45-year-old trees. In Canada's Bay of Fundy, life-giving fog that had sheltered healthy forests of mountain paper birch became acid, which by the late 1980s had rendered leaves brown, branches barren, and killed off five percent of the trees. In southern Mexico acid rain corroded the ruins of the Mayan civilization.

The problem became so obvious and so acute, almost everywhere in the world, that most governments made catalytic converters mandatory on automobiles, and enforced emission standards that required more and more efficient scrubbers to be placed in smokestacks. Levels of oxides of nitrogen and sulfur were reduced in many parts of the world, and the problem of acid rain has been largely forgotten. Acid-burned, dead landscapes have become less common, at least in my part of the world. Even Sudbury, Ontario, has been restored, with grass and trees growing in the surrounding area.

But there is another kind of acid rain that it is not so easy to do anything about. It results from the forest fires that consumed large sections of Canada in 2023 and annually rage across larger and larger areas on every continent except Antarctica. The intentional burning of African forests and savannahs produces such horrendous smoke that is has become the world's leading contribution to air pollution from the burning of vegetation. Primarily it has been farmers and herders who set fires to more and more land to stimulate the growth of crops and grasses. Fires rage for months across thousands of miles of African savannahs and are so extensive that they pump three times more gases and particulates into the air than all the fires set by farmers and settlers throughout South America, including the fires that are devastating the Amazon region. Burning destroys much less rainforest in Africa than in South America, but the savannah fires cover a greater area, burn a greater volume of the dry grass, shrubs, and trees, and are more frequent than anywhere else in the world. The forest only burns once and it is destroyed, but the savannahs and grasslands just become larger and so are burned regularly. The burning fires produce formic, acetic, and nitric acids; acid rain with a pH of 4 falls throughout the year on all the forests of equatorial Africa from Côte d'Ivoire across Central Africa to Gabon, the Republic of Congo, and the Democratic Republic of Congo, and for a good distance westward over the Atlantic Ocean. Red African soils rain out of the sky as far away as Florida, the Caribbean region, and parts of South America. Perhaps some of the fire-born acids blow as far. It is all one atmosphere, and the air knows no boundaries.

In addition to acids, a problem most people think has been solved, the world's rain also contains hundreds of thousands of toxic chemicals, the most studied of which are called per- and polyfluoroalkyl substances (PFAS). These include thousands of different chemicals used in hundreds of types of products, including stain- and water-resistant fabrics and carpeting, cleaning products, paints, fire-fighting foams, cookware (Teflon), food packaging, and food processing equipment. Most PFAS are highly persistent in the environment, and they are so ubiquitous in rainwater that they have contaminated soils worldwide. A 2022 study found that levels of PFAS in rainwater often greatly exceed drinking water standards in the United States, Denmark, and other countries. It found that PFAS contaminate soils above national guidelines and are so ubiquitous that they are emitted in sea spray from the oceans to reenter the hydrologic cycle and rain on us over and over.[30] PFAS have even been found in core samples of snow and ice in Antarctica dating as far back as 1920, with sharply increasing concentrations of PFAS in Antarctic snow in the past few decades. Ironically, production of many of the PFAS chemicals has risen sharply to replace chlorofluorocarbons in an effort to save the Earth's ozone layer.

Chemical Soup

"Better Living Through Chemistry," we were promised with much fanfare after World War II. But whatever else it has brought us, eighty years of more and more Chemistry has brought us an air pollution problem that is so overwhelming that it seems beyond our control.

Only 50 percent of all the pesticides that are sprayed on crops ever reaches the ground, and some that does eventually evaporates back into the air, to be windblown hundreds or thousands of miles until it rains back down someplace else. Pesticide concentrations in average rainwater are about the same as the highest levels found in river water, and sometimes even higher. Pesticides from North American regularly fall on the

British Isles, carried there across the Atlantic by the westerly trade winds. Rain falling on the Shetland Islands has sometimes had the same concentration of pesticides as California's San Joaquin River, which receives irrigation water directly from heavily sprayed fields! Near the equator, trade winds blowing from the east carry pesticide-bearing dust from Morocco across the Atlantic to Barbados.

Air pollution does not stay in one place, but affects every corner of the planet. Just as the air that a person breathes becomes dissolved in the blood and circulates to all parts of the body, so the breathing earth circulates its air to all parts of the planet. High levels of lead have been measured in the surface waters of the world's oceans in many locales, including the high Arctic. And during the 1950s and 1960s radioactive fallout from atmospheric testing used to accumulate in plants, animals, and human beings in Alaska, Scandinavia, and other parts of the Arctic.

Arctic air nowadays suffers from smog every winter and spring, as pollutants travel three thousand miles and more from industrial areas of Europe, Asia, and North America, blanketing the Arctic region with a pervasive haze. Winter air samples in Barrow, Alaska and at Mould Bay and Igloolik in the Canadian North have contained levels of soot similar to that found in fair-sized American cities.

A complete accounting of common air pollutants in industrial societies is beyond the scope of this chapter. They include emissions from automobiles; gases from industrial processes—carbon monoxide, carbon dioxide, ozone, sulfur dioxide, hydrogen sulfide, oxides of nitrogen, chlorine, phosphine, phosgene (a nerve gas and a very common air pollutant), hydrogen cyanide (very widespread); vapors from gasoline and jet fuel; vapors from organic solvents (in dry cleaning fluids, paints, lacquers, varnishes, inks, adhesives, enamels, and many many other substances); pesticides; chemicals in plastics; fluorides (very widespread); arsenic (very widespread, from copper smelting, zinc plating, lead plating, enamelware factories, chemical factories, etc.); metals, including beryllium (in copper alloys and rocket fuels), cadmium, mercury, lead, zinc (the most common metal in city dust), vanadium, nickel, manganese, and chromium; formaldehyde (a raw material, product or by-product in making adhesives, beer, rayon, textiles, disinfectants, dyes, embalming fluids, explosives, paper, pharmaceuticals, soaps, toothpastes, shampoos, air fresheners, deodorants, draperies, carpets, fertilizers, insulation, plywood, particle board, and a host of other products); and so on and so forth. Our air is heavily contaminated, and is poisoning the lungs, blood, and organs of all living creatures.

There is no easy solution, and no easy way to clean up the air, because the dirtying of it is so intimately connected with our every living action. The industries that give us petroleum, plastic, automobiles, paper, textiles, leather, rubber, metals, lumber, glass, food—basically, the industries that give us our food, clothing, and shelter all contribute a lion's share to the problem.

The popular belief that air pollution has been reduced is due to the fact that the Air Quality Index is based on only three of the hundreds of thousands of air pollutants—nitrogen dioxide, sulfur dioxide, and ozone—and only on particulates of a certain size. But all one has to do is look up in the sky and out across the landscape to realize that this is a lie. Both the mountains surrounding my present home city of Santa Fe, New Mexico, and the sun up in the sky, are visible through a haze that is thicker than when I moved here in 2004, and much thicker than the haze around and above New York City in my childhood, when the sky was equally blue overhead and on the horizon, and one could see forever through air that was delicious to breathe.

The haze that envelops the earth now was first reported by scientists at the Smithsonian Institution in 1970.[31] They discovered that 14 to 16 percent less solar radiation fell on Washington, D.C. in 1966 and 1969 than it had in 1907 and 1919. And when they looked elsewhere, they found that on average the sun had dimmed about 6 percent over the entire world in the same half century. In 1973, these researchers, in collaboration with scientists at the National Physics Laboratory in Jerusalem, found 10 percent less solar radiation falling on the top of Mount St. Katherine in the Sinai Peninsula than had been the case in the 1930s.[32] The term "global dimming" was coined by English biologist Gerald Stanhill in the 1990s. He reported that radiation from the sun had been decreasing globally by an average of 2.7 percent per decade for half a century.[33] The intensity of sunlight had fallen by 10 percent over the United States, nearly 30 percent over parts of the former Soviet Union, 16 percent over southern and central Ireland, 22 percent over Israel, 17 percent above the Arctic Circle, and 9 percent over Antarctica. During the first two decades of the twenty-first century, dimming further increased over North America, northeast Asia, China, and Oceania, while Europe brightened.[34] Reports of recent widespread solar brightening are probably not true, because they rely on recent satellite observations which should not be compared with earlier ground observations. The satellite data has been shown to greatly overestimate the amount of solar radiation reaching the ground.[35] The satellite data are useful, however, in estimating trends since the 1980s. None of the satellite data sources show global brightening in recent years, and some of them show significant dimming even over the oceans.[36] Nowhere on earth does as much sunlight fall on the surface of the earth as it did a century ago. What must this be doing to all of life?

It is time that we stopped taking this barrage of chemicals for granted. The same jobs that they do for us today were done without most of them when I was a child. Whether Chemistry has done these jobs better or worse is an arguable point, but even all of its presumed benefits cannot be worth the wholesale pollution of every nook and cranny of our world.

While I was still in college, construction was begun on the Trans-Amazon Highway. Friends of mine were eagerly awaiting its completion so they could go down and explore the green jungle by motorcycle. But it all seemed like such an invasion to me—the road and the motorcycles in a place like that. I felt a deep foreboding about what was going to become of the Amazon rainforest and its people, and I yearned to visit it before it was ruined. I would travel by the power of my own arms and legs, walking and paddling up the river in the traditional way. Wilderness, I thought passionately, is a way of life. Did these people really expect to bring their motorcycles to the jungle and leave the rest of their environment behind? Did they really expect to have a wilderness experience that way?

I never did make it to South America, but in October of 1973, I spent some time in the rainforests of Guatemala. At Tikal in those days, electricity was an iffy three-hour-a-day luxury, the nearest pavement and the nearest telephone were hundreds of miles away, and it felt to me as if the United States and all of its supposed power over the world were totally irrelevant.

The Mayan people had a culture so different from mine that I concluded that American influence here was zero, that it looked as if it would remain zero, and that Americans' ideas of their own self-importance were nothing more than delusions of grandeur! Here were traditional farmers with their own languages, who for the most part did not know Spanish, the official language of their country, who did not even have pack animals, who thought nothing of walking for three days to go visiting, who dressed and thatched their homes in the traditional ways of their ancestors, and who to all appearances were going to live that way for a long time to come. Here in the jungle with them were monkeys and toucans and no obvious connection to the world I came from. The industrial world had no relevance and no impact here, or so I thought.

I was wrong.

CHAPTER 11

Wood

In giving CPR to a drowning person, the first thing to do, one is taught, is to establish an airway, for without oxygen, making the heart beat and the blood circulate will not save the patient. And so we began by inquiring into the state of our atmosphere, for without air life will certainly not survive on this planet. The wholesale pollution of earth's air supply is not only a dire emergency, it is also one of our most clearly defined problems. It is of very recent origin, and since it has definite, identifiable causes, it makes sense to begin talking about solutions. Other problems are much more long-standing, their causes are not so obvious, and for their cures we will have to explore far afield from where most environmentalists stop looking.

One beech tree 120 years old inhales 2,500 gallons of carbon dioxide to make 28 pounds of carbohydrates and liberate 2,500 gallons of oxygen in one sunny day, thus regenerating enough air for the respiration of eight or nine people plus the tree.

Trees represent about ninety percent of the plant biomass of the earth (at least they used to), and they process about half of all the air on the planet (grasses, shrubs, swamps, and marine algae take care of most of the rest). Photosynthesis in a forest is two to three times more efficient than in the same area of grass or cropland. A forested earth recycles more carbon dioxide and oxygen and can therefore support a greater density of life than a deforested earth.

Farmland cannot replace a forest as a carbon reservoir because farm products are consumed within a year and contain much less carbon. Trees live for hundreds of years, and in their wood they store ten to a hundred times as much carbon as farm crops. In temperate climates trees have tremendous root systems that permanently hold together a very deep layer of rich soil full of decaying organic matter that is home to so many animals and plants and their roots that at least half the biomass of the forest is below the ground. When such a forest is cleared to plant crops, all of the carbon in the wood returns to the atmosphere as carbon dioxide, and the soil and the humus layer in it must be protected lest it erode and be washed away as well. Careful agricultural practices can preserve, and even improve, the fertility of the soil, but with a few notable exceptions—Western Europe, South China, Japan, Java, the Philippines—centuries of intensive agriculture have ruined the soil where human beings have settled in large enough numbers.

Woodland is also able to absorb much more water than even well-tended plowed farmland. In an astonishing experiment in New Jersey, forest land absorbed five inches of water per hour for ten hours, while 600 feet away, cleared cultivated land of the same soil type could not absorb more than an inch altogether. The climate and weather in different parts of the world are somewhat at the mercy of global wind patterns, ocean currents, the tidal pull of the sun and moon, the shape of the continents and the geographical features of local terrain. But trees, being storehouses for vast quantities of water, greatly moderate the climate by their ability to store and release heat. Their roots penetrate deep into the earth and conduct heat between ground and atmosphere, and their branches and leaves have enormous emitting and absorbing surfaces. They shelter the ground against solar radiation, both cool and humidify the air by evaporation from their leaves, and are barriers to the winds. Clearing woods dries the air and causes temperatures to become more extreme—treeless land is warmer by day and colder by night than a forest, and suffers hotter, drier summers and more severe winter weather.

In the tropics, clearing a forest to plant grass or crops is much more disastrous. In a tropical rainforest, decay processes are so rapid that practically no humus ever accumulates in the soil. Most of the organic matter is above ground, and when this is cleared off, the soil that is left is so poor in nutrients that it will support grass for only a few years before it turns into parched barren desert. Of course, you can truck in huge quantities of fertilizer, and when the rains stop coming you can dam up some rivers and build huge irrigation projects, but in the end the result will be the same. Where once grew such richness of life that fifty different species of trees could share an acre of land, there will be nothing but a few drought-resistant shrubs that have taken over after the people have gone and the grass has died.

In prehistoric times forests covered some three quarters of earth's land. Today forests—not necessarily in the best of health—cover roughly a third of earth's land, and another third is desert. A lot of the earth's grasslands were once forested, and would be reclaimed by trees today were they not maintained by periodic burning or by continuous cultivation or grazing of farm animals. A certain amount of grassland is perpetuated naturally, where the climate and terrain are suitable, by the grazing of wild animals and by lightning fires. However, the devastation caused by lightning is usually limited by the rain that comes with it. Fires set by people burn without regard for weather conditions and do not come with rain.

The decline of forests is very ancient, and is basically a result of human population pressure. The immediate causes of forest destruction—cutting for firewood or charcoal production; timber extraction; agriculture; cattle grazing; clearing for villages and towns; highways; accidental fires; mining; flooding by dams; improving access to hunting—these are all more or less unavoidable effects of overpopulation.

It is not easy to tell that many desert areas of the world once were covered in thick forest. The climate has changed in some areas. Other desert lands could support a forest even in *today's* climate, but you would never know it were it not for the small remnants of the ancient groves that still stand and flourish. What will Brazil be like a few generations from now? "One day they will wonder, as they already do in so many other parts of the world, where the forests used to stand. And whether it is true that there used to be trees as far as the eye could see."[37]

North Africa and the Middle East have probably never received a great deal of rainfall—at least not since the last Ice Age ended, and the climate before that is difficult to guess—but at one time they were covered by forests. The wooded soils soaked up the rain that did fall, and it percolated down to great depths, forming giant aquifers such as still underlie the

Sahara, although the tree cover is long gone, and the meager rains run off the impenetrable surface in flash floods. The trees were felled in ancient times for agriculture, and the fertile Mediterranean soils supported great civilizations for centuries or millennia, until eventually either the soil was worn out, or overgrazed, or great irrigation systems collapsed or were destroyed by invaders. Israel, the "land of milk and honey," was farmed for thousands of years. The mountains above it in the Sinai wilderness were clothed with dense forests, containing abundant wildlife. The Bible still speaks to us of the "forests of Arabia" (Isaiah 21:12). On the plains of ancient Mesopotamia—present-day Iraq—once grew groves of cypress and palm trees. Ancient Asia Minor—modern-day Turkey—had large oak forests in the time of the Hittite Empire.

The hills of Lebanon were once covered with forests of cedar, of which today some four hundred trees remain. It is astonishing to see a photograph of the tiny green grove of flourishing trees at Les Cèdres, dwarfed by the miles of barren desert around it, a reminder that it is not necessary for this land to be bare. Higher up at 6,000 feet in the northern mountains of Lebanon is the Kammouha Forest, a protected grove of mixed forest, also thriving and also surrounded by barren desert. The trees of Lebanon were felled in Biblical times to supply timber for the ships of the Phoenicians, and later the Romans, and to build the cities of Egypt. King Solomon had "eighty thousand hewers in the mountains" (I Kings 5:15) to get cedars to build his temple and palace.

There are likewise still some isolated olive groves in parts of North Africa that are today for the most part desert. Olives, not native to the area, were originally imported from the east and replaced formerly existing native forest.

The island of Crete was once entirely covered with coniferous forest, of which a small remnant remains in the White Mountains.

Under the sands of present-day Tunisia lie the ruins of great cities and granaries, dating from the days when the region was a breadbasket for the Roman Empire. Roman mosaics depict animal hunts in forests that flourished two thousand years ago and supplied lions for Roman amphitheaters.

The central Sahara, today one of the hottest places on earth, once supported grass, trees, and wildlife. Underneath its sands is a web of valleys and river channels that flowed with water and attracted human settlement until about four thousand years ago. On the Tassili plateau of Algeria, nearly a thousand miles from Mediterranean coast, rock paintings, carbon-dated to between the sixth and second millennia B.C., show hippopotamus being hunted from reed boats. Other rock art depicts rhinoceros, lions, panthers, antelopes, ostriches, elephants, giraffes, hyenas, and camels; the later paintings show herders with their cattle. Fossilized pollen five to six thousand years old proves that cedar trees once grew where now is only sand. 160 ancient cypress trees still stand in an area where it rains perhaps once a decade. These are not desert-adapted trees; their roots descend through sand and soil to depths of up to two hundred feet, where they tap a still-existing underground aquifer. Almost all of these cypresses are more than 2,500 years old.

Much of the primeval forest of prehistoric Australia was probably destroyed by the extensive fires set by its native people. Today's billabongs—the oases of the Australian desert—lie along the courses of long-vanished rivers, and only five percent of the continent is now forested.

The Indus River valley of Pakistan, the cradle of great civilizations, is today a very arid region. To the east in India is the Thar desert, as dry and inhospitable as the Sahara. The remains of Harappan civilization are buried all along the Indus valley, and also under the sands in what is today unrelieved desert. Archaeologists have found more than forty settlements in the empty lower valleys of rivers that once flowed into the Indus but today sink into the desert sands long before reaching it. In Rajputana in northwest India are beds of salt lakes that held *fresh* water until about 2000 B.C. and were surrounded by moist-climate vegetation. Then, over several centuries, they became salty, and some have dried up entirely. Baluchistan, in western Pakistan, now almost a waterless desert, then had many rivers and supported a large agricultural population. The Harappans occupied this whole region from about 3000 B.C. to 1500 B.C. Their cities are notable for their brick fortresses and huge granaries. They grew wheat, barley, melons, sesame, dates, and cotton, traded in wood, metal, and ivory, and built their houses out of burnt brick and timber. At the time of Harappan civilization, the Indus River valley received the same summer monsoon rains that the rest of the Indian subcontinent still receives today.

But sometime after 2000 B.C., the rains stopped coming, and there is an interesting theory as to why.

In the summer the monsoon winds blow northeast over the Indian subcontinent from the Indian Ocean. When moist ocean air blows over the warmer land, it rises, and when it rises it expands, cools, and drops its moisture as rain. The dry conditions in northwest India and Pakistan are apparently due to a thick layer of dust from the desert that cools the land and causes the moisture-laden monsoon winds to sink instead of rise. Falling air absorbs moisture instead of releasing it and creates a desert region below it; therefore the summer air over the Thar desert contains four times as much water vapor as the air over most deserts, yet it does not rain.

The desert perpetuates itself with its own dust. Over each square mile of Rajputana sometimes hangs more than five tons of suspended dust. The daytime sun is hazy red or completely hidden; the nighttime dust veils the stars. The original cause was the growth of human populations. The inhabitants of this land depleted the surrounding forests for timber and fuel, leading to increased rain runoff and soil erosion. Hundreds of years of irrigated agriculture and cattle grazing increased the salinity and decreased the moisture-holding capacity of the soil. The consequent drying up of rivers and streams and the failure of irrigation systems led to the decline of Harappan civilization and created the conditions which have continued and worsened up to the present day. The Thar desert is still expanding its borders at a rate of up to half a mile a year.

It has been proposed that the desert could be reclaimed by simply putting a fence around it to keep out people and goats so the wild grasses could grow; the grass would hold the soil in place, the dust would be kept down, and the monsoon rains might return after an absence of four thousand years. But fencing off millions of hectares of land in densely populated India would not be so easy a thing to do!

Most people have assumed that the desiccation of North Africa, Pakistan, and other presently dry areas of the tropics and subtropics has resulted from some kind of "natural" climate change that occurred after about 3000 B.C., having nothing to do with human activity. It must not be forgotten, however, that in many cases it is obvious that human beings cut those trees down. And in some areas of the world the same process has begun only recently.

> Small low islands in the Pacific are known to have borne dense scrub or forest which has been totally destroyed following the introduction and increase of rabbits, pigs or goats. Small sandy islands of this type are reported to have "split" a rainstorm as it approached from the ocean. Actually, it is probably extreme radiation from the barren sands that evaporates the rain before it strikes the land. On the high islands, clouds passing obliquely over valleys and ridges tend to persist over forested cover and forest plantations but evaporate over anthropically induced savannahs and grasslands. In the French West Indies it has been observed that rain will fall on the forest on each side of a roadway but not on the roadway itself, presumably because of the dryness and radiation over this roadway. Laymen visiting the tropics are astonished at the extreme shade and dank coolness within the forest in contrast to the dry, scorching sunlight of adjacent cleared land. (Frank Egler, American Museum of Natural History)[38]

It is likely that the electrical conductivity of the trees themselves, and of the moist air above them, also has something to do with the falling of rain, as was suggested in chapter 4 in the discussion of the global electrical circuit.

Deforestation and desertification in ancient times were not limited to the tropics. The forests of northern China were destroyed thousands of years ago, and the once deep and fertile soil that nourished early Chinese civilization has been eroding away ever since, 2½ billion tons of yellow subsoil washing away every year and giving the Yellow River and the Yellow Sea their names. We have records from early Chinese history in which officials warned against the consequences of deforestation in the mountains.

Underneath the sands of the Gobi Desert in Mongolia lie dry riverbeds, irrigation canals, ancient millstones, ruined towns, and the remains of the city of Khara-Khoto which, with its Buddhist temples and libraries, its fine buildings and its trading houses, was once the capital city of the

Tangut Empire. Khara-Khoto was visited by Marco Polo and flourished at least until 1368 A.D. Large reservoirs of water today lie just two to three yards below the surface of much of the Gobi Desert, a strong indication that once vegetation grew here and that perhaps it could again.

Buddhist monks seem to have been responsible for a large amount of deforestation in China, Japan, and Tibet. The trees were used for the construction and constant rebuilding of their huge halls and temples; burned as fuel for the cremation of the dead; and cleared for the grazing of their livestock.

The continent of Europe was once almost entirely clothed with forests. In southern Europe, under a dry Mediterranean climate, the forest was more open and composed largely of evergreen oaks and pines. Once cleared, it did not regenerate like the temperate mixed forests to the north; by classical times much of it had already disappeared, leaving behind scrub and, in places, bare eroding soil. Woods that existed in Homer's Greece were already gone in Plato's time.

> What now remains compared with what then existed is rather like the skeleton of a body wasted by disease; the rich, soft soil has all run away leaving the land nothing but skin and bone. But in those days the damage had not taken place, the hills had high crests, the rocky plain of Phelleus was covered with rich soil, and the mountains were covered by thick woods, of which there are some traces today. For some mountains which today will only support bees produced roof beams for huge buildings whose roofs are still standing and there were a lot of tall cultivated trees which bore unlimited quantities of fodder for beasts. The soil benefited from an annual rainfall which did not run to waste off the bare earth as it does today, but was absorbed in large quantities and stored in retentive layers of clay, so that what was drunk down by the higher regions flowed downwards into the valleys and appeared everywhere in a multitude of rivers and springs. And the shrines which still survive at these former springs are proof of the truth of our present account of the country. (Plato, *Critias*)[39]

Other forests disappeared later. Strabo's *Geography* and Pliny's *Natural History* at the beginning of the Christian Era describe the forests on Cyprus, Crete, Sicily, Corsica and other Mediterranean islands, the wooded mountains of southeastern Spain, the forests of the Po Basin, and the wooded hills of many parts of Italy. Wood was so abundant along the Mediterranean Sea that Marseilles and Genoa were famous for shipbuilding. But shipbuilding, agriculture, grazing, industry, warfare, and sheer human numbers decimated the forests of southern Europe, leaving behind the shrubbery and eroding soil that remain today.

Prehistoric central and western Europe were covered by broad-leaved deciduous forests of oak, elm, beech, linden, hazel, and other trees. Pollen analysis has shown that even the dry and now treeless plains of Hungary were once wooded. The temperate forests of Europe fell to axe and fire, and what remains is but a shadow of what once was.

The forests of England were decimated before coal replaced wood and charcoal as the prevalent fuel in the eighteenth century. By that time 65 of the 69 great English forests of medieval times had already disappeared.

At least some, and perhaps all, of the grass steppes in the south of European Russia and Ukraine were once forested. Herodotus, for example, wrote of "the forest country" along the lower Dnieper in Ukraine, and Kiev was founded in the ninth century A.D. on forested land, where now is open steppe. As one looks north to where steppe gives way to mixed deciduous forest, and further north to coniferous forest in northern Russia and Scandinavia, progressively more woodland has survived to the present day, although most of it has been logged over at least once.

The United States was not the untouched paradise many people think it was before Europeans arrived. It was densely inhabited by people who burned both grassland and forest regularly. They burned woodland for agriculture, to create better pasture for animals they hunted, to clear it for ease of travel and for security from their enemies, and for many other reasons. And, like aboriginal people the world over, they were not in the habit of carefully extinguishing campfires like us modern folk who have matches. Fire was precious and to be preserved. When European settlers arrived, the forest cover in the eastern United States was parkland so open that stagecoaches could be driven almost unimpeded from the Atlantic coast all the way through to the Mississippi River.

In the American southwest, what is now treeless desert in Chaco Canyon, New Mexico was occupied in 900 A.D. by the Ancestral Pueblo, whose great cities and irrigated fields were then surrounded by forests of piñon, juniper, and ponderosa pine. They felled those trees for timber and firewood, and when the local forests were gone, they built an elaborate road system to haul spruce and fir logs from mountain slopes fifty miles distant, until their civilization collapsed in the twelfth century.

Rapa Nui (Easter Island), a treeless island in the South Pacific, 2,300 miles west of Chile, was covered by forests of palm, hibiscus, and other kinds of trees when it was settled by Polynesians some two thousand years ago. They cleared the land for agriculture, and to obtain logs for canoes and for use as rollers in erecting the huge statues for which the island is famous. By 1500 A.D., the forests had been so thoroughly destroyed that not a single tree survived, and a civilization that had its own writing system and had built astronomical observatories collapsed into tribal warfare. Today Rapa Nui supports less than a third of its former human population.

The vast forests that have remained intact until modern times, including the tropical forests of South and Central America, Africa, and Southeast Asia, and the northern forests of Canada, Alaska, and Russia are right now being attacked on a grand scale. Damage to the world's forests is now so much accelerated that not only ecosystems but whole continents are suffering the consequences.

The Amazon has become a huge war zone—human against nature and human against human. Forest has been flooded by some of the largest hydroelectric projects in the world; burned off in huge conflagrations to make room for farms and cattle ranches; burned off by the landless poor, who can get title to any land they squat on for five years; torn apart by gold miners; cut down for fuel to smelt iron. In South Pará is the largest open pit iron mine in the world, and 6,100 square kilometers of forest are felled each year to supply charcoal to fuel the iron smelters. Native people and rubber tappers who live in the forest and wish to preserve it are routinely murdered by one and all. Ranchers and squatters are in a perpetual state of war with one another. The forests are the biggest losers. In one six-week period in 1989, 180,000 fires were counted burning in the western Amazon region alone. The burning is still going on more than three decades later. Since 1978 about one million square kilometers of Amazon rainforest have been destroyed across Brazil, Peru, Colombia, Bolivia, Venezuela, Suriname, Guyana, and French Guiana. More than 20 percent of the Amazon rainforest is already gone.

Deforestation in the Amazon. *Photo by Amanda Perobelli.*

Africa is being just as devastated, but the deforestation there has been going on longer. In the century prior to 1980 almost 400,000 square miles of forest and woodland were cleared in West Africa alone.

We have been losing about one and a half acres of tropical rainforest per second for decades. In 1950, about 15 percent of the earth's land surface was covered by rainforest. Today, more than half of that is already gone.

Canada's immense wilderness is also being destroyed, by oil, natural gas, mineral, and hydroelectric interests, and its northern forests have been sold off for decades at dirt-cheap prices, mostly to foreign corporations, to make pulp and paper. In 1989, for example, the province of Alberta gave away virgin forest on one third of its land at $1.25 per tree to Mitsubishi, Daishowa, and other foreign companies whose seven pulp mills denuded the wilderness and polluted every major river flowing toward the Arctic Ocean from Alberta—the Peace, the Athabasca, and the Slave. One thousand square miles of British Columbia's forests are being clearcut every year, endangering the survival of the grizzly bear. Although the Canadian government now asserts that those days are over and that "logging is not deforestation," reality is quite otherwise. A 2019 study found that clearcut forests do not grow back and that current deforestation rates in Canada are nearly fifty times higher than reported by government officials.[40] 3.6 million hectares of forests were clearcut in British Columbia between 2005 and 2017, creating "dead zones" that, combined, are larger than Vancouver Island.[41]

Massive burning of forests is also occurring in the United States, not only on private land but by the National Park Service, the U.S. Forest Service, and every other government agency that manages land. The government has been doing this since 1968, and by now it has most of the population convinced that fire is good for the forest, and that "prescribed burns" need to be set every five to ten years in all the forests of the United States because they contain too many trees and need to be thinned out. The practice is escalating. Whereas all prescribed burns were tens to hundreds of acres in size in the 1970s through the 1990s, some are thousands to tens of thousands of acres today, and amount to millions of acres per year. This is devastating ecosystems from coast to coast, and in Alaska. In 2017, the various government agencies intentionally set a total of 202,250 fires and burned 6,421,972 acres. Fires were set in every state except Hawaii. The agencies that set the fires were the Bureau of Indian Affairs, the Bureau of Land Management, the Department of Defense, the Fish and Wildlife Service, the National Park Service, the Forest Service, many state governments, and some local governments. Among the real parties in interest are oil, gas, mining, and geothermal companies wanting the land for purposes other than producing oxygen and sheltering complex ecosystems; and timber companies that cooperate in "thinning" the forest for its own good before the prescribed burns are set. The U.S. has also trained forest managers in other countries in how to thin and burn their forests, with the result that it has become standard practice in many parts of the world, and is contributing not only to global deforestation but to the increasing numbers of out-of-control forest fires raging on every continent.

It has become a matter of economics—economics and the demands of an ever-growing world population. Or so it seems to a lot of people. But appearances can be deceiving. Modern deforestation is a continuation of a very ancient human practice that, like so many other human practices, has been immensely speeded up and magnified. This is not only because of economics and overpopulation, which are also old upon this earth, but also because the technology is available to do on a much more massive scale that which we have always done.

Human beings have probably been cutting and burning down trees for a million years or so—so long that many grassland, savannah, and forest ecosystems probably depend on it. We have been part of these neighborhoods for a very long time, and our activities have not necessarily been detrimental. But in modern times we are no longer just leaving our mark on the landscape, we are acting on such a grand scale that we are actually denuding the earth.

The consequences are serious and many:

- the destruction of a way of life for forest-dwelling peoples
- the disappearance of a large number of the earth's plant and animal species
- the birth and growth of deserts
- soil erosion by wind and water
- massively increased rain runoff to rivers, and floods in surrounding plains
- reduced evaporation from earth's land areas
- more carbon dioxide in the atmosphere
- less wood available to human beings on a sustainable basis
- a less beautiful earth

In 1982, recovering from illness, I fled the urban jungle of southern California to live in the redwood forest up north. For the next three years I drank deeply of the clean ocean air and luxuriated in the peace and quiet of Mendocino County.

Logging companies—Georgia Pacific and Louisiana Pacific—were the main employers in Mendocino, but I imagined that as long as most of the land was owned by logging companies, it would at least continue to have trees on it, and not houses, farms, shopping malls and factories.

What I had seen ruined in Ithaca in the late 1960s was still to be found here on the Mendocino coast—slow, narrow, meandering dirt roads, crisp delicious air, clean rivers, a sparkling mist-covered seashore with white sand, abundant driftwood and a glistening ocean surf pounding the headlands, a giant silent forest peopled by unafraid wildlife and watered by plenty of pure rain, and a sky so blue whose graceful clouds not a single airplane ever disturbed. The town of Mendocino was an old-fashioned village which preserved its Victorian houses and its wooden sidewalks, where strollers paused to greet their neighbors, where houses, cars, and bicycles were never locked, where blazing stars shone in a pitch-black night sky without competition from streetlights.

But the modern world filtered in here too. Mendocino, because of its purity and its sparse population, was becoming a war zone.

Tourists, who used to make the four-hour trip from San Francisco in their cars on summer weekends, began to come up in their RVs every weekend of the year. Many long-time residents were forced to leave, as much of the rental housing was converted into inns and bed-and-breakfast accommodations. Speculators from Los Angeles bought up property and built hotels and condominiums, and the battle lines between developers and preservationists were drawn. Property values rose sky-high, and the rich people who moved in made war on their hippie neighbors. Zoning regulations and building codes were used to destroy communes. Helicopters searching for marijuana crops terrorized Comptche and other villages. Theft and other crimes became more common, and new and brighter streetlights were installed in town to combat them.

The war on the forests escalated. Tensions grew between loggers and environmentalists as the scale of clearcutting increased, and the last remaining stands of old growth redwoods were going to the sawmills. One company cut down all the trees it could before selling off its vast holdings of forest land to the highest bidder.

I left Mendocino in 1985 and returned to live there again from 1998 to 2003. Jets now flew regularly overhead on their way north and south. The sky was not as clear nor as beautiful nor as quiet as it used to be. The winter rains did not come as regularly as they used to, and wells often ran dry in late summer and fall.

CHAPTER 12

Water

A single body of water is on this earth, filling the great basins of the oceans, pervading the porous rock and soil of the continental crust, flowing in swift streams under sea and over land, whose droplets join together here and disperse there, moving about the earth in the bodies of plants and animals, flying through the air on the wings of birds and in billowing clouds blown by the winds, now vaporous, again liquid, again hexagonal snow crystals, yet again ice, every drop ebbing and flowing to the pull of sun and moon.

Water flows through my body as through the sea. It flows in with my food and drink, circulates in my blood and bathes my cells, and flows out on my breath and in my sweat and urine, to continue on its way downstream. This glass of water I drink has spent time at the bottom of the Atlantic Ocean, in a thundercloud over Malaysia, in a mountain spring in the Himalayas, in the sap of an ancient Lebanese cedar, in the blood of an Amazonian mosquito, and in the sewers of New York City.

The residence time of water in the atmosphere is about nine days; in large rivers two weeks; in the soil two weeks to a year. The turnover time for all the water in a large lake is ten years; in shallow groundwater tens to hundreds of years; in the surface layers of ocean 120 years; in the deep ocean 3,000 years; in deep groundwater 10,000 years and more; in the Antarctic ice cap up to 500,000 years. It is one system, constantly recirculating and coming back around. It is the very foundation life, yet we human beings are poisoning it.

Groundwater

Rainwater that does not run off the ground soaks into it like into a sponge. Some of it evaporates again later, and some percolates down until it reaches the water table, the top of a layer of soil and rock that is saturated with water. Water also seeps down into the porous ground from rivers and lakes. Groundwater flows through the rock and soil just as surface water flows over the land, only much more slowly—a few feet a month or slower. It may return to the surface here and there in cold or hot springs, or flow through the ground into the oceans. Where the water table lies at the surface there are swamps and bogs and marshes.

Of all the water on earth, 97.2% is salt water, 2.1% is snow or ice, and only about 0.6% is fresh water. Of all the fresh water on earth, only 1.2% is in rivers and lakes, which from one point of view can be considered to be groundwater overflow. The fresh water on earth is mainly groundwater, forming a continental ocean beneath our feet as much as three or more miles deep. It is a renewable resource, but only if it is not withdrawn faster than rain and snow can replenish it.

The Ogallala aquifer is a great deep reservoir of water-laden sand, silt, and gravel up to a thousand feet thick underlying a large part of the American Great Plains. The Dust Bowl of the 1930s was turned into America's breadbasket through the mining of this aquifer. First tapped for irrigation in the 1930s, it was extensively exploited after World War II with the development of high-capacity pumps and the sinking of 150,000 deep wells all over the Plains states. The annual overdraft is nearly equal to the yearly flow of the Colorado River, and a reservoir that was filled gradually over millions of years is being drained in a matter of decades. Water tables have dropped hundreds of feet. Up to three-fourths of the groundwater in areas of Kansas is already gone. Springs have vanished. Rivers and lakes have disappeared. Midwestern landscapes that once supported antelope, elk, deer, lynx, grizzly bears, black bears, mountain lions, beavers, turkeys, ferrets, otters, wolves, and ravens now support only corn and cattle, whose thirst is quenched by stealing trillions of gallons of water from far beneath their fields.[42] Total depletion will eventually cause 5.1 million acres of irrigated land to dry up and force a return to dry-land farming. One cure for this situation, proposed in 1982 and updated in 2015 by the U.S. Army Corps of Engineers, is the building of huge canal systems costing billions of dollars to import water from 376 to 1135 miles away in South Dakota, Missouri, and Arkansas, all uphill.[43]

This is an excellent example of how we keep increasing the scale of technology to postpone the day of reckoning, and it is a reckoning that will become ever more catastrophic when it comes. When intensive agriculture turned the Great Plains into a great dust bowl, that was a clear warning that we were already exceeding nature's capacity to provide. But we did not stop. We merely borrowed from the future some more by mining deep underground waters. When they dry up we will steal water from distant lakes and rivers. When *they* dry up we will dam up great canyons and arms of the sea and steal water from even greater distances. These plans are already on the drawing board. One day it will all fail, and when the dust finally returns, and the groundwater is gone, and the rivers and lakes are dry, and the ocean has become an enemy, and there are no more technological fixes big enough to correct the disasters that the previous technological fixes caused, we will finally have to face the consequences of our actions, like an alcoholic who wakes up cold sober after a fifty-year drunk when his supply of liquor finally runs out.

This sort of thing is going on all over the world and is essentially a replay of the building and collapse of so many other agricultural civilizations in semi-arid lands that has occurred during the past ten thousand years, played out this time with much more powerful technologies and coming to a conclusion in only a fraction of the time.

Tucson, Arizona, when I lived there during the 1970s, was getting all of its water from deep wells, which it was rapidly depleting. Although a river with a deep, wide channel runs through the city, it contains no water. The Santa Cruz River runs only three or four days a year during the flash floods that follow summer "monsoon" rains. It hasn't always been this way.

The flowing waters of the Santa Cruz, the San Pedro to the east and their tributaries irrigated the Hohokam civilization for a thousand years, and other farming civilizations before them. This fertile, well-watered land nourished substantial populations. Probably more people lived in the Sonoran Desert in the thirteenth and fourteenth centuries than lived there again at any time up to the twentieth century. And our civilization has been able to surpass the Hohokam in numbers only because we import food to the desert from our Midwest and from Mexico's central plateau.

The Santa Cruz and San Pedro valleys were also the corridors through which Europeans first traveled into the area from Mexico, at that time New Spain. The Spanish explored this fertile land very intensively—at one time eight million sheep and one million cattle roamed Arizona under Spanish rule! Beaver dams were numerous, and trout up to eighteen inches long were abundant in the Santa Cruz, the San Pedro, and Sonoita Creek into the late 1880s. Malarial mosquitoes bred in the marshy valleys through which these rivers slowly wound their ways.

Tucson was a small village, built by the Spanish in the 18th century on the Santa Cruz River at the site of a Pima community. The river flowed year-round, and tall grasses extended into the hills from a streamside forest. Drinking water came from El Ojito Spring, within town limits, and a public bath was built at the spring. In 1857 a dam was constructed across the Santa Cruz to form Silver Lake, and water power turned water wheels for milling flour.

But all those sheep and cattle grazing the desert grasses eventually removed most of the vegetation, and rain began to wash off the land instead of being absorbed into it. Flash floods demolished Silver Lake Dam in 1880. By 1890 severe flooding was causing the formerly quiet rivers to carve steep-walled trenches through which they emptied rapidly and torrentially into the Gila. River flows became intermittent, irrigation ditches were left high and dry, the land dried up and cracked, and its grass cover was replaced by desert shrubs like mesquite, ocotillo, and desert broom. During the drought years of the 1890s thousands of cattle starved to death. Malaria disappeared along with the beavers and the fish. Sahuaro cactus became fewer, and the zone of oak woodland retreated to higher elevations. In town, El Ojito Spring disappeared. Steam-driven flour mills replaced water-driven mills, and underground water replaced surface water for the town's residents. The streamside forest served as fuel for the mills, the water works, and the new railroad, and the trees too were soon gone.

In the twentieth century ranchers used bulldozers and herbicides to get rid of millions of acres of desert shrubs and sagebrush, and planted monocultures of wheat grass and other cereals for their cows to eat. To get the needed water they drilled thousands of wells up to half a mile deep into the desert floor, lowering the water table still more until the rivers no longer flowed at all. In some of the deepest wells the water level dropped 100 feet, and the land surface sank. One well at the University of Arizona mined water that was 25,000 years old, having traveled twenty-five miles through the earth at a rate of five feet per year from the hills where it fell as rain long ago.

A similar situation exists all over southern and central Arizona. Most riverbeds are dry nearly all year and groundwater levels have fallen everywhere. The technological fix here is the $3.5 billion Central Arizona Project, a 336-mile system of concrete-lined canals, tunnels, pumps, and pipelines that have provided Phoenix, Tucson, and surrounding agricultural areas with Colorado River water since 1993. Unfortunately, the waters of the mighty Colorado are also being drained to supply irrigation and drinking water to Wyoming, Utah, Colorado, Nevada, New Mexico, Arizona, southern California—including Los Angeles and the Imperial Valley—and northern Mexico, and the supply is no longer enough for all comers.

All of the major tributaries of the Colorado in Arizona that still flow with water have been so thoroughly dammed and diverted to satisfy an unquenchable thirst that they are tributaries no longer. The loads of silt that used to ride the currents of the Verde into the Salt, and of the Salt, the Agua Fria, and the San Francisco (and once the Santa Cruz and the San Pedro) into the Gila River now pile up in a multitude of large reservoirs. The Gila dries up long before it should drain into the Colorado. The once torrid Colorado flows tame and tranquil year-round, drying up before it ever reaches the sea. Except for occasional flood flows, Arizona rain and snow no longer flows into the Gulf of California. And fertile Mexican deltas that were once nourished by the silt of many great Arizona rivers have been eroding away since the 1960s.

Fifteen hundred years ago the Hohokam settled the fertile lands of what are now southern Arizona and New Mexico, and like their Ancestral Pueblo and Mogollon neighbors to the north and east they irrigated and farmed them for a thousand years. Most likely a large lake once existed in southern New Mexico, fed by rivers flowing through meadows and forests; this once-lush land was inherited by yucca, creosote bush, and cactus after the water supply failed and the farms were abandoned. How many more years, I wonder, will our own irrigated farms last? And what kind of landscape will we leave behind us?

A Thirsty World

Visitors to Mexico City today are often surprised to learn that when founded in 1325 by the Aztecs it was an island in the middle of a large lake. The Spanish, over the course of three centuries, drained most of Lake Texcoco and filled in the valley floor. In recent years Mexicans are draining their underground water as well, and some parts of the city have been sinking up to a foot per year.

During the 1980s, 1990s, and 2000s, Libya drilled more than 1,300 wells, most more than 500 meters deep, to mine 20,000- to 40,000-year-old water from deep underneath the Sahara and pipe it hundreds of miles to Tripoli and other Mediterranean coastal cities, so that the desert could quench the thirst of the seashore! It is the largest underground network of pipes and aqueducts in the world. Some estimate that the aquifer will last only 60 to 100 years. Surface water and oases will then disappear forever, finally destroying the Bedouin way of life and completing a sterilization and depopulation the desert that was begun with deforestation thousands of years ago. Some palm tree oases have already been killed where deep wells have sufficiently depleted the local groundwater.

Israel is mining the sandstone aquifer that underlies the Sinai-Negev Peninsula. This water, which flows to the surface in springs and oases, is also tens of thousands of years old.

In 1987 the United Nations Economic and Social Commission for Asia and the Pacific reported that the following countries were facing problems related to excessive withdrawal of groundwater:

China	Maldives	Indonesia	Vietnam
Japan	Republic of Korea	Fiji	Australia
India	Sri Lanka	Guam	
Kiribati	Thailand	New Zealand	

In the past three and a half decades the problem has become far worse. Serious depletions of groundwater have occurred in Pakistan, Mexico, Saudi Arabia, Yemen, Australia, Israel, Jordan, Syria, South Africa, Namibia, Turkey, Bangladesh, Nepal, Chile, Argentina, and California's Central Valley.[44]

In 2022, it was reported that out of 107 countries using groundwater for irrigation, about half are draining it faster than it is replenished.[45]

Endangered Springs

My body steamed in the hot sulfurous water as cold November air blew softly on my head, the stars blazing in the blackness above me. There was no sound except for the burbling hot and cold creeks which joined together to produce the magical pool in which I soaked. Somewhere not too far away, I knew, a few huge majestic birds, the last of their kind, nested down for the night. My own troubles began to seem far away, as the unbearable stresses of medical school drained out of me, washed downstream by the earth's own medicinal waters. For this was my sanctuary too.

Located in the Los Padres National Forest, Sespe Hot Springs were accessible by either a leisurely eighteen-mile walk from the end of the road along Sespe Creek, or else by a four-hour hike four thousand feet down a mountain, and a *very* strenuous return trip up the same mountain. This large roadless area was strictly protected from any development by virtue of its being a nesting sanctuary for the California condor, the largest and rarest soaring bird of North America.

The hot water, almost boiling, flowed continuously out of the rock into a small basin that you could set pots of food in to cook—no need for a campfire here—and then it flowed about a hundred feet until it mixed with the water of a cold creek. Several pools deep enough to bathe in had been enclosed by rocks, each pool having a different mixture of hot and cold water and a different temperature for bathing. An old wooden shack with wooden benches, two tiny windows, and spaces between the floorboards, had been erected years before over the hot creek as a sulfurous team bath!

At night, if the air was chilly, you lay on the ground, because the earth beneath your sleeping bag was warm. The stars were unbelievable. And far off in the distance, above the horizon about sixty miles away, the white glow of Los Angeles was a reminder that somewhere civilization still existed.

Sespe, to me, was an especially magical hot springs. The air always felt electrically charged. So did the highly mineralized water, so cleansing and stimulating to the body.

On April 19, 1987, the last remaining wild California condor was captured and taken to the Los Angeles Zoo. Small captive breeding populations there and in San Diego represented a last-ditch effort to save this great vulture from extinction. With the removal of the birds from the wild, however, their former habitat now had no condors to protect, and owners of the 13,820-acre Hudson Ranch were free to subdivide it for homes and recreational facilities. Other development projects and activities within the condor's former range included ski resorts, oil drilling, hunting, and agriculture. Forests were being cleared to build reservoirs, roads, power lines, and wind farms. Before the capture of the last condor,

Amos Eno of the National Audubon Society said, "The endangered condor is holding back the dike on a number of devastating projects. It will be extremely difficult, in our view, to maintain habitat in suitable form without wild condors."[46]

In 1987 captive Andean condors from South America were released in the vicinity of the Sespe Condor Sanctuary; they "represented" the California species in the wild while the breeding program continued. In a victory for environmentalists, the Hudson Ranch was purchased by the federal government and became part of the Sanctuary. On June 19, 1992, 31.5 miles of Sespe Creek were officially declared a wild and scenic river, and included in the newly designated Sespe Wilderness, protected from dam building and any other development. Today, eighty-nine California condors once again make their home in and around the Sespe Condor Sanctuary to protect Sespe Hot Springs from exploitation and extinction.

As geothermal development accelerates, hot springs worldwide are facing extinction. Hot springs and geysers are the surface flow of hot groundwater, which is just groundwater that has been heated by contact with hot rock in a volcanically active area. The mining of geothermal fields will exhaust most of them in thirty to one hundred years, depending on the size of the reservoir and the rate of extraction. The hot springs above these fields will cool off and eventually disappear. The world's largest geothermal energy project, at The Geysers, California, once promoted as a limitless source of cheap energy, began to run out of steam after just thirty years of operation. It remains in operation today by the injection of 20 million gallons of treated wastewater from Santa Rosa and Lake County every day one to two miles deep into the earth to access the remaining heat. Here in New Mexico, my favorite wild hot springs used to be Spence Hot Springs near the village of Jemez Springs. But Los Alamos National Laboratories, using hydraulic fracturing techniques, blasted a hole measuring close to one-third of a cubic kilometer two miles deep in the earth about 4 miles distant from Spence Springs in an experiment to prove the feasibility of getting geothermal energy from hot dry rock. And from 1991 to 1995 they operated a test geothermal power plant there that injected over one hundred gallons of water per minute into the hole under a pressure of about 4,000 pounds per square inch to extract heat from the two-mile-deep rock.[47] When I used to soak in Spence Hot Springs the water temperature was at least 105 degrees. Today it is about 95 degrees and is now considered only a warm spring.

If geothermal energy is not inexhaustible, neither is it clean. If you multiply the sulfurous smell of a hot spring a thousandfold, you have air pollution. Hot water under pressure is also a powerful solvent, which is why hot springs always contain a lot of sulfur and other minerals and salts, and often are much saltier than the sea. When this water is pumped to the surface in large quantities, it can cause serious chemical and thermal pollution of freshwater rivers and lakes if it is allowed to run off. Reinjecting geothermal brines back into the ground from which they came prevents surface pollution but may hasten the eventual cooling of the entire field.

Iceland uses more hot groundwater for heating than any other country; because of the presence there of so many active volcanoes, it is blessed with an underground heat source that may well be inexhaustible. Here in the United States, however, the small islands of bubbling paradise I've grown to love may soon be gone forever. Their fortunes, more and more, will be linked to the fortunes of endangered species like the magnificent California condor, who, like the hot springs, can only survive if they are left alone.

Toxic Wastes

Groundwater is vulnerable to a constant rain of pollutants that percolate down from the earth's surface. Farms, factories, oil wells, and copper mines; cesspools, sewer lines, landfills, and gas tanks—all are sources of chemicals and contamination that find their way earthward. Salt, applied to roads in winter, enters the groundwater. So do the solvents used in hundreds of thousands of businesses to clean metal parts. So do pesticides sprayed on farmland. So do radioactive isotopes generated by nuclear power plants. Many of the 1.2 million abandoned oil and gas wells in the United States, and unknown numbers in other countries, continue to bring salty, radioactive, noxious liquids *up* from a mile or more beneath the ground to contaminate fresh drinking water closer to the surface.

Most horrible of all are toxic wastes that are *intentionally* pumped into wells drilled deep into the earth. As regulation of surface disposal became stricter, deep well injection became the most popular method of disposing of hazardous wastes in the United States. Since the 1980s, the *majority* of my nation's toxic and radioactive wastes—more than 30 trillion gallons of it[48]—have been injected down deep wells directly into the ground, much of it under pressure, to depths of from a few hundred to over ten thousand feet. The oil and gas industry has been pumping "brines" from their drilling operations into deep wells for over eighty years. The drug, chemical, pharmaceutical, refinery, steel, and other industries have been pumping toxic wastes down deep wells for over sixty years.

Altogether there are about 1.7 million municipal, industrial, commercial, agricultural, and domestic wells in the United States that have been injecting some kind of fluids below the surface of the earth. Most are on farms and oilfields, and their owners have not been very particular about where they pump the stuff. Sites selected by industry for injection of their most hazardous wastes are supposed to be porous, brine-containing sandstone or limestone beds between impervious layers of rock. Sometimes explosives have been used to blast out underground disposal caverns. It has been claimed that these areas are isolated from deep freshwater aquifers, but in practice groundwater contamination has been widespread. Most of it is hidden, unacknowledged, until, from time to time, it erupts to the surface in spectacular accidents, such as occurred at Erie, Pennsylvania in 1968 when an overpressurized well erupted, spewing its toxic contents into the air like a geyser. Or in 1975 in Beaumont, Texas, when five million gallons of dioxins, herbicides and other chemicals spewed into a nearby drinking water aquifer from an injection well used by Velsicol Chemical Corporation. Earthquakes have been caused, such as the ones that shook Denver while the Rocky Mountain Arsenal was disposing of nerve gas and other chemicals in the 1960s. More than 1,300 earthquakes, up to magnitude 5.3, were caused by that injection well between 1962 and 1968. Earthquakes shook a nuclear power plant at Perry, Ohio in the

1980s after the industrial disposal of pesticide wastes down a deep well a few miles away. Between then and now, earthquakes related to wastewater injection have occurred in Ohio, Colorado, Oklahoma, Arkansas, Texas, and West Virginia.[49]

This stuff is down there permanently and irretrievably. Since groundwater flows so slowly—as slowly as five or ten feet per year, and not more than a mile a year through the most porous ground—it is not self-cleansing except over geologic time.

In 1984 the United States amended the Solid Waste Disposal Act to regulate, for the first time, the disposal of toxic wastes in deep injection wells. But the law contains so many exceptions and loopholes that underground disposal of wastes remains the most popular way to get rid of them. And Class V injection wells, of which there are about 1.5 million in the U.S., inject wastes that are classified as "non-hazardous" into or above underground sources of drinking water. Their uses include draining cesspools, municipal wastes, and septic systems used by apartment buildings and strip malls, and since all modern households, businesses and municipalities use and discharge toxic chemicals, much of the drinking water in the U.S. is being contaminated.[50]

The oil and gas industry alone disposes of about 3 billion tons of brine per year down 170,000 injection wells, amounting to about ten tons per person per year, or about 60 pounds per person every day.[51]

It is clear that effectively outlawing deep well disposal would only increase surface disposal and incineration of these poisons. The stuff itself has not been outlawed anywhere, and it has to go somewhere—if not below ground, then above ground, or into rivers, lakes, and the ocean, or up into the air. It is all one system, and the sheer quantities of poisons our civilization is generating are overwhelming the self-purifying capacities of air, water, and soil alike. As John Hernandez of the U.S. Environmental Protection Agency said of toxic chemicals, "Where in the world are we going to put that stuff is a mystery to me."[52]

The EPA states that about 2.5 billion tons of solid, industrial, and hazardous wastes resulting from the manufacture and use of goods throughout the economy are disposed of each year in the United States.[53] But it also states that 7.6 billion tons of "non-hazardous" industrial solid wastes are produced annually.[54] The problem is in the definition of "non-hazardous."

Although the United States regulates hazardous wastes more severely than the majority of the world's nations, the legislation is so anemic as to have virtually no effect on the torrents of poisons flowing out of our industrial cauldrons into our communal water supply. Only a few hundred different chemicals are regulated in any way whatsoever, out of the 50 million chemicals that are produced. Adding the 6 billion tons of mining wastes and 300 million tons of municipal wastes to the 7.6 billion tons of solid industrial wastes and 3 billion tons of oil and gas wastes that our nation produces annually, one calculates that at least 280 pounds of solid and liquid wastes are generated in the United States per day for every resident of my country. And such is the modern way of doing things that virtually every drop of it is contaminated with hazardous substances. If you are a Canadian or of some other nationality, the numbers are somewhat different, but not dramatically so. And they are not likely to change much in the near future as long as we continue to "modernize" the world's farms and factories, and as long as we continue to live the way we do and consume the kinds of products we are accustomed to.

Hi-Tech Poisons

They are among the newest of our technologies, and they are among the most addictive and destructive. It is no coincidence that the rapid growth of the computer industry in the 1980s took place at the same time as a frightening deterioration in our world's environment. The connection between the two is *speed*. Human transactions which used to take hours or days now take microseconds. The industrial economy thrives on the increased speed; the earth's ecology is incompatible with it. Machines can be made to run as fast as one wants; living creatures, which can only go at their same age-old tempo, cannot compete in the same world with them.

The electronics industry, on which we have recently made our entire transportation and communication network completely dependent, is also a very polluting industry. Enormous mounds of computer paper are chewing through Canadian forests. Computer screens, wireless or wired, emit harmful electromagnetic radiation. So do televisions, cash registers, photocopiers, automobile ignition systems, and every other kind of electronic equipment, anything with a screen, and anything with a digital display. The radiation emitted by computer and TV screens covers the entire electromagnetic spectrum, and is very difficult and expensive to shield against, much more expensive than any manufacturer has ever spent. And now that virtually all electronic equipment is or is becoming wireless and is *intentionally* emitting radiation, very few people even remember that electronic equipment *itself* emits radiation. I will take up electromagnetic pollution in more detail in chapter 18.

More relevant to this chapter is the fact that the manufacture of electronic parts is extremely polluting to groundwater. The production of computer chips requires the etching of silicon wafers using acids, solvents, and toxic gases like arsine and phosphine. Other chemical and toxic metals play a large role in attaching chips and other components to printed circuit boards. The final circuits are cleaned with more industrial solvents.

The manufacture of computer chips requires more than 1,000 different chemicals and metals, most of them toxic.[55] The electronics industry has become one of the largest producers of hazardous wastes in the United States and a major polluter of groundwater. When most computer chips were manufactured in Silicon Valley, well water in that region of California was rendered permanently unfit to drink. Now that most chips are made in Taiwan, South Korea, Japan, and China, everyone has forgotten this. The Silicon Valley Toxics Coalition, founded in 1982 to expose this problem, no longer exists. But groundwater continues to be contaminated wherever computer chips are made, including in the United States. The Intel Corporation, which makes chips in the U.S., Israel, Ireland, Malaysia, and Germany, used and contaminated 16 billion gallons of water worldwide in 2021. Since Intel produces only 6 percent of the chips in the world, this industry pollutes more than a quarter of a trillion gallons of water globally each year.

Dams

They've constipated the planet.
— Sandy Healy

Just how much they've constipated the planet is difficult to realize. Dams and reservoirs were so much around me as I was growing up that I took them for granted as a normal part of the earth's geography, like rivers, mountains, and valleys. They provided me with all of my drinking water in New York City. They provided me with my favorite swimming holes in Ithaca. They calmed the waters in most of the canoeable rivers in my state.

Dams are as old as civilization itself. Deforestation, agriculture, and large irrigation systems have formed the background for most of the great human civilizations, ancient and modern. Ancient Sumer, Babylonia, India, China, Sri Lanka, Cambodia, the Roman Empire, the Inca and Aztec empires, and the nations of the American southwest all were watered by large irrigation works.

Today's dams dwarf those of the ancients in both size and number. In Egypt the Aswan High Dam on the Nile is seventeen times more massive than the Great Pyramid of Cheops. The Akosombo Dam on the River Volta submerged 4 percent of the country of Ghana. 434 dams have been built or are under construction in the Amazon Basin, and 463 more are in various stages of planning.[56] The National Inventory of Dams lists 91,815 dams in the United States. There are about one million dams on the continent of Europe.[57]

Large dams are neither simple nor benign. They do not create lakes, just large tubs of water. A lake is a living breathing system, complete with bugs, plants, fishes, beavers, bears, birds, and whoever else grew up there, all in some sort of equilibrium with one another established over a long period of time. It is true that in nature sudden earthquakes, floods, and landslides occasionally create instant lakes, but these are rare, and when they do happen, they are just as disastrous as our huge dams are. Our dams are worse, because they permanently prevent fish and other animals from swimming up and downstream. Our dams are worse, because the water plants that take over a new reservoir and thrive on the rotting underwater vegetation get in the way and clog up our irrigation canals and our turbines. We kill them off as weeds so they can never do their job of cleaning up the mess and preparing the way for a greater diversity of life to return.

Our dams drown forests and flood farmland, displace human beings and drown wild animals, cut off fish from their spawning streams and suck them into turbines where they are pulverized. The sudden filling of a large lake causes earthquakes. Soil is no longer carried and deposited downstream, and the river channel and banks erode rapidly while the new reservoir fills up with silt. Floodplains no longer receive the sediments that accompany periodic flooding, and irrigated farmland, while it receives a lot more water, becomes less fertile. The water table rises, and the ground becomes waterlogged. Stagnant water behind the dam and in the irrigation canals breeds mosquitoes and disease. Dissolved salts become more concentrated by evaporation from the reservoirs and canals, and instead of being washed into the ocean they accumulate in the irrigated soil, poisoning the vegetation.

Hoover Dam on the Colorado River, built in 1936, was the first dam over 500 feet high ever built in the world. The filling of the reservoir behind it caused the first dam-related earthquakes. Since then, some three hundred more such huge dams have been erected all over the world, on every continent. The enormous power of these engineering projects to alter the earth has been taken much too lightly, probably because they seem like just bigger versions of what human societies have been building for so many thousands of years. But it is precisely their huge scale that is so devastating. They don't weave themselves artfully into local ecosystems like the dams beavers build out of sticks and mud. They are not as forgiving as the earthworks that diverted water onto the fertile fields of my ancestors for so many generations, and whose cumulative impact on the land became evident only after hundreds or thousands of years. Modern dams are huge monsters of impudent concrete and steel whose imposing walls halt the flow of the very largest rivers, drown entire watersheds, and poison the lifeblood of wild systems.

A technology that destroys whatever it touches cannot be used wisely. Its power is such that if it is used at all on this earth, nothing is safe from its reach. Nothing, it is rightly said, is sacred anymore.

Japan stilled the waters of its last wild river, the Nagara, in 1993. China and the five countries of Indochina have constructed 60 hydroelectric dams that have tamed the 2700-mile-long Mekong River, at a high cost to native forests, wildlife, and villages. In Sarawak on the island of Borneo, the drowning of a culture and the flooding of hundreds of square miles of rainforest is underway; three large hydroelectric projects, including the largest dam in Southeast Asia, have already been built and a fourth is under construction. French Guiana flooded 120 square miles of dense unbroken rainforest behind a dam on the River Sinnamary. The electricity it generates supplies the launching site for the European space Agency's satellites. Chile has built three dams along the Biobío River in one of the least contaminated parts of the planet, invading vast expanses of virgin forests and decimating the Mapuche-Pehuenche people. India has dammed the sacred headwaters of the Ganges, submerging 100 Indian villages behind an 850-foot-high wall that has stopped the flow of the Bhagirathi, the Ganges' major source. A mammoth development project, underway in another part of India, including 30 major dams, 136 medium dams, and nearly 3,000 smaller dams in the Narmada River valley, will, if completed, flood over 2,100 square miles of forest and farmland and displace more than a million people. China built a 607-foot-high dam that drowned one of the great scenic wonders of the world, the spectacular Three Gorges on the Yangzi River. It also flooded nineteen towns and counties, 100,000 hectares of fertile farmland, and the homes of about 1.4 million people, many of them subsistence farmers, fishers, and craftspeople. Its completion was the last straw for the baiji, a river dolphin that had lived in the Yangzi for 20 million years. The dam went into operation in 2003, and the baiji was declared extinct in December 2006. The Chinese paddlefish was declared extinct in 2022. The Yangzi softshell turtle, the largest freshwater turtle in the world, was declared extinct in 2023.

The fate of a few of the world's most famous rivers shows the devastation that very large dams always cause:

The Amazon

The Amazon River runs 4,000 miles to the Atlantic Ocean and fifteen of its tributaries are over 1,000 miles long. This great river basin holds about one fifth of the world's fresh surface water. Since these rivers run mostly through flat, low-lying country, building a high dam means flooding immense areas of jungle. The Balbina Dam on the Uatumá River flooded 900 square miles of jungle to produce just 250 megawatts of electricity, about enough to supply half the needs of nearby Manaus. A bit more efficient, the Tucuruí Dam on the Tocantíns River flooded 1,100 square miles of jungle and produces 8,370 megawatts. The Samuel Dam, on a tributary of the Madeira, was built on land so flat that engineers had to build thirty miles of dikes to contain a lake of 200 square miles. The Amazon Basin is now home to about twenty-five million people, demanding goods and services of every kind. The population continues to rise as dams such as Tucuruí provide a cheap and abundant supply of electricity for industrial development. The hundreds of dams already built in the Amazon Basin are flooding hundreds of thousands of square miles of forest and threatening migratory fish, turtles, and river dolphins with extinction.

The history of dams in rainforests has been a most unhappy one. The first such dam of substantial size was built in Suriname, creating Lake Brokopondo in 1964 and submerging 600 square miles of virgin forest. The rotting vegetation under the lake gave off hydrogen sulfide gas, and the stink was so bad that workers at the dam had to wear gas masks for two years. It also acidified the water and corroded the dam's cooling system, which needed replacing. By 1966, the lake was half covered with water hyacinth and ferns, impeding navigation and clogging up machinery. The government of Suriname spent $2.5 million spraying Lake Brokopondo with 2,4-D to destroy the weeds. Native people and wildlife found their drinking water first clogged, then poisoned.

This type of experience has been repeated over and over. Within three years of the damming of the Congo River in Africa, water plants had clogged up 1,000 miles of the river. In Colombia the Anchicayá Reservoir silted up almost completely in just ten years. In the Philippines the Ambuklao Reservoir, which was supposed to pay for itself in sixty years, filled up with silt in just thirty-two.

Ghana's Akosombo Dam displaced 70,000 independent farmers from 740 settlements. Parasitic diseases became a much more serious problem—river blindness, spread by flies, and schistosomiasis, carried by snails. Huge chunks of the West Africa coast have been swept away by the powerful Guinea current, as the river no longer replaces sand that is washed out to sea. The coastal town of Keta fell into the sea, twenty miles down current from the Volta estuary.

The Nile

The River Nile, longest in the world, watered and nourished Egyptian civilizations for many thousands of years. Egyptian agriculture, unlike that in other ancient lands, did not require a large system of irrigation canals. Once a year the Nile flooded a wide swath of land along the river valley, spreading a fresh layer of fertile soil over the earth. The annual flooding prevented salts from accumulating in the soil, because as the water drained through the soil it carried them back into the river and out to sea. And so agriculture in the Nile Valley did not suffer the same fate as in so many other ancient civilizations around the world. The fertility of Egyptian soil was renewed, year after year, century after century for millennia—until the Aswan Dam.

The Nile was dammed at Aswan in 1902. The dam was raised in 1912 and again in 1932. The existing Aswan High Dam, completed in 1970, created behind it a lake 340 miles long and up to 22 miles wide, flooding farms and villages and displacing 100,000 Nubian people from their homeland in Egypt and Sudan. The dam has brought the dubious benefit of eliminating the annual flooding and transforming 2.5 million acres from seasonal to year-round irrigation. Farmers now harvest two or three crops each year instead of only one, and 950,000 acres of desert have been reclaimed for agriculture. Hydroelectric turbines produce up to 10 billion kilowatt-hours of electricity per year.

Such benefits have fueled industrialization and population growth to such an extent that the growth of cities has eaten up as much farmland as has been reclaimed from the desert. The surrounding water table has risen, and so has the salt content of the irrigated earth. Vast amounts of pesticides and fertilizers are being used to grow crops on deteriorating soil. Shorelines and the riverbed below the dam have eroded, and some beaches and other areas of the Nile delta have washed into the Mediterranean Sea. The Egyptian government has periodically dosed its canals and irrigation drains with massive quantities of herbicides in an attempt to destroy water weeds, but they always come back. The dead and rotting weeds only add to the nutrient supply that caused the problem in the first place. The sardines in the river, which once supported a large fishing industry, are gone, and the lake has been stocked with other types of fish. Sediments, unexpectedly, have mostly been deposited above the lake instead of in it; this has created unforeseen problems upstream, but this reservoir is expected to survive for over a thousand years.

The Euphrates

Ancient source of fertility for the Fertile Crescent, nourisher of Mesopotamia's great civilizations, one of the four rivers in the Biblical Garden of Eden—this mighty river has been thoroughly dammed up, tamed, conquered, destroyed. Both Syria and Turkey can now turn the river on and off at will. During January and February of 1990 no water at all flowed through Syria and Iraq as Turkey filled the reservoir behind Atatürk Dam, its third on the Euphrates. Turkey is one of the most dammed countries in the world, with more than 860 active dams and more planned.[58]

Farewell to Québec

One of the biggest, most devastating hydroelectric developments in the world was initiated in 1971 in James Bay territory in Canada's province of Québec. Dams, dikes, and reservoirs planned in the watersheds of the La Grande, Great Whale, Nottaway, Broadback, Rupert, Laforge, and Eastmain Rivers would alter waterways in an enormous area extending from James Bay in the west to the Newfoundland border in the east; from within 225 miles of Montreal in the south to Ungava Bay, 750 miles further north. The projects that have so far been built have had huge human and environmental consequences.

Phase I of the James Bay Project, completed in 1985, involved the building of 9 dams and 206 dikes, diverted 7 large rivers, and created five reservoirs. Virgin forest equal to half the area of Lake Ontario was drowned. The Caniapiscau River was made to flow backwards, and the reservoir created by the Caniapiscau Dam became the largest "lake" in Québec. That reservoir flooded a caribou calving area, and in October of 1984 the newly swollen river drowned ten thousand migrating caribou on their annual trek from Labrador to their winter grazing lands near Hudson Bay. Phase II, completed in 1996, created three more large reservoirs and drowned more virgin forest. Phase III, completed in 2012, diverted half the total water flow of one of Québec's largest rivers, the Rupert, and created the 600-square-kilometer Eastmain Reservoir. Formerly roadless areas have been opened up to mineral exploration and clearcut logging. Increased plumes of fresh water into Hudson Bay have decreased the bay's salinity and caused abrupt freezing along its coastline in winter, entrapping eider ducks and beluga and killer whales. The flooding has released large amounts of mercury from the bedrock in some areas and contaminated the fish and all who eat them. Roads and modernization have drastically changed the nomadic societies of Cree and Inuit who have lived by hunting, trapping, and fishing for thousands of years. Where in 1971 there was only virgin forest and free-flowing rivers, twenty years later there were already 3500 miles of power lines, 930 miles of roads, five permanent villages, and five airports. Many hunting and trapping grounds, and some of the richest wildlife habitats, were now underwater.

Hydro-Québec is taking a vast territory notable for its running water and turning it into a vast territory of stagnant reservoirs, virtual toxic sinks for pollutants in the air. It is complete and utter madness. (Daniel Green, Société Pour Vaincre la Pollution, 1991).

Today Hydro-Québec, courtesy of dams not only in James Bay territory but throughout the province, generates 36.767 billion watts of power from 63 hydroelectric stations, courtesy of 628 dams and 28 large reservoirs. It is the third largest hydropower company in the world, behind only China's Yangzi Power Company and Brazil's Centrais Electricas Brasileiras. These developments have been built *by* Hydro-Québec, but they have been built *for* the whole world—that is, you and me. Most of the power is exported to New Brunswick, Ontario, and the northeastern United States, or used by foreign-owned aluminum and magnesium smelters operating along the Saint Lawrence River. Québec is the world's fourth-largest producer of aluminum even though there are no aluminum mines in Québec. This electricity-intensive industry is attracted here primarily by the cheap and abundant supply of electric power the province is selling.

At times, doing the research for this section, I have felt overwhelmed by the sheer numbers of dams and reservoirs everywhere I looked all over the world. They appear to be a very basic part of civilization itself. As world civilization has become rapidly more mechanized; as worldwide transportation and communication systems have become every larger; as world economies have become larger and more interconnected; as world populations have grown ever more rapidly—so we have built larger and larger dams and reservoirs on more and more rivers.

Civilization cannot survive without water, and we modern people use one hell of a lot of water. To produce a ton of copper, 70,000 gallons of water are used in mining the rock and 30,000 gallons more in refining it. To make a ton of steel requires 60,000 gallons. It takes 110 gallons of water to produce one pound of rayon; 200 gallons to make one pound of rubber; 30 gallons to make one pound of paper. In our homes, Americans each use an average of 138 gallons of water per day for flushing toilets, washing ourselves, our clothing, and our dishes, filling our swimming pools and watering out lawns. The food we eat is grown with an immense amount of irrigation. To grow a pound of wheat requires 50 gallons of water; a pound of potatoes, 80 gallons; a pound of rice, 500 gallons. To produce eggs for breakfast consumes 120 gallons of water for each egg; a one-pound steak for dinner, 3,500 gallons. The total average daily use of water in the United States adds up to some 1,000 gallons per person.

At the same time, civilization has become much more energy-intensive. The average person in the United States uses 10 kilowatts—a hundred times as much non-food energy as food energy. The worldwide average is already up to 2.3 kilowatts.

Such profligate waste of water and energy on a planetary scale is doing in our rivers and wreaking havoc on living systems. Rivers are not only complex habitats themselves, they are important movement corridors between other ecosystems. It is all well and good to build fish ladders whenever you put up a dam, but if there are eight or nine dams on the same river very few fish can make it to the top of the last ladder to spawn in their streams and even fewer of their offspring survive the return trip to the ocean without being crushed to bits in the turbines.

The Opposition

I have found, in my seventy-three years of living, that it is not so easy to stop another person from doing something that he or she wants to do. How much harder it is to stop a whole gang of people from doing what they want to do.

When the damage that any group of people can do is limited enough, not only can we all cope, but some sort of balance can be reached between people of different points of view. But when the available technology is so powerful and devastating as it has become, and its damage so irreversible, no such balance is any longer possible. The earth is now at the mercy of whoever wishes to use, or abuse, that power.

Despite substantial and widespread opposition in every case, and sometimes in spite of organized international protests, the enormous dam projects on the Nagara River in Japan; on the Narmada River in India; on the Yangzi River in China; and in James Bay Territory in Canada were all pushed forward by a coalition of government and industry. The inequality in power between those for and against development is due to this inescapable fact: while a reprieve secured by the opposition may always be reversed at some future date, the damage caused by the dam is always permanent.

In one of the most heroic protests I can remember, thirty-year-old Californian Mark Dubois chained himself to a rock in May 1979 in a hidden part of the Stanislaus River's upper canyon, vowing to drown with the canyon if the New Melones Reservoir was filled. The engineers did halt the filling of the reservoir, and Mark Dubois unchained himself and went back to his private life. But despite this reprieve and the opposition of California's Governor Jerry Brown, Congress refused to add the Stanislaus River to the national Wild and Scenic Rivers System, and the reservoir was eventually filled, permanently destroying one of the most heavily used whitewater rivers in the country.

But perhaps dams need not, after all, be so permanent. No protester has ever blown one up, as a few radical groups have advocated. But each year the fifty-year operating licenses of some two hundred aging American dams expire. A 1986 law requires the Federal Energy Regulatory Commission to give consideration to fish, wildlife, and environmental quality before relicensing them. And since then, a number of dams have actually been torn down to restore the environment. The two large dams on the steep Elwha River, for example, in Washington's Olympic National Park were finally removed in 2014. They were built in the early twentieth century without fish ladders, and they wiped out nine races of native salmon, some of which used to grow to 100-pound size. Some two dozen species of birds and mammals who in some way depended on the salmon for food virtually disappeared from the Elwha Valley. The river runs free again today, allowing salmon once more to spawn and support life.

One of my favorite hot springs bubbles out of the rock along the Elwha River several miles above the second dam. On my hikes up the river to soak my body years ago I never thought much about the dams that were there and what they might have done to the local ecology. I hope one day to return to the river that now flows free.

As this book goes to press, the largest salmon restoration project in history is restoring the Klamath River in California to its natural state.

The removal of the first of the four dams on the river was completed in September 2023, and the other dams will be removed during 2024, allowing salmon, steelhead trout, sturgeon, Pacific lamprey, and many other native fish species to return to the river for the first time in over a century.

The Biggest Dammed Reservoirs in the World

Name of dam	Year built	River	Country	Capacity (million cubic meters)
Owen Falls	1954	Nile/Lake Victoria	Uganda	204,800
Kariba	1959	Zambezi	Zimbabwe and Zambia	185,000
Bratsk	1964	Angara	Russia	169,270
Akosombo	1966	Volta	Ghana	148,000
Daniel Johnson	1968	Manicouagan	Canada (Québec)	141,852
Guri	1986	Caroni	Venezuela	138,000
Aswan	1970	Nile	Egypt	132,000

The Flush Toilet

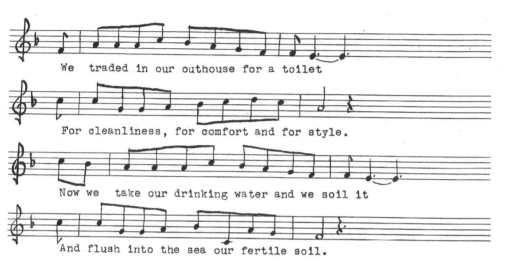

We traded in our outhouse for a toilet
For cleanliness, for comfort and for style.
Now we take our drinking water and we soil it
And flush into the sea our fertile soil.

The Sea

The oceans are dying. The pollution is general.

— Jacques-Yves Cousteau
September 1970

Human beings have been altering earth's waterways for thousands of years. We have been poisoning them in a big way for less than seventy. When I was a young boy most of the rivers and lakes in my country flowed with pure, drinkable, and delicious water. The lakes started to be poisoned when motorboats came into use. It takes just one motorboat, with its oil and gasoline, to contaminate an entire lake. Today it is a rare lake that if it is big enough to have one, does not have dozens. The trend spread to Canada later, but with a vengeance. In 1982 sailboats were still the preferred Canadian watercraft. In that year Canadians spent $80 million on sailboats and only $35 million on powerboats. By 1988 the picture was reversed: $43 million for sailboats and $325 million for powerboats. The silence and purity of Canada's wilderness were gone.

We have trashed our rivers by using them as sewers for our factories, our farms, and our homes. Detergents played a big role in getting us accustomed to the change. In the 1950s and 1960s we watched as our rivers and streams became covered with detergent foam. Manufacturers finally changed the formulations so that didn't happen anymore, but the damage was done: nobody expects rivers to run clean anymore. It has become acceptable to have sewers for rivers, and to depend on factories to filter and chlorinate our water for us before we drink it. But our brothers and sisters in the wild—the bears and the squirrels, the ducks and the geese, the lizards and the snakes—nobody filters the poisoned water for them. The beavers and otters have got to live in our waste, and the fishes have got to breathe it besides.

Even mountain streams that are not tainted with chemicals are contaminated today with giardia, an intestinal parasite that was unknown in America's wilds before the 1960s.

The Mediterranean Sea is suffering under the impact of raw, treated, and partially treated sewage from 175 million coastal residents and 270 million tourists each year; of factories along its shores discharging mercury, lead, detergents, and solvents; of supertankers on their way to and from the Middle East that in the course of normal operations discharge each year half a million tons of oil into the sea.

The level of the Caspian Sea is rapidly dropping because its biggest tributary, the River Volga, has been so extensively tapped for irrigation. In the southern Caspian, near Iran, oil pollution has turned the sea bed into tar.

Irrigation has caused the Aral Sea to virtually dry up. It has changed from the world's fourth largest inland lake to two bodies of water, north and south, separated by dry land in between and containing together only one-tenth the former amount of water. 300 bird species, huge herds of antelope, and all native fish species are gone.

Lake Baikal, the earth's oldest, deepest, and largest freshwater lake, is home to more than a thousand species of plants and animals found nowhere else in the world. It became the rallying point for the Russian environmental movement, which succeeded in shutting down a polluting pulp mill on its southern shore. However, untreated sewage from homes and hotels continues to be discharged into the lake.

The Great Lakes of North America used to receive the effluent of factories that manufactured thousands of toxic chemicals and metals. While much manufacturing has since relocated overseas, that has not stopped the pollution of these waters by agricultural chemicals.

Eventually, it all drains into the oceans. And the cities and factories along the coastlines of the world add to the mess. So do shipwrecks and oil spills.

A phenomenal number of ships have gone down during the last four hundred years. Apparently there are about ten thousand shipwrecks in the waters of my home state, New York. There are an estimated three million shipwrecks in the world. The ocean floor must be paved with them! In modern times every ship that goes down carries a load of oil with it, and a sunken supertanker is a tragedy of gigantic proportions.

The news media report on the very biggest, most devastating oil spills. We heard about:

- the *Torrey Canyon*, which ran aground in 1967, spilling 30 million gallons of oil into the English Channel;
- the oil well blowout that shocked Californians in 1969 by blackening Santa Barbara beaches with 2 million gallons of oil;
- the *Metula*, which spilled 16 million gallons of Persian Gulf crude when it ran aground in the Strait of Magellan in 1974;
- the *Argo Merchant*, which ran aground off Nantucket Island in 1976, spilling 7.6 million gallons of oil onto Cape Code and into the fish habitat of Georges Bank;
- the 1977 Ekofisk well that blew out, spewing a million gallons a day into the North Sea for a week until it was capped;
- the *Amoco Cadiz*, which sank off the Brittany coast of France in 1978, releasing its 63-million-gallon cargo;
- the Mexican Ixtoc oil well blowout in 1979 that spewed half a million gallons a day into the Gulf of Campeche for almost a year;
- the eight Iranian oil wells, bombed by Iraq in 1983, that gushed 80 million gallons into the Persian Gulf;
- the *Exxon Valdez*, which caused environmental tragedy in Prince William Sound, Alaska when it ran aground in April 1989, spilling its 11-million-gallon cargo of crude oil;
- the U.S.-Iraq war of January to February 1991 that destroyed the Persian Gulf with perhaps the largest oil spill the world has ever known;
- the *Braer*, which ran aground in January 1993 spilling its 20-million-gallon cargo of crude off the coast of the Shetland Islands;
- the *Deepwater Horizon* oil well blowout in 2010 that gushed for three months, spewing 200 million gallons of oil into the Gulf of Mexico.

No area of the world has escaped. Even Antarctica is regularly contaminated by oil spills. In 1989 the *Bahía Paraíso*, an Argentine Navy supply ship, went down in Bismarck Strait, spilling 250,000 gallons of diesel fuel and killing thousands of seals, penguins, and other birds in their breeding grounds; and at American bases 50,000 gallons of jet fuel, gasoline, and heating oil spilled onto the ice near McMurdo Station, and 20,000 gallons of fuel oil spilled on to the ice at the South Pole. In 2007 the MS *Explorer*, an adventure travel ship, sank in Antarctic waters spilling 50,000 gallons of oil. At a single Australian Antarctic station there were 38 diesel spills reported between 2008 and 2018.[59]

During the twelve months following the *Exxon Valdez* disaster there were *over ten thousand* oil spills of all sizes reported on land and water in the United States alone.

Who is to blame? As long as we all use so much oil, a percentage of it will continue to wind up polluting our land and our rivers and our lakes and our oceans.

As a conservative estimate, about three million tons of oil from ships, tankers, refineries, petrochemical factories, offshore oil wells, and disposal of industrial wastes winds up in the oceans every year. Much more oil is discharged by ships in cleaning their ballast tanks, pumping their bilges, and other standard operations than is spilled in shipwrecks. About 55,000 oil wells have been drilled in the ocean floor off United States coasts. Each drilling platform during its average twenty-year lifetime has between 1 and 3 large spills (1,000 barrels or more), 25 medium spills, and 2,000 small spills of less than 50 barrels.

That isn't all. The oceans receive an unknown large amount of oil that drains into them from polluted rivers. In fact, spent crankcase oil from automobiles may be a greater source of ocean pollution than oil drilling and tanker transport combined. And the 90 million tons of oil that evaporates into earth's atmosphere every year during transport, storage, and use eventually rains back down on land and sea. The total quantity of hydrocarbons being added to the oceans each year from petroleum may be as large as the amount from the growth of living organisms.

We also send twelve times as much iron into the oceans as do geological processes; twelve times as much copper; forty times as much antimony. We have at least doubled the total amount of lead in the oceans in the past century.

Plastic in all its forms pollutes the seas very extensively. Hundreds of thousands of birds and marine mammals die each year by getting stuck in plastic garbage or swallowing it and choking. Birds strangle in the holes of six-pack holders. Turtles choke trying to eat floating garbage bags, perhaps mistaking them for jellyfish. Birds and marine animals of all kinds die because pieces of indigestible plastic block or ulcerate their digestive tracts. Whales have been found dead with plastic garbage bags lining their stomachs.

It is still a fact that ocean liners, merchant ships and naval vessels of every nation dump their garbage overboard into the ocean. This included up to a quarter of a billion plastic containers per year for a couple of decades until the world community finally realized that the floating stuff was cluttering up the oceans and not disappearing. Since December 31, 1988, it has been illegal by international treaty to discard plastics into the ocean, but enforcement in international waters is problematical, and not all nations have signed the treaty.

Up to 14 million tons of plastic enters the oceans annually from coastal communities.[60] Commercial fishing fleets annually lose between 500,000 and 1,000,000 tons of plastic nets, lines, ropes, and buoys. Lost fishing nets continue to entangle fish, and also seabirds that prey on them, and seals, and whales. Birds become fatally entangled in lengths of fishing line, wrapped around their feet and wings.

Garbage washing up on shore along with the driftwood has become a familiar sight to beachgoers. One three-hour long cleanup of 157 miles of Texas shoreline in September 1987 netted 31,773 plastic bags, 30,295 plastic bottles, 15,631 plastic six-pack rings, 28,540 plastic lids, 1,914 disposable diapers, 1,040 tampon applicators and 7,460 plastic milk jugs. Thirty-five years later the Ocean Conservancy's International Coastal Cleanup continues, every September, around the world. To date, 350,000,000 pounds of plastic garbage have been collected. It is a losing battle. "I have been on beaches in Hong Kong, Saint Helena in the South Atlantic, and Indonesia where you can watch plastics and debris in the barrel of each wave crash onto the beach. Literally, the trash starts getting replaced as soon as you pick it up," says Nicholas Mallos, who oversees the annual cleanup.

I personally had to wash "beach tar" from my feet many a time on the beautiful Southern California coast when I lived there in the early 1980s, and I sat on tar-stained rocks on remote beaches of Mexico's Yucatán Peninsula in the early 1970s.

We all know that the coastal waters of our world are polluted. One would like to think, however, that the vastnesses of the open ocean, way out there in the middle of nowhere, thousands of miles from any land—one expects that those great blue deeps are ever clear and pristine. So few of us ever venture out across the seas anymore in anything but the cabin of an airplane, or perhaps high on deck a speeding ocean liner, that they remain distant and mysterious to us, perhaps not part of the modern world. What, then, of Jacques Cousteau's dismal assessment? Was he really right when he said the oceans were already dying in 1970?

The deepest part of the North Atlantic's western basin is called the Sargasso Sea, named for the Sargassum weed that floats in mats upon its surface. It is an area of relatively calm, warm blue water the size of the United States, encircled by great ocean currents. To the south is the westward-flowing North Equatorial Current carrying water from Africa to the Caribbean. To the west and north is the Gulf Stream, a swift 95-mile-wide river one mile deep that brings warm water from the Gulf of Mexico to bathe Northern Europe. To the east are the undersea mountains of the Mid-Atlantic Ridge.

Fishes that live in the Sargasso Sea are camouflaged with mottled yellow-brown skin and branched and feathery growths that mimic their weedy homes. Humpback whales spend time in these waters. Millions of eels leave their freshwater streams in Europe and North America and swim for thousands of miles to the depths of the Sargasso Sea where they spawn and die—a migration opposite to that of Atlantic salmon. A small gray, black, and white seabird called the cahow, thought extinct for three hundred years, still breeds on the islands of Bermuda and lives on Sargasso weed the rest of the year.

The Sargasso Sea is the wilderness of the Atlantic Ocean, remote from ocean currents and rarely visited by ships. It once contained the cleanest water in the world.

> When I went under sail across the Sargasso, from the West Indies to Africa, 58 years ago, lowering my dory almost daily, it was then the purest and most pellucid water in the world, actually more transparent, as determined instrumentally, than any spring, or pool of

melted snow, or mountain tarn, and completely devoid of continental dust. (Robert Cushman Murphy, 1970)[61]

But when Thor Heyerdahl sailed across the Atlantic in his papyrus boat in 1970, his crew found the Sargasso Sea so polluted they were reluctant to wash in it. A continuous stretch of 1400 miles was filled with masses of asphalt-like oil.

> We saw visible pollution floating past our reed bundles every day. . . . We sailed past tarlike clots of oil 43 days out of the 57 the crossing lasted. Earlier, in 1947, we had sailed in perfectly pure water from Peru to Polynesia on the Kon-Tiki raft. We found no pollution then, although we sifted the sea water by towing behind us a fine-meshed plankton net.[62]

A survey of the Sargasso Sea in 1972 revealed an average of nine thousand particles of plastic per square mile of ocean. Sargassum weed in the 1970s was found to be far less abundant than it was in the 1950s.

> On a cruise through the Sea in July 1969 when we tried to sample the neuston, the organisms inhabiting the upper few centimeters of the ocean, we were notably unsuccessful. Quantities of oil-tar lumps in various sizes up to 6 centimeters clogged the net with every tow. After two to four hours the net became so fouled that it had to be cleaned with solvent. On another day we struck so much oil we had to discontinue work. Some tows during this particular cruise netted more oil lumps than Sargassum. (John and Mildred Teal in *The Sargasso Sea*)[63]

The polluted Atlantic has been displaced in the public's mind by an even greater disaster in the Pacific. In 1997, Captain Charles Moore, returning from Hawaii to California after competing in the Transpacific Yacht Race, found his boat sailing for thousands of miles through a giant soup of plastic debris. More than twice the size of the United States, it has come to be known as the Great Pacific Garbage Patch. In popular conception, it is a pair of calm vortexes, or gyres, in the North Pacific where garbage accumulates after being transported there by surrounding ocean currents. Outside of those patches—so people think—the Pacific Ocean remains clean and pure. But it is not so.

> The assumption has been that plastic litter in the North Pacific clusters into "garbage patches"—one located midway between Hawaii and California and the other between Hawaii and Japan. Both occur in "oligotropic" areas where life-forms are comparatively scant. This might have meant polluting plastics were safely sequestered from more "productive" parts of the ocean that teem with marine life and fishing vessels, somewhat shielding the food chain from contamination. What we find in the transition zone shatters all assumptions. Our TZ trawls are the worst—the most plastic choked—we've ever seen . . . We're finding the highest levels of pollution in a highly productive zone.[64]

On his voyages in the Pacific, Moore found more plastic than plankton in the ocean, by a large margin. "In 1999, the ratio of dry weight of plastic to plankton was six to one. In 2008, it was forty-six to one, and in 2009, it was twenty-six to one, which is surprisingly high, considering these were the most plankton-choked samples we'd ever taken." Deepwater lanternfish were eating large amounts of plastic instead of plankton.[65] The Ellen MacArthur Foundation states that if current trends continue, there will be more plastics by weight than fish in the oceans by 2050.[66]

The Sargasso Sea in the Atlantic, and the two gyres in the North Pacific, are representative of what is going on in every ocean. "Forty percent of the world's oceans lie within subtropic gyres," points out Moore, and "all are similarly polluted with plastic trash"[67] And plastic is not confined to these calm areas. One study found that 80 percent of the plastic washing ashore in the Netherlands had been pecked by seabirds. Another found that 90 percent of blue petrel chicks at South Africa's remote Marion Island had plastic in their stomachs, fed to them by their parents. Other studies have found plastic in the stomachs of 95 percent of puffins, 93 percent of blue petrels, and 80 percent of northern fulmar.[68] After reviewing fifty years of studies of 186 species of seabirds, a group of Australian scientists concluded that 90 percent of all individual seabirds today have eaten plastic.[69] As many as 580,000 pieces of plastic per square kilometer of ocean have been observed, and the highest concentration of plastic, they reported, is not in the Atlantic or Pacific but in the Tasman Sea between Australia and New Zealand. Even pre-production plastic pellets, which are the feedstock for every type of plastic product, and which are a major international commodity that is shipped everywhere, are continually spilled into the world's oceans. Their strong resemblance to fish eggs has made them virtual staples in the diets of several seabird populations.[70] Remote Kamilo Beach, near the southern tip of Hawaii's Big Island, has so many trillions of plastic flecks on it that some say they outnumber grains of sand on the beach.[71]

Researchers in southern California have concluded there is so much plastic everywhere at sea, "contaminating coastal, deep-sea, near-shore, and open ocean pelagic habitats," that if you are a fish anywhere on the planet you are not only ingesting plastic, but being poisoned by it, since every type of plastic is nothing but soup of toxic chemicals.[72]

A whale that stranded itself and later died on an Australian beach was found to have nearly six square yards of compressed plastics in its gut. A gray whale stranded on a Puget Sound beach had in its abdomen sweatpants, a golf ball, surgical gloves, small towels, plastic fragments, and twenty plastic bags. The stomachs of two pygmy sperm whales that were stranded in the northern Gulf of Mexico were found to be "completely occluded by various plastic bags."[73]

When 400 million tons of the stuff are manufactured annually, trying to clean up the mess it is making is futile. As Moore says, "all the creative and well-meaning schemes focused on cleaning up the gyre will work about as well as trying to bail out a bathtub with the tap still running."[74]

When you add microplastics (see chapter 19) into the mix, the problem becomes even more insoluble. Microplastics are plastic particles less than 5 millimeters in size. The popular conception is that microplastics are simply what is left when plastic bags, containers, and other pieces of trash break down over the decades, and that if we were all more careful in disposing of our trash the problem could be solved. This conception is wrong. The International Union for Conservation of Nature published a monograph in 2017 titled *Primary Microplastics in the Oceans: A Global Evaluation of Sources*, by Julien Boucher and Damien Friot. They concluded that between 15 and 31 percent of all the plastic in the oceans *originates* as microplastics, and that the largest sources of this universal pollution are the laundering of synthetic textiles and the abrasion of automobile tires while driving! Textile fibers that are abraded and shed in washing machines are discharged in sewage water and eventually end up in rivers, lakes, and the oceans. And all car and truck tires, which are about 60 percent synthetic polymers, gradually wear away on the world's roads. Tire dust is constantly spread by the wind and washed off the road by rain and, again, eventually ends up in the oceans.

They Nuked Paradise . . .

The United States, Great Britain, and France contaminated large areas of the Pacific Ocean by exploding atomic bombs. Great Britain stopped in 1958, the United States in 1962, and France in 1996. France exploded 41 nuclear bombs in the atmosphere and 152 underground at the two tiny atolls of Moruroa and Fangatoufa. By 1980 the base of the Moruroa atoll was used up—full of giant holes like Swiss cheese—so they then drilled bomb shafts in the bottom of the lagoon. Large amounts of nuclear wastes that had accumulated on Moruroa were washed out to sea by a typhoon on March 11–12, 1981, including some forty pounds of plutonium. The islands and the lagoon were all sterilized. Coral reefs in the area are all dead. And human leukemia, lymphoma, brain tumors, and thyroid and lung cancers remain prevalent throughout French Polynesia.

About two hundred nuclear reactors travel the world's oceans today on submarines, icebreakers, aircraft carriers, and other vessels. Accidents involving American and Soviet ships, bombers, and rockets left at least fifty nuclear warheads and nine nuclear reactors scattered on the ocean floors. And the sea floors under Arctic waters are littered with more than 17,000 pieces of radioactive debris. Dumped into the Kara and Barents Seas by the former Soviet Union, they include fourteen nuclear reactors, spent fuel rods, nineteen ships containing radioactive wastes, and three nuclear submarines.[75]

Strip-Mining the Seas

Heedless of the fact that all of the ocean's fish are eating, breathing, and swimming in water contaminated with petroleum, microplastics, plutonium, and thousands of toxic chemicals, the world's eight billion people still want to eat fish, and fishing fleets from every nation are using more and more sophisticated technologies to satisfy them. This will not stop unless either the demand for fish ceases, or the technologies to satisfy that demand cease, or until there are no more fish in the ocean. And an international team of scientists led by Boris Worm concluded in 2006 that the third possibility is the most likely: if world consumption of fish continues at its present rate, wrote the authors, the oceans will be essentially devoid of fish by the year 2048.[76] That study was updated in 2016, with similar conclusions.[77] In 2003, reviewing half a century of ocean studies, Worm and his colleague Ransom Myers found that 90 percent of all large fish—tuna, swordfish, marlin, cod, halibut, skate, flounder—that swam in the oceans in 1950 were already gone.[78]

As fish stocks became depleted, larger and larger nets with smaller and smaller mesh size were used. During the 1980s and early 1990s, between 1,000 and 1,500 Japanese, Taiwanese, and South Korean ships systematically strip-mined the Pacific Ocean to supply an increasing global demand for seafood. Each ship carried a nylon monofilament up to thirty *miles* long and thirty to fifty feet deep that drifted at night trapping or entangling every living thing that swam into it—fish, marine mammals, diving birds, and turtles indiscriminately, so long as they were larger than the size of the mesh. A fleet of fifteen ships could sterilize 900 square miles of ocean at a time. In 1989 the United Nations General Assembly passed a resolution calling for a moratorium on drift nets longer than 2.5 kilometers in international waters, and in 2002 the European Union banned the use of drift nets of any size. But the bans are not well-enforced, and the practice has continued.

Another method of strip-mining the seas, which is practiced worldwide, is bottom trawling. This is the practice of dragging a large steel beam and metal chains along the *bottom* of the ocean, capturing every type of living thing in a thousand-foot-long net that is attached behind it, and smashing everything it does not catch. This destroys sea beds, coral reefs, and everything else the trawl drags over.

Computer technology, in combination with sonar, has accelerated the strip-mining of the oceans even more by giving fishermen 3-D images of their targets. Michel Derosière, fishing in the English Channel, described it this way:

> It's as if the water had been drained away and I can look right down and see exactly what the sea bed looks like . . . It's fantastic. It means I can update the data in real time—so we can fish areas we used to avoid, whatever there might be down there, even the most treacherous shelves or rock formations. And we always know exactly where we are—to within a meter.[79]

"It is a revolution for us," said Icelandic fisherman Halli Stefanson. "You can get 17 tons as a result of two minutes' fishing."[80]

Today China catches more ocean fish than any other country, followed by Indonesia, Peru, Russia, and the United States. The markets are worldwide.

CHAPTER 13

Peace Walk (1991)

In the spring of 1986, I heard there were a large group of people walking across the United States, walking for an end to nuclear weapons in the world, for an end once and for all to the threat of nuclear annihilation I had grown up with. They called themselves the Great Peace March for Global Nuclear Disarmament.

I left New York at the beginning of July bound for the Midwest. I stopped for a few days at the annual Rainbow Gathering in Pennsylvania, where I met some walkers who were "on vacation" from walking. Together, Jay "Cloudwalker" Stolzberg, Laurie McWhorter, and I hitchhiked back to the March, arriving in the quiet tree-lined streets of Anita, Iowa on July the tenth. Six hundred people in their blue, green, yellow, and orange tents were camped in a park beside a lake enjoying a day of rest, and that evening folk singer Pete Seeger entertained us and lent his energy and support to the cause.

The next day I walked with them ten miles eastward, and the following day twenty. By week's end I had my walking legs, and for the next five months I was a member of a community that moved an average of fifteen miles a day and rested one day a week.

We were a strange meld of modern industrial society with food-gathering nomadism. Our tents were made of nylon, and we were accompanied by so many cars, trucks, and buses that there was a vehicle for about every five people. But we walked, and the vehicles moved at a walker's pace rather than the other way around. Most of us carried no clocks or money, we lived outdoors with the sun, moon, and stars for a roof, and we sat and slept upon the ground. We depended partially on donations for our food, which we shared communally in the manner of a nomadic tribe. We bathed where and when we could, which by American standards was not often.

And I saw, and heard, and smelled, and felt what was happening to the earth around me, one step at a time, across the Iowa prairies to the Mississippi River, across the farm country of the Midwest, across the Allegheny Mountains and the forests of Pennsylvania. I slept outdoors in my native New York City for the first time in my life, and I walked down the industrialized east coast to my nation's capital, the District of Columbia.

The air was fresh-scented and crystal clear over the cornfields of Iowa. It was a deep blue, even down to the horizon in every direction, as air should look. But from the Mississippi River all the way to the Atlantic Ocean hung a pall of pollution which colored the air an off-white or dirty-yellow—fading to clear blue overhead, but noticeably yellow low on the horizon.

The worst day was walking over the Goethals Bridge from Staten Island, New York into Elizabeth, New Jersey. The chemical stench was indescribable, the air a most unhealthy hue, and the scene that greeted my eyes on the other side of the bridge where the oil refineries and chemical factories stand—it is like a huge bombed-out, sterile landscape, filled with ugly twisted metal and concrete towers and smokestacks belching flames and smoke of many colors—it was like walking into hell. I could scarcely comprehend that human beings actually work eight hours a day in a place like that, looking at that scenery and breathing in that air.

But perhaps the hardest part of my five months of walking along this country's roads, wide and narrow, was stepping over all the dead animals that line the shoulders of every road. They lie there, thousands upon thousands of them, killed by speeding cars, buses, and trucks, small animals and large, common species and endangered ones, many more than are killed by guns. I stepped over dead cats and dogs, and I stepped over many varieties of dead birds, snakes, and turtles. There were squirrels, skunks, and raccoons, and there were some animals I did not recognize. I saw a few deer, and even a rotting cow. And there would have been human beings lying there too if they were not so efficiently carted away as soon as they're killed.

And I vividly remember walking alongside the fields of the Midwest, and for several days, carpeting the roads upon which we walked were thousands and thousands of dead and dying yellow butterflies.

Many if not most of us who walked across the United States in 1986 still dedicate ourselves to peace and environmental concerns. Our lives have been changed, each person differently according to our nature. I have readjusted to sleeping in a bed, sitting in chairs, and living indoors, but I don't feel that I fully belong to this way of life. Clocks and money were hard to come back to. I have a terrible time buying shoes anymore because my feet are wide and flat from a couple of thousand miles of steady walking.

I have not since seen so many butterflies as I saw that summer while we were walking. They seem to have all but disappeared here in the Northeast compared to the number I remember brightening the skies of my youth. I often wonder where they have all gone, and whether the flowers miss them too.

CHAPTER 14

Plants and Animals

Remember, if you can, a time and a place where you once looked about you at the bluest morning sky you ever saw, perfect blue from horizon to horizon. Clouds floated by, in their varied roundnesses resembling sheep, or mountains, or dragons of your imagination. A breeze blew softly across your face, and you smelled the sweetness of the air, a mixture of delicious scents—fresh, dark soil, flowers, pine needles, perhaps lingering moisture from last night's rain. The air is buzzing with the wings of small insects and the songs of grasshoppers or crickets, the call of distant birds, one to the other, the flapping of wings as a flock of them passes overhead. The soil is cool beneath your bare feet. You are thirsty. You walk down to the stream to drink really good-tasting water.

Imagine, all about you, life in its abundance. The stream is thick with fish, and beavers make their homes in it, and otters. Minks inhabit its shores, and fishers stalk their porcupine prey from the high branches of the forest. Great numbers of martens, weasels, and wolverines share these woods. Black bears fish in the stream and snack on numberless varieties of fruits and berries. The forest is home to deer, moose, elk, caribou, and buffalo, and to foxes, wolves, cougars, and bobcats.

Imagine rivers so thick with fish they could stop a boat during spawning runs that lasted a month at a time. Beaches and offshore islands overflowed with nesting birds, a dozen species or more sharing the same space. Flocks of birds were so enormous they blocked the sun as they flew overhead in their millions—sometimes in their billions.

The Gulf of Saint Lawrence was visited by large schools of dolphins and whales of many sizes and varieties. Its shores were peopled by black, brown, and white bears. Enormous numbers of walruses lived here. Penguins frequented the entire Atlantic coast from Greenland to Florida. Parrots graced the summer skies with their green, yellow, and orange plumage. Four kinds of seals played in their millions up and down the seacoast in the Gulf and River of Saint Lawrence, and one variety made Lake Ontario their home.

On the Arctic tundra, in late spring and early fall, migrating caribou formed an endless undulating sea of flesh as far as the eye could see for days at a time. On the Great Prairies the buffalo were equally many, and pronghorn antelope traveled these plains in vast herds. And in their season of ripeness strawberries grew so thick upon the prairies that the animals' hooves seemed as if covered with blood.

Such was the density and variety of life that greeted the arriving Europeans on this North American continent more than four centuries ago.

Now imagine a somewhat changed land. The forests, from the Atlantic Ocean to mid-continent, are thicker, and the prairie grasses further west are interspersed with trees. Buffalo and musk oxen are no longer the largest animals on the continent. Elephants roam the land, along with camels, horses, giant wolves, lions and saber-toothed tigers. With them are bear-sized beavers, giant armadillos and large ground sloths. And vultures soar above them with wing spans of up to twelve feet.

Such was the variety of life that greeted the first Indians who arrived here. Such abundance did the earth once provide on every land and in every ocean. Only in Africa has a semblance of this living wealth survived.

To make sense of the recent history of extinctions on the earth requires an understanding of the geography of the continents and of human migrations that is often made difficult by language and by modern forms of transportation. We speak of seven "continents"—Antarctica, Australia, South and North America, Africa, Asia, and Europe—that are vastly different from one another in both size and degree of isolation. The airplane has been the great equalizer, and the ease of air travel hides these differences. An even bigger deception, however, is the belief that the airplane has connected previously unconnected territories—the belief that the "one unified world" is a modern invention.

The whole world is more or less accessible to human beings, over land by foot and over water by boat. Australia was colonized more than 130,000 years ago,[81] so people must have had ocean-going boats for at least that long. The Indonesian island of Flores was colonized by humans at least one million years ago.[82] There has never been a land bridge connecting Java to Flores, which means people had to have traveled thirty kilometers over water, so people were capable of traveling at least short distances by sea a million years ago. So few of us have ever crossed the oceans in anything but an airplane or a steamship that we aren't aware that it can be done fairly easily on even a wooden raft or a boat made of reeds. The trip across the ocean was probably not made very often until the time of Columbus, but it was made often enough that new technologies and population growth had effects the world round.

Europe, Asia, and Africa form one very large land mass that until recently had no real barriers to migration of humans or other animals. The Sahara has become an arid barrier only during the last 4,000 years, and the Suez Canal divided Africa from Asia only in 1869. Before that the Isthmus of Suez was no barrier to the movement of animals east and west. During the Ice Ages the Gulf of Suez was often dry land, and the passage at Suez was even wider.

North America has had an intermittent land connection to Siberia in the far North. During the Ice Ages human beings and other animals were able to move back and forth along this land bridge. But the connection was often interrupted, the journey was long and cold, and the land link has not existed at all for the last 12,000 years. North America is a relatively isolated continent—less so for humans and more so for other creatures—and it is much smaller than the land mass of Europe-Asia-Africa.

South America is not much more than half as big as Africa. Connected to North America only by a narrow isthmus hundreds of miles long, it is even more isolated from Europe-Asia-Africa.

Australia, among inhabited continents, is the most isolated. It is less than half as large as South America, has no land connections with the rest of the world, and historically has been the least visited by human beings.

It is a fact that most of the largest animals and birds are gone from the continents of Australia and North and South America, and that on the Eurasian-African land mass most of the large Pleistocene (Ice Age) animals that remain are in Africa south of the Sahara, and in Southeast Asia. A big argument has raged among scientists about what caused these mass extinctions—a changing climate, or hunting by humans? It appears to me that in their zeal to prove a theory, neither side has been very objective in their examination of the evidence.

On every continent and island where it occurred, the large wave of extinctions was preceded by the arrival of human beings who hunted them. The dates for the extinctions given by Paul Martin are 13,000 years ago for Australia, 11,000 for North America, and 10,000 for South America.[83] Large animals lived on in the West Indies until the arrival of human beings 4,000 years or so ago. On the more isolated islands of Madagascar and New Zealand, the Pleistocene fauna did not begin to disappear until 800 or 900 years ago. The extinctions of smaller animals all over Polynesia followed human colonization in relatively recent times. On uninhabited islands like Mauritius and the Galápagos, the entire ecosystem survived intact until European settlers arrived or, in the case of the Galápagos, until it became a regular port of call for navy boats, whaling vessels, and pirates.

When all these animals became extinct, there were no comparable extinctions in the plant world, and no marine organisms that died out. Only large animals that were hunted by people. And the extinctions left empty niches in terrestrial ecosystems which have mostly gone unfilled. The climatic shift hypothesis explains none of this. The ice has retreated many times without harming big animals.

But Paul Martin and others who have argued the case for "Pleistocene overkill" have found it necessary to believe, for example, that human beings did not arrive in America until 12,000 years ago, and that all the large herbivores and their predators died out within the space of about 1,000 years. There is plenty of evidence to the contrary on both these points. People have likely been in North and South America for at least 300,000 years.[84] And the animals probably didn't die off all that suddenly.

For an idea of what might have happened to other animals around the world, let's look at how it went for the Old World lion. Fossil evidence shows that in Ice Age times lions lived throughout almost the whole of Europe, Asia, and Africa. They ranged west as far as England and Wales, east through Holland, France, Germany, Switzerland, Austria, Czechoslovakia, and Russia into at least the western part of Siberia, south through virtually all of Africa, and southeast through Asia Minor, Arabia, the entire Indian subcontinent, and Sri Lanka. They vanished from England 40,000 or 50,000 years ago, and from Germany sometime after 30,000 years ago. They were still in Greece in the fifth century B.C., but by the end of the first century A.D. they were extinct in Europe. In the Middle East lions were common in Biblical times. The last Israeli lion was killed at Lejun, near Megiddo, in the thirteenth century A.D. Lions survived in Arabia until the twentieth century. The last two lions of Iraq were captured on the Khabur River about 1913, and after the First World War lions were gone forever from Turkey. In Asia the last refuge of the lion appears to be the Gir Forest Sanctuary in Gujarat State in India, where a few hundred remain. In Africa, the Cape lion of South Africa became extinct in 1865, and the Barbary lion north of the Sahara in 1922. The remaining lions of Africa are under heavy assault from human civilization, as are all of that continent's wild animals.

What happened to the animals, in both the New World and the Old, doesn't seem so mysterious to me. As various human cultures developed more powerful technologies and their populations increased, they began to exterminate the local wildlife. This began in the Old World about 100,000 years ago and has taken quite a long time because the Eurasian-African land mass is so enormous—the animals had 31.5 million square miles in which to roam. So did human beings. Later Paleolithic technology didn't catch on everywhere at once. The last big refuges for both wildlife, and for humans who preferred a simpler life, were in Africa, India, and Southeast Asia—not surprisingly, because those are our origins. We evolved in those ecosystems, and those ecosystems evolved with us in them. Africa has the most extensive savannahs in the world, and therefore the largest refuges for big animals, partially because of fires set periodically by human beings over hundreds of thousands of years. These fires maintain the open spaces where the grasses grow and prevent the forest from reoccupying the land.

For a million years, sharing Europe's forests with human beings were hippopotami, rhinoceros, elephants, leopards, lions, wolves, bison, aurochs, horses, boars, pandas, cave bears, hyenas, musk oxen, and many types of deer, elk, and antelope. In the north were woolly rhinoceros and woolly mammoth. In the sky flew giant vultures and giant swans. Some of these animals are extinct today. Others have retreated to Asia or Africa, where they live still. Horses, wolves, pigs, sheep, goats, and aurochs (cattle) have been domesticated, and they are doing very well indeed.

In the Americas, fluted spear points appeared about 12,000 years ago. North American has only 8.5 million square miles, not 31.5 million, and South America is even smaller. Without as much room to run, the wildlife here were exterminated fairly rapidly. By 10,000 years ago most of the largest animals were becoming uncommon throughout the Western hemisphere, but some probably lived on quite a bit longer, here and there in reduced numbers. The 18th-century explorer Charlevoix encountered among the Algonquins of North America a tradition of "a large elk, beside which the others seemed like ants. He has, they say, legs so high that eight feet of snow do not embarrass him: his skin is proof against all sorts of weapons, and he has a sort of arm that comes out of his shoulder, which he uses as we do ours."[85] Apparently elephants had lived in Canada in the not-too-remote past.

When Europeans showed up with their guns, they began to slaughter the rest of the native wildlife and replace them with domesticated European cattle. This time it was not just the largest land animals that suffered. Buffalo were slaughtered, but so were prairie dogs, prairie grasses, passenger pigeons, and walruses. All life is in danger—not just in size, but in quantity and in variety—plants and animals, terrestrial and marine alike.

In Australia, an even smaller land mass, the woomera, or spear-thrower, appeared about 6,000 years ago.[86] Ten-foot-tall kangaroos, hippo-sized wombats, large koalas, giant emus, marsupial lions and other large animals disappeared at about that time. Here too, European civilization, though it arrived later, is threatening the remaining wildlife.

New Zealand was first colonized by humans perhaps 2,000 years ago. New Zealand is rather small, and its unique fauna were killed off rather fast. In less than 2,000 years, up to twenty-seven species of flightless wingless birds, from turkey size to thirteen feet tall, became extinct. The smaller kiwi is the only living reminder of the birds who once peopled these islands.

Madagascar has been settled for some 10,500 years, and in that space of time the pygmy hippopotamus, two kinds of giant tortoise, aardvarks, a dozen kinds of flightless birds, and fourteen varieties of lemurs became extinct. The elephant bird survived until about 1700 A.D. At ten feet in height and 1,100 pounds it was probably the largest (though not the tallest) bird that ever lived. It laid two-gallon eggs that were over a foot long, and may have been the inspiration for the fabled roc of the Arabian Nights.

On Madagascar can also be seen how quickly forest can be changed into grassland. All of the area now prairie in the western highlands was tree-covered only recently. Groves are still preserved in holy places, on inaccessible slopes and where the land is unsuitable for grazing or agriculture.

The island of Mauritius in the Indian Ocean was home to the most famous extinct animal of all: the dodo. Indonesian sailors probably visited Mauritius in the first millennium B.C., but the island had no human settlement until the Portuguese and then the Dutch arrived in the sixteenth century A.D. Into an environment that had never known mammals, the settlers brought dogs, cats, monkeys, swine, and rats, and they hunted some of the native animals to extinction. The dodo, a large flightless dove, lasted only a hundred years under the assault—the last one died in 1681. In addition to the dodo, two kinds of owls, a red rail, a broad-billed parrot, two types of giant tortoise, and two varieties of snakes have been exterminated in less than 400 years under, successively, Dutch, French, British, and independent rule. The last Round Island boa died as recently as 1980.

Mauritius is also home to the *Calvaria major* tree, of which no new seedlings germinated after the extinction of the dodo. By the 1970s only thirteen were left living, and all were over 300 years old. It seems the tree's fruit must pass through the gizzard of a large bird before the seed can germinate. After turkeys were imported to Mauritius and force-fed *Calvaria major* fruits, the first new seedlings in 300 years started to grow.

Like a family member that dies, every species that goes extinct leaves a big hole in its ecosystem. Daniel Janzen and Paul Martin listed fourteen genera and thirty-seven species of native trees and large shrubs of Costa Rica whose seeds were probably dispersed by some of the fifteen genera of large animals that have become extinct during the last 10,000 years.[87] In today's forests peccaries, tapirs, agoutis, and small rodents eat the fruits and disperse the seeds of these trees. But a large proportion of the fruit rots in the trees or on the ground beneath them without being tasted by anybody. The horses, llamas, elephants, bears, and giant sloths aren't around anymore to eat them. These are large fruits and nuts that resemble those eaten by large mammals in Africa today, and which are readily eaten by introduced horses and cattle. On African wildlife preserves it is a rare event when the animals do not consume all of the fallen fruit crop—but where the elephants have been killed off in Uganda, the fruits of the *Balanites wilsonia* lie rotting on the ground just like so many Costa Rican trees' fruits do.

Janzen and Martin suggested that, in the United States, the Kentucky coffee bean, the honey locust, the Osage orange, the pawpaw, and the persimmon may once have grown more densely and had a wider range before the native large mammals, their seed dispersers, were killed off.

Where I lived in Mendocino, huckleberry bushes grew prolifically in the pygmy forest. Every fall when the tasty huckleberries were ripe, they would sit there in their millions uneaten upon the bushes, unless I went out myself and gathered them. This was very strange to me, because in my youth at Lake Oscawana, New York, if I wasn't quick enough at the blueberry bushes, the birds would beat me to every last blueberry. Perhaps the answer is similar. Whoever used to feast on huckleberries on the Mendocino coast doesn't live there anymore.

We humans have been very successful at killing off most of the land animals bigger than ourselves and driving the remainder onto a few wildlife preserves in Africa.

By nature's standards human beings are not very big. Living lions grow to be 10 feet long, and the largest tigers are over 13 feet. The white bear and the grizzly bear reach a length of 10 feet. Indian rhinoceroses can be 6½ feet high at the shoulder, 14 feet long and weigh over 2 tons. African rhinos are even larger, bearing horns up to 5 feet in length. Hippopotami grow up to 15 feet long, 5 feet high at the shoulder, and 5 tons. Bison grow up to 12 feet long and 7 feet high at the shoulder. The Alaskan moose stands 8 feet high with antlers 6 feet across. The African elephant is 11 feet high, and the giraffe 18 feet. Bactrian camels are 9 feet long and 7 feet high. Nile crocodiles grow to be 21 feet long and weigh a ton. The Komodo dragon of Indonesia reaches a length of over 13 feet. Galapagos tortoises with shells 4 to 5 feet long can weigh up to 500 pounds. Pythons of India and Southeast Asia reach a length of 30 feet, and South American anacondas can be still longer. Even birds up to 13 feet tall have existed only recently.

In the seas live clams 4 feet long, weighing 500 pounds, and, near Japan, crabs 13 feet from claw to claw. Through the earth's waters swim seals up to 15 feet long weighing 4 tons; walruses 11 feet long; and manatees 13 feet long. Stellar's sea cow, which was exterminated from the Bering Sea about 1767, was 20 to 30 feet long and weighed up to 7 tons. Bluefin tuna grow to be 14 feet long and weigh 1,500 pounds. Ocean sunfish can weigh 2 tons. There are even river fish 14 feet long—the pirarucu of the Amazon. The basking shark can measure 35 feet in length and weigh 15 tons. Squids 60 feet from head to tentacles inhabit the ocean deeps. In the coastal waters off England swim worms 90 feet long. In the north Atlantic the arctic jellyfish reaches a diameter of 8 feet and has tentacles up to 120 feet long. Sailing on the surface of the ocean are Portuguese men-of-war whose tentacles may reach 165 feet in length. The largest animals in the world—the blue whales—grow to be 100 feet long and weigh more than 150 tons.

The plant kingdom is peopled by individuals just as large. Practically all of the earth's trees are much larger than we are. Any that are our size we consider to be dwarf or pygmy trees. The plant kingdom has its whales too. Only recently large areas of Vancouver Island, California, and Chile were covered by trees 200 feet tall and 15 to 20 feet in diameter. And we have been exterminating trees as ruthlessly as we have been killing animals, allowing a few of them to remain standing in forest preserves.

The dinosaurs died out some 65 million years ago, leaving the birds and the crocodiles as their descendants and closest living relatives. But they didn't die out because they were too big. Dinosaurs came big and small, and the average dinosaurs were not all that much larger than the average mammals that replaced them.

We humans are such fearful animals that we have been exterminating all of the largest life forms and then pretending they existed only in the remote past, and died of natural causes. There will come a time not too far in the future when no more giraffes, or elephants, or rhinoceros, or hippopotamus, or lions, or tigers, or leopards, or buffaloes, or musk oxen, or bears, or ostriches, or crocodiles will roam this earth anymore, and scientists will argue among themselves about the cause of their disappearance.

Oxygen

The extermination of large animals took place about 10,000 years ago. The extermination of small animals is taking place now, due partly to humans crowding out all other animals, and partly to the incredible efficiency of modern hunting and trapping technologies. But there is another technology that is little more than a century old, that we have welcomed into our lives, and that is so universal that there is no escape from it today anywhere on earth. It kills by interfering with the foundation of life itself.

Every living thing—as we all know—must eat and breathe in order to live. The combination of fuel and oxygen feeds the fires of life. And if you starve someone, or suffocate them, they will die. Which is what we are doing to all living creatures today.

Our bodies do not burn our food with an actual flame. The combustion is much slower: inside every living cell of every plant, animal, and insect are hundreds to thousands of even tinier structures called mitochondria. And embedded in the inner membrane of each mitochondrion are a series of enzymes, called an electron transport chain, whose function is to both facilitate and moderate the combustion. It is there that the electrons generated by the digestion of the food we eat are shunted to the oxygen we breathe. The result is a slow process of combustion that sustains us from cradle to grave. If you interfere with the flow of electrons in the mitochondria of any living being, you are both starving and suffocating that being. And if you interfere with the flow of electrons in the mitochondria of *every* living being, you will exterminate all of life on earth. We are doing exactly that today with what is euphemistically called "non-ionizing radiation."

It began in 1897 with Marconi's first radio station, intensified during the twentieth century with the invention of television and radar, and has culminated in the placement of wireless communication devices in the hands of almost every person on earth. These cannot function unless every square inch of the planet is intensely irradiated at all times. Today every living thing on earth swims in a sea of radio-frequency (RF) radiation that is millions to billions of times stronger than the RF radiation from the sun and stars with which we evolved. This radiation, as all computer engineers know, interferes with electric currents wherever they are, and is the reason money is spent hardening electronic equipment against such interference. What has been ignored for over a century is the interference the radiation causes with electric currents in all *biological* systems, which did not evolve with it and are *not* hardened against it. This includes the electric currents in our nervous systems, our brains, our hearts, and our acupuncture meridians, as well as the tiny electric currents in the inner membranes of the mitochondria in all of our cells, which regulate our metabolism.[88]

If you are a large animal, the slowing of your metabolism has many types of effects: it lowers your body temperature; you gain weight; you are prone to diabetes, heart disease, and cancer. These effects have been observed not only in humans but also in zoo animals, domestic animals, and wild animals.[89] But the smallest animals, with the highest rates of metabolism, are being killed outright.

Insects
In 1904, the bees on the Isle of Wight off the southern coast of England—the location of the world's first permanent radio station—began to die. By 1906, when the fifth radio station on the 36-kilometer-long island went into operation, 90 percent of all the honey bees on the island had disappeared. The radiation was depriving them of oxygen, and they could not even fly. "They are often to be seen crawling up grass stems, or up the supports of the hive, where they remain until they fall back to the earth from sheer weakness, and soon afterwards die," wrote biologist Augustus Imms of Christ's College, Cambridge.[90] Swarms of healthy bees, imported from the mainland, were dead within a week. In the succeeding years, "Isle of Wight disease" spread like a plague along with radio technology into Australia,

Canada, the United States, South Africa, Italy, Brazil, France, Switzerland, Germany, and the rest of the world. In the 1960s it was renamed "disappearing disease," and today it is called "colony collapse disorder." Fully half of all honey bee colonies in the United States were lost to this disease in the year ending April 1, 2022, and again in the year ending April 1, 2023.

The problem, as anyone with eyes and ears can see and hear, is not confined to honey bees. There are far fewer butterflies, moths, bumble bees, crickets, and other insects than there were a few decades ago. In much of the world people driving cars no longer have to scrape the bodies of insects off their windshields frequently, if at all. In 2017, scientists reported a 75 to 80 percent decline in total flying insects in sixty-three nature protection areas in Germany.[91] In 2018, another team of scientists reported a 97 to 98 percent decline in total insects caught in sticky traps in a Puerto Rican rainforest.[92] In 2019, scientists from Australia, Vietnam, and China reviewed seventy-three reports of insect declines from across the globe and concluded that 40 percent of all insect species on earth are threatened with extinction.[93]

Without insects, a multitude of other creatures—birds, bats, reptiles, amphibians, small mammals, and fish—are starving to death.

Birds

The disastrous effects of radio waves on birds were first noticed during the 1930s by people who raced homing pigeons, and by military officers who used carrier pigeons for communication. Charles Heitzman, a father of the pigeon-racing sport in the United States, and Major Otto Meyer, one-time head of the United States Army's Pigeon Corps, were both alarmed by the large numbers of pigeons losing their way during the heyday years of the expansion of radio broadcasting.[94]

In the late 1960s a team of Canadian researchers deliberately irradiated chickens, pigeons, and seagulls with radio waves at a strength similar to what most human beings are irradiating their brains with today. The birds would scream, defecate, try to escape, and collapse to the floor within five to twenty seconds—but not if they were first defeathered. Chickens that had been plucked showed no reaction at all to being irradiated until about the twelfth day when their regrowing feathers were about one centimeter long. These scientists then experimented just on feathers, and proved that bird feathers make fine receiving aerials for microwaves.[95]

In 2000, when cell phone towers and antennas were proliferating in Spain, wildlife biologist Alfonso Balmori began to study their effects on birds in his city of Valladolid. The effects he found were dramatic and universal. White storks that built nests less than 200 meters from a cell tower often fledged no chicks at all, and on average fledged half the number of baby storks as those whose nests were further away. House sparrows were incredibly more numerous in less irradiated areas—forty-two sparrows per hectare where the electric field was only 0.1 volts per meter, down to only one or two sparrows per hectare where the electric field was over 3 volts per meter.[96] Antennas had been installed in the "Campo Grande" urban park in Valladolid during the 1990s. By 2003, half the park's fourteen resident bird species had either seriously declined or vanished despite the fact that air pollution improved. Most kestrels had disappeared, and green woodpeckers, short toed treecreepers, and Bonnelli's warblers, all previously common, had disappeared entirely and were not seen again. A 2007 study by Jenny De Laet and James Denis Summers-Smith found more than a 90 percent decline in house sparrow populations in London, Glasgow, Edinburgh, Dublin, Hamburg, Ghent, Antwerp, and Brussels.[97] The United Kingdom added the house sparrow to its Red List of threatened and endangered species in 2002 after the bird's population in British cities fell by 75 percent in just twelve years.[98] A 2008–2009 study by zoologist Sainudeen Pattazhy in Kerala, India, found that house sparrows were virtually extinct there. "Continuous penetration of electromagnetic radiation through the body of the birds affects their nervous system and their navigational skills. They become incapable of navigation and foraging. The birds which nest near towers are found to leave the nest within one week," he wrote. "One to eight eggs can be present in a clutch. The incubation lasts for 10 to 14 days. But the eggs which are laid in nests near towers failed to hatch even after 30 days."[99] Ornithologist Mohammed Dilawar, in Delhi, reminisced that until March 2001, house sparrows were always in and out of their home. A decade later they were completely gone. As he told a reporter, "We left for a while to return to see, the commonest bird had flown the nest."[100]

The decline was so severe in Switzerland that the Swiss Association for the Protection of Birds declared the house sparrow "bird of the year" for 2015. Pattazhy's observations about bird navigation were not just conjecture. Scientists at the University of Oldenburg in Germany, in an effort to determine why migratory songbirds were no longer able to orient themselves toward the north in spring and toward the southwest in autumn, placed European robins in aviaries which they surrounded with grounded aluminum screening that kept out radio waves. They began this study in the winter of 2006–2007. "The effect on the birds' orientation capabilities was profound," wrote the authors of the study. Only when the aluminum screening was grounded did the birds orient normally in springtime.[101]

In the last couple of years seabirds of all types have been dying by the millions in their breeding grounds on the coasts and islands of many nations, as new towers and antennas have been erected to provide new and better cell phone service even in the most remote places on earth.

Photo by Hanne Wilhelms

It is being blamed on avian influenza. However, on those coasts and islands not yet irradiated by new towers, birds are *not* dying.[102] And influenza, whether of birds or humans, is caused by electricity and not a virus.[103]

Bats

Bats, especially insectivorous bats, are under assault from several directions. They are starving to death because of lack of insects to eat. Their fast metabolism is being slowed, which impairs their ability to fly. And radiation in the ultrasonic range is directly interfering with their echolocation,[104] which impairs their ability to navigate and hunt.

In July 2006, when John Ackerman and two other cavers entered Bat River Cave in southeastern Minnesota, the walls of the bat gallery—a chamber with an underground waterfall—were thick with three species of bats. In the winter of 2010–2011, a count was made: 4,112 bats were hibernating in the cave. On February 9, 2020, Ackerman and four other people went back into the cave: only 82 bats were left.[105]

The Pennsylvania Game Commission reports a 99 percent decline in bat populations in that state.[106] Bats are dying in forty states and seven Canadian provinces.

"This is our worst ever year for starving bats found in the wild," said Hazel Ryan of the Kent Bat Group in England in the fall of 2019.[107]

John Allen used to see bats regularly where he lives, at Maori Island in Grovetown, New Zealand, but they have disappeared because there is no food source for them. "It's absolutely incredible," he says. "There are no insects flying around at night now. You can leave your windows and doors open with the lights on and nothing will fly in."[108]

Swedish scientists found a 59 percent decline in bat populations in Sweden between 1988 and 2017.[109]

Bats have been on earth for more than 50 million years. With 1,400 species, they are the second largest order of mammals, next to rodents. The disappearance of bats is leaving a giant hole in the global ecosystem.

Amphibians

Amphibians were here when the dinosaurs were here, and they survived the age of mammals. If they're checking out now, I think it is significant.

— David Wake, Director of the Museum of Vertebrate Zoology, University of California, Berkeley, 1990

They are ancient animals with abilities to survive beyond belief. They live both in water and on land. They can breathe through their skin. They can regenerate limbs and organs. They don't get cancer. They have been around for 365 million years and have survived four mass extinctions during the history of life on Earth. Although they do not have a high metabolism, they are disappearing more rapidly than any other class of animals.

Even before cell phones, the proliferation of radio and TV towers, radar stations, and communication antennas in the 1960s, 1970s, and 1980s began killing off these most hardy, well-adapted, and important forms of life.

- The northern leopard frog, *Rana pipiens*—the North American green frog that croaked from every marsh, pond, and creek when I was growing up—was already extremely rare by the end of the 1980s.
- In the Colorado and Wyoming Rocky Mountains, boreal toads used to be so numerous that, in the words of Paul Corn of the United States Fish and Wildlife Service, "You had to kick them out of the way as you were walking down the trail." By 1990 they were difficult to find at all.[110]
- Boreal chorus frogs on the shores of Lake Superior, once innumerable, were extremely rare by 1990.
- In the 1970s David Wake could turn up eighty or more salamanders under the bark of a single log in a pine forest near Oaxaca, Mexico. In the early 1980s he returned and was able to find maybe one or two after searching the forest all day.[111]
- Until 1979 frogs were abundant and diverse at the University of São Paulo's field station at Boracéia, Brazil, according to Stanley Rand of the Smithsonian Tropical Research Institute. But when he returned in 1982, of thirty common frog species, six had disappeared entirely and seven had decreased in number drastically.[112]
- In 1974 Michael Tyler of Adelaide, Australia, discovered a new frog species that brooded its young in its stomach. It lived in a 100-square-kilometer area in the Conondale Ranges, 60

kilometers north of Brisbane, and was so common that he could collect a hundred in a single night. By 1980 it was extinct.[113]

- The golden toad lived only in a 320-acre stunted forest in Costa Rica's supposedly pristine, protected Monteverde Cloud Forest Preserve. In the early 1980s Marc Hayes of the University of Miami typically counted 500 to 700 males at one of the species' breeding sites. After 1984 that site never had more than a dozen males. At another site Martha Crump observed a thousand males in 1987, but only one in 1988 and another single male frog in 1989.[114] Today the species is extinct.[115]

In 1990, when I began researching this magical class of vertebrates, there were not many amphibians left in all of Europe. Out of more than five thousand known species worldwide, about a dozen were doing well.

By the time I wrote *Microwaving Our Planet* in 1996, every species of frog and toad in Yosemite National Park had become scarce.[116] Seventy-five species of the colorful harlequin frogs that once lived near streams in the tropics of the Western Hemisphere from Costa Rica to Bolivia had not been seen in a decade.[117] Of the fifty species of frogs that once inhabited the Monteverde Cloud Forest Preserve, twenty were already extinct.

Similar population crashes were occurring in North, Central and South America, Europe, and Australia. Only in Africa and Asia, when I wrote that book, were amphibians doing well. That has since changed. On March 15, 2023, a team of nineteen American scientists published a paper titled "Continent-wide recent emergence of a global pathogen in African amphibians."[118] Amphibians, say the authors, were doing fine on the dark continent until about the year 2000—which coincidentally is when telecommunications companies began lighting up that continent with cell phone signals in earnest.

The single most rapid and catastrophic crash in amphibian populations occurred in the year 1988 in the Monteverde Cloud Forest Biological Preserve in Costa Rica, a location that has long puzzled scientists because it is strictly protected and supposedly untouched and pristine. That is what I once thought as well. But right in the middle of this two-square-mile preserve, on top of a hill called Cerro Amigos ("Friends Hill"), is an antenna farm called Las Torres ("The Towers"). This is what it looks like today:

What is completely neglected in the sciences of biology, medicine, and ecology is our electrical connection to earth and sky. As I discuss in chapter 9 of my book, *The Invisible Rainbow*, we are all part of the global electrical circuit that courses through the sky above us, flows down to earth on atmospheric ions and raindrops, enters the tops of our heads into our bodies, flows through our meridians, exits into the earth through the soles of our feet, travels along the surface of the earth, and flows back up to the sky on lightning bolts during thunderstorms. Those of us who are most vital and have the strongest connection to earth and sky—healthy, vigorous young adults and pregnant women—died in the largest numbers in the 1918 flu, which was caused not by a virus but by the use of enormously powerful VLF radio stations by the United States when it entered the First World War.

Salamanders, toads, and frogs have more vitality than other forms of life. The density of the strings that connect them to earth and sky—their meridians—is greater. It is why they rarely (and salamanders never) get cancer: both their external and internal communication systems are too strong for their cells to escape control. It is why frogs can partially regenerate lost limbs, and salamanders can regenerate them completely. It is why salamanders can even regenerate their heart—and do it within hours—if half of it is cut out—an astounding fact discovered by Dr. Robert O. Becker and written about in chapter 10 of his classic book, *The Body Electric*.[119]

It is also why amphibians are dying out. Animals with such a strong connection to Earth's orchestra—who are so attuned to it that they have survived for 365 million years—cannot withstand the chaos that we have superimposed on it during the past half century and more—the chaos that we have injected into the living circuitry with our radio and TV stations, our radar facilities, our cell phones and cell towers, and our satellites.

It is why, in 1996, when parades of cell towers were marching from coast to coast in the United States, and sprouting at tourist destinations, mutant frogs were turning up by the thousands in pristine lakes, streams, and forests in at least thirty-two states. Their deformed legs, extra legs, missing legs, missing eyes, misplaced eyes, misshapen tails, and whole body deformities frightened school children out on field trips.[120]

It is why developing frog embryos and tadpoles exposed by researchers in Moscow in the late 1990s to a personal computer developed severe malformations including anencephaly (absence of a brain), absence of a heart, lack of limbs, and other deformities that are incompatible with life.[121]

It is why, when tadpoles were kept for two months in a tank on an apartment's terrace in Valladolid, Spain, 140 meters from a cell tower, 90 percent of them died, versus only 4 percent mortality in an identical tank that was shielded from radio waves.[122]

Where Have All the Songbirds Gone?

That was the title of Joseph Wallace's heart-rending article in the March/April 1986 issue of *Sierra* magazine. So many of the warblers, tanagers, flycatchers, and other colorful singers that once graced the forests of North America—so many of the birds that used to chirp and sing outside my window from early spring to late autumn—were becoming fewer year by year, and have dwindled even more today. The causes, said Wallace, were to be found both in Latin America, where the forests in which they spent the winter were disappearing at an incredible rate, and up here in North America, where so many patches of forest were growing smaller and becoming separated by fields and roads. Today they are increasingly also being killed by the radiation from cell towers and antennas.

Forgotten Fliers

It is not just loss of habitat and microwave radiation that has been decimating bird populations. It is also guns.

The Passenger Pigeon

> The migratory or wild pigeon of North America was known by our race as o-me-me-wog. Why the European race did not accept that name was, no doubt, because the bird so much resembled the domesticated pigeon; they naturally called it a wild pigeon, as they called us wild men.
>
> This remarkable bird differs from the dove or domesticated pigeon, which was imported into this country, in the grace of its long neck, its slender bill and legs, and its narrow wings. Its length is 16½ inches. Its tail is eight inches long, having twelve feathers, white on the under side. The two center feathers are longest, while five arranged on either side diminish gradually each one half inch in length, giving to the tail when spread an almost conical appearance. Its back and upper part of the wings and head are a darkish blue, with a silken velvety appearance. Its neck is resplendent in gold and green with royal purple intermixed. Its breast is reddish brown, fading toward the belly into white. Its tail is tipped with white, intermixed with bluish black. The female is one inch shorter than the male, and her color less vivid.
>
> It was proverbial with our fathers that if the Great Spirit in His wisdom could have created a more elegant bird in plumage, form, and movement, He never did.
>
> When a young man I have stood for hours admiring the movements of these birds. I have seen them fly in unbroken lines from

the horizon, one line succeeding another from morning until night, moving their unbroken columns like an army of trained soldiers pushing to the front, while detached bodies of these birds appeared in different parts of the heavens, pressing forward in haste like raw recruits preparing for battle. At other times I have seen them move in one unbroken column for hours across the sky, like some great river, ever varying in hue; and as the mighty stream sweeping on at sixty miles an hour, reached some deep valley, it would pour its living mass headlong down hundreds of feet, sounding as though a whirlwind was abroad in the land. I have stood by the grandest waterfall of America and regarded the descending torrents in wonder and astonishment, yet never have my astonishment, wonder, and admiration been so stirred as when I have witnessed these birds drop from their course like meteors from heaven.[123]

— Chief Simon Pokagon of the Pokagon Pottawattami band, 1895

The bird Chief Pokagon so beautifully described may once have been the most numerous bird in North America, with a home range from Hudson Bay to the Gulf of Mexico, and from the Great Plains to the Atlantic Ocean. Its brooding areas were often twenty to thirty miles long and three to four miles wide, every tree filled to its limits with nests. As recently as 1870 a flock one mile wide and 320 miles long, containing perhaps two billion birds, passed over the Ohio River at Cincinnati. Yet these pretty doves were hunted so mercilessly—for their meat, for their feathers and for pure sport—that fifty years later they were extinct. The last living passenger pigeon passed away in the Cincinnati Zoo in September of 1914.

The Carolina parakeet was a relatively small parrot, twelve inches long, weighing about ten ounces. Until the late 1800s these green and yellow birds, with orange-yellow heads, were extremely common in the eastern forests of North America. Like the now-endangered spotted owl of the Pacific Northwest, Carolina parakeets built their nests within hollow trees of mature forest, and they suffered from the rapid destruction of their homes.

These birds were also hunted for their meat and feathers, and for sport, and farmers killed them because they were seedeaters. Like all parrots, these were highly social birds. When one was wounded or killed, the rest of its flock would noisily swoop or hover over the fallen one. In this way hunters were able to wipe out an entire flock after shooting down one bird.

The last known Carolina parakeet died captive in the Cincinnati Zoo in February of 1918. It was not a good decade for birds in Cincinnati.

Eskimo curlews once linked ecosystems throughout the Americas in their long migrations. They spent the early summer in their breeding grounds beyond the tree line in the Alaskan and Canadian Arctic. They flew eastward in July to feast on the ripe purple berries which carpet Labrador and Newfoundland at that time of year. Throughout August great flocks of them left the Atlantic coast from Newfoundland to New York heading out over the open ocean on a 3,000-mile journey to the northeast coast of South America. There they resumed their journey southward over land and spent the summer (our winter) on the grassy Pampas of central and southern Argentina. In late February or early March they departed northward again, this time towards Central America, and a few weeks later their flocks darkened the skies over the Gulf coast of Texas. In the spring they feasted on grasshoppers and worms all over the American prairies before resuming their northward flight to their Arctic nests.

The Eskimo curlew was thirteen to fourteen inches in length. Its dark brown feathers were edged with white, and it had a long whitish neck and long legs. This wading bird was a harbinger of spring on the tundra plains bordering the Arctic Ocean. There the Inuit called it pi-pi-pi-uk in imitation of its distinctive soft whistle. This same bird was welcomed as a messenger of their spring 10,000 miles away by native Patagonians.

White men hunted them relentlessly in Labrador and New England in late summer; in South America during our winter; and on the prairies from Texas to Canada in spring. By the end of the 19th century only tiny remnants of the once great flocks remained. No Eskimo curlews have been seen in Argentina since 1939. Single birds were sighted during the 1950s and 1960s in the Northwest Territories, in Texas, and along the east coast, and one was shot in the Bahamas in 1963. None have been seen since.

Plants and Animals

Photo by Ann Tate, airbirdpix.com

The Eskimo curlew's close relative, the little curlew, still does its job of nesting in eastern Siberia and wintering in Southeast Asia, Australia, and New Zealand, but in the Western Hemisphere this link of north and south no longer lives. Curlew berries still grow abundantly all over Newfoundland and Labrador, but nobody is there to eat them. And insects eat farmers' newly planted crops every spring all over the American prairies. Curlews are not there to eat them either.

The Last Refuge

"Farewell to Africa," said Robert Jones in a forty-seven-page article I could hardly bear to read. That was in 1990 when I began my research for this book.[124] Areas of Kenya where only recently had roamed great herds of antelope, zebra, elephants, and rhinoceros, were now populated only by wheat, onions, and cattle, grown to support a human population that had tripled in thirty years, and that today has almost tripled again. The national parks and wildlife reserves, even on the great Serengeti Plain of Kenya and neighboring Tanzania, are only sparsely populated with animals, and the few that are left are mercilessly pursued by minibuses and hot-air balloons full of tourists with cameras and video camcorders. The future of Kenya's wild animals depends on large private ranches—fenced to keep out poachers—and on the success of government wildlife policies. Richard Leakey, who quit as director of Kenya's Wildlife Service in January 1994, summed up the situation this way:

> The era of free-ranging game in Africa is finished. It's no more realistic to have wild animals running loose here than it is in your American farmland or suburbs. Wildlife, if it's to survive in Africa, must pay its own way from now on. If we can demonstrate that revenues from wildlife—from tourism, hunting, hides, meat, and the like—are greater than from the plow, then we can save it. Otherwise, *na kwisha kabissa*—it will be finished completely. Only farmland will remain.

National Geographic devoted a large portion of its December 1990 issue to the southern African nation of Botswana, which the editors called "perhaps the last place where the great herds of game animals synonymous with Africa stand a good chance of arriving at the next century's doorstep intact." Northern Botswana's wetlands and rivers, its dry woodlands and its savannahs succored some of Africa's last great free-roaming herds of Cape buffalo, zebras, and antelopes. It was illegal to hunt elephants and a herd of 60,000 was the largest remaining on the continent; it had been illegal to hunt them since 1983. The world's largest inland delta, flooded by the Okavango River, nourished egrets, herons, storks, pelicans, ibis, ducks, geese, eagles, and many other kinds of birds that fed together on the abundant fish, frogs, and mollusks. Hundreds of thousands of flamingos came to feed when the heavy rains came to replenish the dry land. Said Frans Lanting, "Here I found an Africa I thought no longer existed. Here was a place where the antelope, the zebras, and I could drink freely from the same waters."[125]

Botswana has set aside 17 percent of its land as national parks or wildlife reserves. But the real reason Botswana is still a refuge is its scarcity of people. With only five people per square kilometer, it is one of the most sparsely populated countries in the world. And only one twelfth of its population lives in the wild northern regions. In 2016 Botswana was home to 130,000 elephants, more than one-third of all the elephants in Africa.[126]

Yet even in Botswana the future of wildlife is in jeopardy. The population of Botswana has doubled in thirty years. The northern roads have been paved, and the Okavango delta is now only a seventeen-hour drive from metropolitan Johannesburg. Planners look to the Okavango River as a resource on a thirsty continent, and cattle owners eye its green pasturage. The delta used to be protected by the tsetse fly and the sleeping sickness it carries, deadly to cattle and humans but harmless to wildlife. The tsetse has disappeared, eradicated by insecticides. Safari leaders complain of the demands put on wildlife by modern trekkers. "The regrettable part, as I see it after 45 years of hunting, is that so much is wanted in such a short time. A safari was once three months; now it is three weeks. It's not that people can't afford the money. It's that they can't afford the time."[127] Safaris sometimes promise twenty-one trophy animals in twenty-one days, and shooting from the back of a truck is common. The ban on hunting elephants was lifted in 1994 after the elephant population had doubled. The ban was reinstated in 2014, and lifted again in 2019.

Unfortunately, although we civilized people have speeded up our lives, our animal neighbors have not, and cannot.

In Botswana's Kalahari region also lives an endangered, almost extinct culture of people. It is one of the oldest cultures on earth. But none of the native San live the hunter-gatherer life anymore. Their homeland is divided by cattle fences, and the nomadic way has become impossible. The San have been unable to farm the veld whose natural wealth once nourished them so abundantly. Most of them live malnourished and diseased at the Botswana government's expense, addicted to tobacco and alcohol.

A World Park

If there are two areas about which, environmentally, the whole world seems to be coming to agreement, it is these:

1. not to kill whales anymore, and
2. to set aside Antarctica as a world park.

The ranks of the various species of great whales have been decimated from centuries of hunting on the high seas—hunting for their meat and for the large quantities of fuel oil that could be rendered from their blubber. In the 1800s the 100-foot blue whale had a world population of about a quarter million. Today there are perhaps 2,000.[128]

But, however much reduced in numbers they are, no species of whale has yet been hunted to extinction. The International Whaling Commission continues to uphold a moratorium on all commercial whale-hunting, agreed to in 1986. But three countries are still reluctant to abandon a long bloody tradition. Iceland, Norway, and Japan still engage in commercial whaling, using technology against which the animals have no defense. Whales, like fish, are located using sonar, and killed with trigger-released exploding harpoons. Subsistence hunting of whales also continues. Inuit and other native hunters in Alaska, northern Canada, Greenland, and Russia still kill belugas, which are small and relatively common, as well as limited numbers of bowhead, minke, fin, humpback, and gray whales. Subsistence hunting of humpback whales also occurs in the Caribbean, and of sperm whales in Indonesia. Indonesia's Sunda Islands hunters still pursue their prey in the traditional way with wooden sailboats and bamboo-shafted harpoons. Aboriginal hunting in the Arctic, however, has adopted more modern technology. Snowmobiles, outboard motors, Caterpillar tractors, cell phones, and exploding harpoons have made subsistence hunting in the Arctic as potentially threatening to these largest of mammals as the commercial whaling the world is outlawing.

As Africa is the last refuge for large animals on the land, so Antarctica is the last haven on earth for wildlife in the sea. The numbers are diminishing, but in addition to hosting whales, Antarctica is still the summer breeding ground for some 20 million seals of 6 different varieties, 12 million penguins of 5 species, and about 40 other kinds of birds numbering perhaps 100 million, all feasting on the prodigious amounts of plankton, krill, and fish that thrive in its offshore waters. Every species has its own particular niche; the terrain on which it builds its nest, the food it eats, the time of day it is active, and the territory it forages or hunts is a little different from everyone else's, allowing all to cooperatively share what is the last large intact ecosystem in the world.

Seven nations—Chile, Argentina, Norway, Great Britain, New Zealand, Australia, and France—once staked claims to parts of the southern continent, but they put those claims to the side in 1961 when they signed an International Antarctic Treaty. It has been ratified by fifty-six nations, and declares:

> It is in the interest of all mankind that Antarctica shall continue forever to be used exclusively for peaceful purposes, and shall not become the scene or object of international discord.

Would that the word "Earth" were substituted for the word "Antarctica." But it is a start. The treaty forbids military activity and bans nuclear explosions and radioactive waste disposal. The signatories have met annually and agreed upon measures to protect the plant and animal life of the region. And since 1977 there has been a moratorium on mining and oil drilling on the continent. In 1989 the General Assembly of the United Nations overwhelmingly approved a resolution calling for Antarctica to be declared a global wilderness.

But there are still some dire threats to the well-being of this southernmost neighborhood of the earth:

- Pesticides and heavy metals have been found in Antarctic birds since the early 1960s.
- The Southern Ocean is heavily polluted by plastic.
- The coming and going of modern ships brings with it the certainty of occasional oil spills.
- Argentina, Chile, Uruguay, Australia, and Finland have cell towers at their Antarctic stations.
- There is still a yearly springtime hole in the ozone layer over the continent. Photosynthesis in the ocean has been inhibited down to a depth of 300 feet. All life here, from birds to seals to plankton, is being damaged by increased ultraviolet radiation.
- Litter and raw sewage from a summer human population of about 5,000 scientists and more than 100,000 tourists has become a blight on the continent. I find it sad that hundreds of millions of birds, seals, and

whales can live together in the Antarctic forever, but let humans come in and the neighborhood starts to deteriorate immediately. Twenty-nine countries have built seventy research stations all over the continent, and they don't sound like places I would want to visit. There does not seem to be the concept anymore that wilderness ought to be quiet. The Antarctic Peninsula, for example, has sixteen stations, with ten more on nearby King George Island; airplanes, helicopters, trucks, and bulldozers are in constant operation throughout the summer. Nearly every base has its own helipad, landing strip, harbor, and garbage dump. To me this is no longer wilderness. I am sure that none of the scientific studies of Antarctic wildlife take into account the effects of continuous noise on the animals' lives.

The plants and animals in the oceans are showing increasing signs of ill health. The first species of reef-building coral went extinct in the eastern Pacific in 1983. Today one-third of the 845 species of coral on earth are at elevated risk of extinction.[129] The health of the oceans as a whole is dependent on the phytoplankton—all the tiny different drifting plants that inhabit the upper regions of water and convert the sun's energy into living matter. They are the ultimate source of food for all the fish in the sea, and a 2010 study concluded that these organisms have been declining by about 1 percent per year since at least 1899.[130] This means the earth has lost 71 percent of its phytoplankton in last 124 years. The implications are ominous. Land life and sea life are inextricably linked, and if the drifting plankton are dying, our own life-support systems are in grave danger.

No one knows how many kinds of plants and animals share this earth— most naturalists estimate between five and fifty million different species. Half of them, more or less, live in tropical forests, and about two percent of these are losing their homes each year. 179 species of birds in Indonesia are under threat of extinction. Of 266 kinds of freshwater fish native to the forest rivers of Malaysia, 140 were already extinct in 1983.[131] Those scientists who have dared to guess how many species on earth become extinct every year have come up with wildly differing estimates—from as few as 3,000 to as many as 150,000. The late Terry Erwin, former curator of Coleoptera at the National Museum of Natural History in Washington, D.C., said:

> In fully describing the distribution of insects in time and space in the tropics, we should think in terms of more than 30 million, or perhaps 50 million or more, species of insects on Earth. A large number of species are tied only to certain forest types that are found on very small patches of soil deposited differentially through time by the vast and meandering Amazon River system. The extermination of 50 percent or more of the fauna and flora would mean that our generation will participate in an extinction process involving perhaps 20 to 30 million species. We are not talking about a few endangered species listed in the Red Data books, or the few furbish louseworts and snail darters that garner so much media attention. No matter what the number we are talking about, whether one million or twenty million, it is massive destruction of the biological richness of Earth.
>
> We are rapidly acquiring a new picture of Earth, and it is crammed with millions upon millions of nature's species on the verge of being replaced by billions upon billions of hungry people, asphalt, brick, glass, and useless eroded red clay baked by a harsh tropical sun.[132]

The 400th anniversary of Columbus' voyage to this continent was commemorated by a fair in Chicago in 1893. It elicited these words from Chief Simon Pokagon, whose father had sold Chicago to the United States sixty years before for three cents an acre:

> Where these great Columbian show-buildings stretch skyward and where stands this "Queen City of the West" once stood the Red Man's Wigwam; here met their old men, young men and maidens; here blazed their council fires. But now the eagle's eye can find no trace of them. Here was the center of their wide-spread hunting grounds, stretching far eastward, and to the great salt Gulf southward, and to the lofty Rocky Mountain chain westward. All and

about and beyond the Great Lakes northward roamed vast herds of buffalo that no man could number, while moose, deer, and elk were found from ocean to ocean; pigeons, ducks, and geese in near bow-shot moved in great clouds through the air, while fish swarmed our streams, lakes and seas close to shore. . . .

The cyclone of civilization rolled westward; the forests of untold centuries were swept away; streams dried up; lakes fell back from their ancient bounds and all our fathers once loved to gaze upon, was destroyed, defaced or marred, except the sun, moon, and starry skies above which the Great Spirit in His wisdom, hung beyond their reach."[133]

> — excerpted from *The Red Man's Rebuke*,
> printed on white birch bark in 1893

CHAPTER 15

Space

The Great Spirit did not hang the sun, moon, and starry skies quite high enough beyond our reach, and the changes we are now making in the heavens above may turn out to be far more permanent than anything we are doing down here on earth. For sixty-five years the United States and Russia have been sending rockets and satellites into orbit around the earth for military, research, and communication purposes, and more recently many other countries and private corporations have been getting into the act as well, to the tune of 223 launchings worldwide in 2023. "Out of sight, out of mind" has been the prevailing attitude, as exploding rockets have sent clouds of debris hundreds of miles long, and outer space has become the world's biggest garbage pit. Since 1957, there have been more than 640 breakups, explosions, collisions, and other satellite-destroying events. This has resulted in the creation of more than 5,500 tons of space debris. About 17,000 satellites have been sent into orbit, some 13,000 of which are still up there. There are about 20,000 other radar-trackable objects floating around and colliding with one another in the ocean of outer space, 1,000,000 objects from 1 cm to 10 cm in size, 130,000,000 objects from 1 mm to 1 cm in size, trillions of paint flakes and fragments of metal, ceramic, and plastic, and uncountable numbers of aluminum oxide dust particles. The debris includes expended rocket stages, rocket panels, fragments of exploded satellites, nuts and bolts, hand tools that have slipped from the gloves of space-walking astronauts, and bodily wastes that were flushed from the Space Shuttle. There are even old nuclear-powered satellites that have been abandoned in their orbits. In the late 1950s the

United States exploded atomic warheads in low space orbits, and for thirty years the United States and the Soviet Union exploded nearly ten million pieces of debris into low and high orbits in anti-satellite and ballistic missile tests. Back in 1963 the United States Air Force launched half a billion tiny copper needles into orbit to create a mirror for radio waves. Some of those needles are probably still up there.

And since space is unregulated, the problem is getting rapidly worse. More than 1,000,000 additional satellites are in various stages of planning, and governments and corporations are lining up to launch them.[134] Companies, universities, and government agencies in Algeria, Angola, Argentina, Armenia, Australia, Austria, Azerbaijan, Bangladesh, Belarus, Belgium, Bhutan, Bolivia, Brazil, Bulgaria, Canada, Chile, China, Colombia, Czech Republic, Denmark, Djibouti, Ecuador, Egypt, Estonia, Ethiopia, Finland, France, Germany, Ghana, Greece, Guatemala, Hungary, India, Indonesia, Iran, Ireland, Israel, Italy, Japan, Jordan, Kazakhstan, Kenya, Kuwait, Laos, Latvia, Lichtenstein, Lithuania, Luxembourg, Malaysia, Mauritius, Mexico, Moldova, Monaco, Mongolia, Morocco, Myanmar, Nepal, Netherlands, New Zealand, Nigeria, North Korea, Norway, Oman, Pakistan, Papua New Guinea, Paraguay, Peru, Philippines, Poland, Portugal, Qatar, Romania, Russia, Rwanda, Saudi Arabia, Singapore, Slovakia, Slovenia, Solomon Islands, South Korea, South Africa, Spain, Sri Lanka, Sudan, Sweden, Switzerland, Taiwan, Thailand, Tunisia, Turkey, Turkmenistan, Uganda, Ukraine, United Arab Emirates, United States, United Kingdom, Uruguay, Venezuela, Vietnam, Zimbabwe, the European Space Agency, the European Organisation for the Exploitation of Meteorological Satellites, the European Telecommunications Satellite Organization, and the Arab Satellite Communications Organization have already been launching satellites and constellations of satellites into low, medium, high, and geostationary orbits.

There is now so much junk up there that it interferes with astronomical observations. Both satellites and space debris now streak all of the photographs taken at Mount Palomar and other observatories. The orbiting junk is also an increasing hazard to orbiting satellites and astronauts. In 1984, Space Shuttle astronauts retrieved the Solar Maximum Mission satellite after it had been orbiting at 350 miles up for four years. Its aluminum exterior had an average of six holes per square foot from high-velocity impacts with orbiting flakes of paint. Many satellites will eventually be rendered useless from constant sandblasting by orbiting dust. But no one bothers to retrieve such satellites anymore: they are either left up there in perpetuity or de-orbited and burned up in the atmosphere. It is planned obsolescence on steroids. Communication satellites that are in geostationary orbits at 22,300 miles up will orbit the earth forever but are designed to last only seven years. Then another one is sent up to replace it. Two, because a second one is always launched as a backup. Weather satellites in geostationary orbit are designed to last only five years. The thousands—soon to be tens of thousands—of communication satellites in medium, low, and very low orbits are designed to last five to ten years, necessitating an almost daily procession of rocket launches from spaceports around the world to replace them, and old satellites raining back down, forever into the future. The upper stages of all the rockets, together with the used up satellites, burn up in the atmosphere on their way down, dispersing their incinerated metals and plastics throughout the air we all breathe.

Satellites are also interfering with stargazing, both by humans, who watch the stars for pleasure, and birds and other animals who use the stars for navigation. Not only are artificial satellites already reflecting so much sunlight that they are brightening our night skies by as much as 10 percent, but they are also already outnumbering the visible stars. "With the naked eye, stargazing from a dark-sky location allows you to see about 4,500 stars. . . . Once Starlink approaches 12,000 satellites in orbit, most people in Canada will see more satellites than stars in the sky," wrote Canadian astronomers in 2021.[135]

Space junk will not disappear any time soon. A marble-sized object in a low orbit, 180 miles up, takes only days or months to re-enter the atmosphere. But at 300 miles, orbital decay takes at least a year; at 500 miles, 30 years; at 800 miles, 300 years. Geostationary orbits 22,300 miles above the earth hardly decay at all—that junk is up there permanently, forever.

Near-daily rocket launches are also spewing black soot, water vapor, aluminum oxide, nitrogen oxide, and chlorine into all the layers of the atmosphere. Some of these substances destroy ozone, while others contribute to global warming. Water vapor increases the conductivity of the upper atmosphere, which may increase the current in the global electric circuit and therefore the frequency and violence of thunderstorms worldwide.

Thousands—soon to be tens of thousands—of communication satellites are also irradiating the entire surface of the earth with microwaves, including every square inch of the oceans. The radiation at the surface of the ocean from these satellites already exceeds the level that has been proven to cause whales to strand.[136]

When I was young the only fast-moving objects in earth's night skies were shooting stars. They were things of mystery, upon which one made wishes. The heavens are a tapestry of unsurpassable beauty, to be gazed at while lying on one's back in a pine forest, gazed at for hours as the dome of stars above turns and shifts slowly and inexorably, shifts with the rhythms of the seasons, year in and year out. In a whole human lifetime the moving tapestry hardly changes at all, yet one is never bored with it. The sun, moon, and planets are like travelers among the stars, as we are travelers here on our earthly landscapes. Once in a great while a comet visits us, lighting up our night skies with its luminous tail, a messenger from realms unknown and far away. There is mystery above us, and all of our science and technology cannot take that mystery away, it can only cause us to forget, for a while perhaps, that it is up there. And all of those satellites that we are sending up into the skies, all those moving lights can only hide that mystery, and distract us from it, from the mystery to be found in the star-studded unknown, and in the unknown depths of our own souls. It is still out there, waiting only for one to shut out the lights, lie on one's back in a pine forest, and look at it.

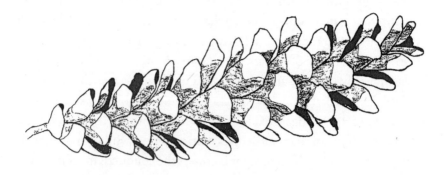

THE FLAMES OF PROGRESS

A Parable

There lived a family in a special house whose floors were paved with silver and gold, and wildflowers, and edible plants. Its walls were built of beautifully grained wood, that the people made into furniture, and burned for fuel, and the boards always grew back just as beautiful as before. Generation after generation were born there, and married there, and had their children there, and died there.

One day someone invented a new and bigger kind of stove. It burned up boards faster than the house could grow them, and the flames reached so high that the great house caught on fire. The people coughed and wheezed and wondered why the air was so smoky. And every day they cut some more boards from their burning house to feed the flames that consumed it.

CHAPTER 16

Living on the Edge

After World War II there came into being a new medical disorder caused by the wholesale poisoning of our living environment. "Environmental illness," it was called, or "twentieth-century disease." In reality it was the environment that was ill, and the twentieth century that was diseased. And in the twenty-first century the disease has become a pandemic. The toxins and electromagnetic fields have become impossible to escape. But people are so mesmerized by the way of life woven by those toxins that they do not even acknowledge the connection.

Although I grew up in New York City, I spent summers in the country from the age of five, and I fell deeply in love with the sights, sounds, and smells of the natural world. When these were gradually replaced by the sights, sounds, and smells of technology, my spirit was disturbed. As I grew older the world grew noisier and noisier. It became hard to find a place to live that wasn't under an airplane route. It soon became even harder to escape the noise of cars and trucks. Roads were built absolutely everywhere, and cars didn't shut up even at night. How could people be so oblivious to the world around them, I wondered? Didn't it hurt their ears like it hurt mine? How did they all sleep at night?

In 1978, I began medical school. On the first day, I found myself and eighty-nine new classmates standing in an unventilated basement room with twenty-two human cadavers pickled in formaldehyde. My colleagues used to joke about the smell, but I couldn't wait until my course in gross anatomy ended.

Southern California has a lot of nature, and the rest of the first two academic years were not quite so stressful. There were many lectures in large auditoriums, and I spent a fair amount of time in the woods at places like Sespe Hot Springs. I cut a lot of classes. But once the clerkships began there was no escape. We lived and worked in the hospital, putting in 100- or 120-hour weeks and being awake on call all night once or twice a week. We worked seven days a week throughout the year, with no vacations except two weeks at Christmas time—and we paid dearly for the privilege, both in tuition and sometimes in health. The amount of chemicals and radiation we were exposed to was phenomenal! Disinfectants, insecticides, antibiotics, drugs of all kinds, detergents, radioactive isotopes, you name it, it was there, in the air, on the floor, on the walls, in the patients. Every patient was x-rayed, regardless of their complaint. It was the price of admission to the hospital. Every obstetrics patient had a "Doppler" ultrasound exam to listen for the baby's heartbeat. There came a time when I would turn on the Doppler to listen to my patient's baby, and I would get a sharp pain in my own stomach.

Then came the surgery rotation. I assisted at a few surgeries. But there came a time when an hour in the operating room would leave me with pains in my hips for days. So I made a deal with my professor: in exchange for being excused from the OR, I would write a research paper on a topic of my own interest. I wrote a fourteen-page article with sixty-six references entitled "Effects of Radiant Energy on Living Organisms: A Review of the Literature." I wrote:

> The mechanization of medicine results in exposure of hospital patients and staff to radiated energy of various types. In particular, the practice of surgery often involves exposure to nuclear isotopes, X-rays, microwaves, ultrasonics, lasers, ultraviolet and extra-low-frequency (60-cycle) radiation. These may be divided into ionizing and nonionizing forms. The biological effects of ionizing radiation are relatively well-known; I will confine myself here to the nonionizing effects of all kinds of radiation. An enormous body of literature exists on this subject; however, it is an interdisciplinary field with which purely medical researchers seem to be largely unfamiliar.

An understanding of the interactions of living organisms with their electromagnetic environment has required the combined expertise of physicians, biologists, chemists, engineers and physicists in many of their subfields.

I found that significant electromagnetic sources in hospitals included electrocautery units, diathermy machines, timer units, thermostats, treadle operated switches, neurosurgical stimulators, ultrasonic devices, image intensifiers, fluorescent lighting, and mobile communication systems. With the computerization of almost everything and wireless connections to almost everything, the list is much longer today. But operating rooms are still the worst environment, and the most intense fields come from the electrocautery units used by every surgeon in every operation to coagulate bleeding vessels. Electrocautery works by sending microwaves directly through the patient into the floor, generating heat at the point of contact with the blood vessel. When I was in school there were still one or two older surgeons on staff who claimed that this was sloppy surgery. They insisted on tying off blood vessels with sutures the old-fashioned way and would have no part of devices that fried them closed and filled the operating room with smoke and the stench of burning flesh.

One day in February of 1982, on the pediatrics ward, at about the end of my third year of medical training, I collapsed. My chest hurt, I was very short of breath, and I felt like I was about to die. During the next two weeks I lost fifteen pounds. I finally made the decision I had been unable to make for three and a half years: I quit medicine. It wasn't worth dying for.

AND THE WAY OF LIFE TO WHICH OUR CIVILIZATION HAS BEEN SO COMMITTED FOR SO LONG—THAT WAY OF LIFE IS NOT WORTH OUR DYING FOR.

I joined my girlfriend in the woods of Mendocino and began to clean out my body in an environment of pure air, clean water, quiet, and minimal radiation. I began to learn what it is like to live without electricity, and how much more peaceful an unwired house feels.

The first few days were a shock to my ears! I had never, in all my life, heard such quiet before. No airplanes flew overhead. No cars roared past my window. No hum of electric motors surrounded me in my home. All was shockingly still. And gradually I began to hear the music of the forest. I heard the ocean surf gently in the distance as it pounded the cliffs with every incoming wave. I heard insects talk to one another, and trees rub their branches together. The breezes made their own soft noises. Deer were almost soundless, but squirrels chattered rather loudly, and birds sang to one another. And with the long Mendocino winter came the comforting patter of rain upon the roof. Flickering candles make their own little noises. And the crackling and roaring of a winter's fire in the fireplace is almost as comforting as the light and heat that it puts out.

My sleep was soon disturbed by a noise that did not belong in the Mendocino woods—a low rumbling sound that came and went without warning, as if someone were switching on and off a great motor somewhere far away. This was not a locally generated noise. I heard it in several places in coastal California, and I later heard it while camping at 9,000 feet elevation in the Sierra Nevada mountains near Likely, California, and in my rented cabin in the forest near Powassen, Ontario. I spoke to a few people in the Pacific Northwest who thought they were hearing ships at sea, and an electronics expert who thought it was the "Woodpecker," an over-the-horizon radar system in Russia which had been interfering with amateur radio communications since 1976. Or was it the Navy's ELF (extra-low-frequency) generator in Michigan's upper peninsula, which broadcast its signals around the earth to submarines under the sea? Or one of the military, communication, research, and weather satellites above us that were blanketing the earth with radar and microwaves? The PAVE PAWS radar on Beale Air Force base broadcast its surveillance signals westward toward Mendocino 125 miles away. An American over-the-horizon radar sat just fifty miles northwest of Likely. And a large missile base was located at North Bay, Ontario just twenty miles from Powassen. I didn't know what particular source was the cause of a certain noise inside my head. I only knew that it didn't belong there, and that the earth was no longer a quiet place for human or beast.

It took me three years to fully recover my health and endurance after leaving medical school. I make special efforts to remain healthy. I have never owned a cell phone, and do not use wireless technology of any sort.

I never use pesticides of any kind. I take no medications, and no supplements. I avoid exposure to the toxic chemicals that surround me. I eat only organic foods, and I get spring water delivered to my home in glass bottles. My home contains no plastic bags, and my refrigerator few plastic containers. I do not wear synthetic fibers. My woolens are untreated and not moth-proofed. I wash my clothing and dishes with pure unscented soap, my comb is made of natural rubber, and I wash my hair with egg yolks. My toothbrush is made of soft natural bristle. You will not find a supply of poisonous chemicals underneath my kitchen or bathroom sink.

Choosing to live this way has had several important consequences: (1) I have become very attuned to what chemical toxins and electromagnetic fields do to my body; (2) I am healthy and physically fit; (3) my way of living supports the natural world of animals and plants around me. What I find extraordinary is how people can stand to go on poisoning themselves and their world as they do.

But most people still don't know. The pervasive electrochemical haze we all live in is invisible, and neither newspapers, magazines, books, television programs, nor social media tell people about it. Nobody even stops to wonder where the modern pandemics of obesity, diabetes, heart disease, and cancer come from. Or why there are so few bees pollinating their flowers, or insects smashing into their windshields, or birds flying in the sky. Neither biologists, nor ecologists, nor doctors, nor physicists, nor chemists, nor engineers learn about toxicity in school. The next few chapters, therefore, are in the nature of a documentary on some aspects of the environmental problem that are so central to it, and so omnipresent that few people acknowledge them as part of the problem at all.

CHAPTER 17

Chemistry

The Second World War was a sea change in the human relationship with the earth. Tools were now available that took apart the atom and disassembled and reassembled molecules. The devastating effects caused by these technologies upon the physical, cultural, and psychological lives of all people are unavoidable. The tools that killed so many millions of people in war would now, if used in peace, destroy the earth.

The marketing phrase Americans used to hear after World War II, when the chemical industry was still in its infancy, was "Better Living Through Chemistry." It gave us such seemingly useful items as insecticides, synthetic fertilizers, antibiotics, detergents, synthetic fabrics, and plastics. The chemical industry is now grown up to be a tyrant, ruling every aspect of our lives, and it has gone absolutely wild. Over 50 million different chemicals have been created, and more than 350,000 of them are on the market. They are in your food and your water as deliberate additives. Your house is painted with them and insulated with them. Your clothes are dyed with them. Your bed is permeated with them. You feed your car with them. You clean your floors and your oven with them. You wash your hair, your clothing, and your dishes with them, and you brush your teeth with them. Unless you are very lucky, your teeth are filled with them.[137]

As a result of chemistry, we in North America now produce more than 280 pounds of more or less toxic wastes per person every day. I don't have figures for other countries, but since so many of the goods we now consume are made elsewhere, they can't be doing much better.

The government and the media focus incessantly on the problem of toxic wastes. But the real sources of the problem—in our laboratories, our factories, our stores, and our homes—have been universally ignored.

CHAPTER 18

Radiation

The universe we live in is organized by two kinds of long-range forces that can be likened to the two great forces of evolution, cooperation and competition. Electromagnetic fields broadcast information about one's size, shape, structure, and state of motion. With electricity and magnetism one says exuberantly, "I am here!" With gravity one says stubbornly to all the universe, "Follow me! Follow me!"

We organic beings are exquisitely sensitive instruments, more finely tuned than any radio. We follow the earth that we stand on wherever it goes, and we experience the lesser pull of the moon and the sun. And electromagnetic fields surround us carrying information from the earth, sun, moon, planets, and stars, and from the other living things that share our world with us. Biologists are accustomed to thinking only in terms of chemistry, but chemistry is just close-range electromagnetic interactions.

We all live in the magnetic field of the earth, which averages about 0.3 gauss near the equator and 0.6 gauss near the poles, and varies locally. We also live in a vertical electrostatic field of about 130 volts per meter in fair weather; the earth beneath our feet is negatively charged, and the ionosphere above our heads positively charged.

Besides these strong static fields are weak low-frequency fields which are generated by lightning and resonate in the atmospheric cavity formed between the ionosphere and the earth's surface. The observed resonant frequencies, called the Schumann resonances, are 8, 14, 20, and sometimes 26 and 32 cycles per second. These tiny electric and magnetic fields are no stronger than 0.001 volts per meter and 0.0000001 gauss, yet we shall see their importance to all earthly life.

We all participate in the global electrical circuit. Ionospheric air currents, driven by solar and lunar tidal forces, cut across the earth's magnetic field generating electric currents of 100,000 amperes some sixty miles above us. Thunderstorms and lightning bolts—100 or so strike the earth every second—bring electric charge down to the ground. Gigantic currents course through the earth beneath us. Our bodies are electrical conduits; electricity makes the return trip to the ionosphere through us and by the slow movement of charged particles through the air.

Cosmic rays also rain down upon us constantly, showering us with positive and negative ions and gamma radiation. Some of them originate in the solar wind, others come from the galaxy or beyond. Other ions and gamma rays come up from the ground, emitted by uranium, thorium, radium, and other radioactive elements.

Most animals have evolved special sense organs—eyes—to see the most abundant form of electromagnetic energy that comes our way, the light from the sun and stars. Sharks, rays and catfish also have organs that sense the electric fields of their prey, and electric eels use electricity like we use radar. Human beings—even deaf people—can hear pulsed microwaves, and in a quiet enough house one may hear the electric fields generated by 50- or 60-cycle current. But all organisms are affected by all the radiant energy around them. Plants respond to light even without eyes. Bees orient the sheets of honeycomb within their hives to the earth's magnetic field. Migrating eels in the Sargasso Sea guide themselves by the tiny electric fields generated by water currents flowing through the earth's magnetic field. And it has been proven that electric fields of less than 1 volt per meter—much weaker than the field near most electrical appliances—affect circadian rhythms, behavior, and cell chemistry in many animals including humans. Magnetic fields as small as 1 milligauss have been shown to affect many of life's functions in organisms as diverse as algae, snails, fruit flies, birds, and human beings. The coloration of fiddler crabs, and the oxygen consumption of potatoes, carrots, shellfish, and rats vary with fluctuations in the intensity of cosmic radiation.[138]

The nervous systems of vertebrates are sensitive to extremely small electromagnetic fields broadcast at precisely the Schumann resonant frequencies. Mood and behavior in chickens and monkeys, as well as in human beings, can be altered by small fields of 8 or 14 cycles per second.[139] Brain waves are tuned to these frequencies. In a state of relaxed wakefulness or meditation we produce alpha waves of between 8 and 14 cycles per second. Slower than 8 characterizes sleep. The same pattern is observed in all vertebrates.[140] And so it appears that in meditation or relaxation one tunes oneself to the primary resonant frequencies of the earth.

The electromagnetic fields we generate join us with our environment, and our own good health depends on the ability of our bodies to maintain them. The electrical system of our heart must function precisely, as must the electrical system of our brain. The positive-to-negative polarization of our head back to front, and of our body from spine to limbs, must be maintained if we are to remain conscious and alert. Wounds will not heal if the proper electrical currents are not present. But even this is a crude understanding. Acupuncture theory maps out precise pathways through which qi flows inside our bodies, and the points where it flows in and out, to and from the earth and the air around us. Qi must flow in the proper direction with the proper strength at the proper time of day in each channel within our bodies for the maintenance of good health. And this varies according to climate, season, and location on the earth. Electricity is part of, if not identical to, qi.

Through our eyes and through our skin we absorb the full spectrum of sunlight into our bodies. Without ultraviolet light our bones grow soft and deformed. Without blue light we become jaundiced. Red blood in our arteries and blue blood in our veins absorb different parts of the sun's nourishing spectrum.

Until the twentieth century everybody on the earth—plants, animals, and bacteria—lived out their lives, growing, eating, sleeping, thinking, playing, hunting, mating, in rhythms that were subject only to their own body's cycles and those of their natural environment. The electromagnetic field of the earth was steady and dependable. The sun shone steadily, and so did the moon and the stars, and their rhythms were dependable. Day followed night. Each place on the earth has its own permanent characteristics. All of the variable fields in one's neighborhood came from one's neighbors, fellow humans, animals and plants.

We are now playing with our life force without respect for life. In the past century, especially in the last few decades, we have blanketed the earth with so many new powerful and violent electromagnetic fields that we have made communication difficult all over the planet—communication between cells, communication between organisms one to the other, communication between organisms and their environment. By creating a network for machines to communicate with one another, we are interfering with sensitive biological systems. Like butterflies in the path of speeding automobiles, our organs, our nervous systems, our immune systems, and our growth and repair processes are being battered by artificial electric and magnetic fields of every size and shape from all directions.

The sun is not the greatest source of light for many of us. We live under incandescent, fluorescent, or LED lighting that gives us a very different spectrum than the sun, and that showers us with light by night as well as by day. Even the great outdoors suffers from light pollution in most parts of the world. Astronomers have trouble seeing the stars; so do plants and animals in their outdoor homes.

Crisscrossing our world are millions of miles of electric wires carrying 50- or 60-cycle-per-second alternating current. 4,000 to 35,000 volts runs through the wires that accompany most of our roads and railways, and lines carrying 110 or 220 volts bring electricity into our homes, offices, and factories. In the late 1930s high-tension wires carrying 230,000 volts from electric generating stations were strung across the United States. In the 1950s lines carrying 345,000 volts became common. In the 1960s 500,000-volt lines were in place. Today a growing network of wires carry a million volts and more.

These high-tension lines have been strung at a cost of massive deforestation and air, noise, and electromagnetic pollution. Long swaths of land 150 to 200 feet wide have been cleared of all trees and vegetation and saturated with herbicides, poisoning the land and fragmenting the continent's forests. Birds in flight often crash into power lines and their supporting structures. Other birds die from electrocution; this is a significant cause of eagle mortality in the western United States. Corona discharge causes the wires to hum and crackle continuously. The surrounding air glows blue, and is filled with ozone, oxides of nitrogen, and free radicals. And electromagnetic radiation is broadcast over the landscape from what amount to giant antennas that we have erected from coast to coast, and from the Arctic to the Gulf of Mexico.

The electric field generated by a 345,000-volt wire is typically 10,000 volts per meter near ground level directly beneath the wire, and the magnetic field can reach 700 milligauss. Seventy-five feet away, at the edge of the utility company's right-of-way, the electric field is still 1600 volts per meter, and the magnetic field 100 milligauss. Five hundred feet away it may exceed 3 milligauss, still biologically powerful. Presumably a similar magnetic field exists high above the wire where birds fly. No matter where we live, these fields surround us, outside and inside our homes.

In cities the ambient magnetic field is often more than 1 milligauss, with fields of 3 milligauss not uncommon. Even in rural areas remote from power lines, 60-cycle electric and magnetic fields are likely to exceed 0.001 volts per meter and .00001 milligauss, ten and one thousand times the natural background, respectively. No one is totally free of it, not human, not bird, not plant, and not fish.

Even the oceans do not escape our electronic reach. Very-low-frequency (VLF) radiation is capable of penetrating deep into the seas, and enormous VLF antennas have been erected in 16 countries to communicate with submarines.

Higher on the electromagnetic spectrum, we have filled our environment with radio waves of all frequencies, from AM and FM radio and VHF television broadcasting, ship and aircraft radios, police and fire radios, burglar alarm systems, ham radios, radar stations, emergency communication systems, military communication systems, cell towers on earth and in the oceans, Wi-Fi, the 15 billion mobile devices in people's hands, the 10,000 satellites orbiting the earth, and the soon-to-be-trillion Internet-of-Things antennas that are being installed in practically every machine and consumer product. The density of radio waves around us is already millions to billions of times greater, everywhere on earth, than the natural flux reaching us from the sun and stars. All of the world's animals and plants in air, land, and sea are becoming roadkill on the information superhighway. And in their efforts to save endangered species, scientists are driving them further toward extinction by attaching radio collars to their last living representatives. They are being used to track birds, dolphins, whales, turtles, sharks, polar bears, musk oxen, camels, wolves, elephants, albatrosses, penguins, snakes, amphibians, bats, fish, butterflies, and dragonflies. And as everyone who has bothered to study the issue has discovered, this causes weight loss, reduced activity, reduced social interactions, reproductive failure, and vastly increased mortality.[141]

Continuing higher up the electromagnetic spectrum, we come to infrared, visible, and ultraviolet light. Although many human beings see the sun only rarely nowadays, sunlight is still by far the greatest source of energy falling on the surface of the earth and is of exceeding importance to life. Animals have eyes to see it with. Green plants manufacture all of their food from water, air, and sunlight. Our cells and their components are approximately the size of the light waves they absorb.

Still higher in frequency are X-rays and gamma rays, with wavelengths the size of atoms and smaller. They are called "ionizing" radiation because they have sufficient energy to remove orbital electrons from atoms. X-rays are used very widely in medicine and dentistry. Radioactive substances, emitting ionizing alpha, beta, and gamma rays, are also heavily employed in medicine. Each year about 3.6 billion diagnostic X-rays and more than 50 million nuclear medicine procedures are performed globally.

Environmental levels of ionizing radiation have been increased by the explosion of more than 2,000 atomic bombs, and by the by-products of the nuclear weapons and nuclear power industries. Unlike other electromagnetic pollution, this radiation will not go away instantly when we turn off the power. We have managed to sterilize or lethally contaminate fair-size chunks of the earth—Moruroa in the South Pacific; some of the Marshall Islands; parts of the Nevada desert; part of the Hanford Nuclear Reservation in Washington State; a portion of the south Ural Mountains in Russia; the area around Chernobyl in northern Ukraine; parts of Kazakhstan; parts of the islands of Novaya Zemlya in the Russian Arctic; an area of both land and ocean in Fukushima, Japan. Nuclear power stations dot the surface of the globe, each one in service for twenty or thirty years until it becomes too radioactive to operate, at which time it must be entombed forever and replaced with a new one. 130 nuclear submarines ply the depths of the oceans for twenty or thirty years until they become lethally radioactive, and their parts become brittle, and they must be entombed somewhere and replaced with new ones.

The by-products of the nuclear industry are all around us, closer than most people think. Smoke detectors contain americium 241. Cobalt-60, supplied by Bruce Power's Bruce B nuclear power plant in Ontario, is used to sterilize medical equipment, including gloves, masks, and syringes, throughout the world. Radioactive metals are finding their way into an array of products that are made with recycled materials: kitchen cheese graters, reclining chairs, women's handbags, tableware, fencing wire and fence posts, shovel blades, elevator buttons, airline parts, and steel used in construction have been found to be radioactive. Cesium-137, a waste product of nuclear weapons manufacture, is being used to irradiate some food—meat, shellfish, eggs, fruits, vegetables, grains, and spices—so it will last longer on grocery shelves. About 120,000 tons of food are irradiated annually in the USA for human and animal consumption, including one-third of all spices sold in this country. In Asia the amount of irradiated food was estimated at 285,223 tons per year in 2010. Food is irradiated in more than fifty countries. The United States, China, the Netherlands, Belgium, Brazil, Thailand, and Australia irradiate foods commercially. Australia, India, Thailand, Vietnam, and Mexico irradiate food for export.[142]

Ionizing radiation causes cancer. But so does non-ionizing radiation, which is emitted by every wireless device and every radio antenna. Cancer is primarily a failure of cell-to-cell communication. It is a disease, not of mutation, but of uncontrolled growth and de-specialization. We send out electromagnetic signals to all of our cells instructing them as to their function and rate of growth. Anything which sufficiently disrupts our bodies' own internal system of communication can cause cancer. Cancer cells are simply normal cells which have gotten out of control, and which are expressing genes that are supposed to be shut off. Muscle cells may forget to be muscle cells, and they may cease cooperating with their neighbors. Non-ionizing radiation also interferes with tiny electric currents *within* cells—the currents in our mitochondria, in a system called the electron transport system, which transports the electrons generated by the digestion of the food we eat to the oxygen we breathe. This starves cells of oxygen, slows their metabolism, and makes it more difficult for them to hear the signals that are telling them to stop growing and what kind of cell to be.

In the womb, internally generated electromagnetic fields guide the growth of a developing fetus. External fields that disrupt this communication can lead to birth defects or miscarriage.

> As we live and breathe and pulsate with the earth,
> like willows waving and whispering with every smallest breeze,
> so do our cells dance to rhythms beat on their tiny membrane drums
>   by one orchestra, and one choreographer.
> The musicians still play, though we choose not to listen,
> and like cancerous cells that become strangers and grow untutored
>   within their body,
> so are we becoming unschooled strangers upon our earth.

CHAPTER 19

Plastic

The lakes of petroleum, the beds of coal, and the fossilized fields of stone that lie deep beneath the ground in so many parts of the world have a very important function. They are the remains of plants and animals who lived long ago and have so far escaped their destiny of final decay and oxidation. As long as they remain buried deep underground, oxygen remains in the atmosphere for us to breathe. But the human race is pumping that petroleum out of the ground, burning some of it, and spreading the rest in a toxic film all over the surface of the earth. We are making plastic out of it, and the plastic is everywhere. From the time they are born, our babies are given plastic diapers to wear, plastic toys to play with, plastic bottles to drink from, and plastic pacifiers to suck on. Plastic pipes carry drinking water into our houses. Ion-exchange systems use plastic membranes to purify that water. We wrap and store our food in plastic and drink our beverages from plastic containers. Hospitals store all their blood and intravenous fluids in plastic bags. Even organic farmers irrigate their fields using plastic pipes and plastic hoses and protect their seedlings with black plastic mulch. Our cars are half plastic. Our computers and cell phones are thirty to forty percent plastic.

These materials are so much around us that we don't notice the harm they are doing to us and to our planet. Plastics are neither durable, repairable, recyclable, nor biodegradable. They are difficult to sterilize, and they are permeable to air, water vapor, oils, and environmental pollutants. They are themselves pollutants, containing as they do poisonous and cancer-causing chemicals that continuously ooze out of them. And the microplastics that they shed and break down into are in our water, our air, our soils, and our bodies. Microplastics have been found in deep oceans; in Arctic snow[143] and Antarctic ice; drifting in the atmosphere and falling with rain over mountains and cities. Scientists reported that an average of 365 particles of plastic per day were falling out of the air on every square meter of land at a remote location in the Pyrenees.[144] Another team sampled the air in Grand Canyon, Bryce Canyon, Craters of the Moon, Joshua Tree, and seven other national parks and wilderness areas in the western United States. They reported that 4 percent of the dust in the air over these areas is microplastics, and that on average 132 microplastic particles rain on every square meter of these parks every day. From this sampling they estimated that between 1,000 and 2,500 metric tons of microplastics fall out of the air on protected areas in the western United States, including national parks and wilderness areas, each year. That is equivalent to between 120 and 300 million plastic water bottles. These conclusions must be taken with a grain of salt, because the researchers are in just as much denial as the rest of society: the devices with which they collected and filtered their air sample were made of plastic![145] But the results of a team of Hong Kong and Canadian scientists about microplastics in the atmosphere are so enormous that they could not have been due to contamination by plastic equipment. They collected an astounding half a million polyester microfibers released from a domestic tumble dryer during a single fifteen-minute cycle and concluded that the average household could release between 90 and 120 million synthetic microfibers into the air in a year.[146]

Microplastics have been found in honey, sugar, beer, milk, table salt, rice, vegetables, seaweed, seafood, bottled water, tap water, energy drinks, and soft drinks.[147] They have been found in human placentas[148] and in stool samples from every volunteer from Europe, Japan, and Russia.[149] 40 percent of the dust in homes worldwide is made up of microplastics.[150]

Our modern attitudes about garbage are traceable to the widespread use of plastics. These are the first building materials ever used by human beings that cannot be repaired. The habit of throwing away instead of repairing them has spread to everything we make. We throw away glass bottles after just one use even though they are still as good as new and are easily sterilized. Nobody repairs ceramics and porcelain anymore. Few people darn their socks and patch their clothing—except occasionally to be stylish. We discard whole light bulbs when only a tiny filament needed replacement, and we throw away whole pens when only the ink was used up. Our dumps are full of refrigerators, stoves, and other appliances that perhaps needed only some new wires, or to be sanded and polished. People discard their automobiles after just a few years' use. Beautiful old houses and large buildings that might have stood another one or two hundred years if cared for, are torn down and replaced with new structures that will not last as long.

Our world did not begin to use large amounts of plastics and synthetic fibers until after World War II, so that the wholesale replacement of natural materials has occurred during my lifetime. The approximate dates of first commercial manufacture for some polymers are:

phenolics (Bakelite)	1908
acrylics	1931
polyvinyl chloride	1936
nylon	1938
polystyrene	1938
polyethylene	1942
polyesters	1942
polytetrafluoroethylene (Teflon)	1943
polyurethane	1953
polypropylene	1958
polyethylene terephthalate (PET)	1978

Plastic grocery bags weren't even introduced until 1978. By 1982 they had captured only 5 percent of the American market, but by 1988 already 70 percent of all grocery bags used in the United States were plastic. Currently the United States uses almost two billion barrels of petroleum each year to manufacture 124 billion pounds of all kinds of plastics. These plastics could in theory be made out of corn instead of petroleum, but this would require about two-thirds of our corn crop.

Globally, one million plastic bottles are purchased every minute, and up to five trillion plastic bags are used every year.

Additives in Plastics

Many polymers are inert and stable over immense periods of time. Others begin to break down almost immediately. Polyvinyl chloride, for example, begins to release hydrochloric acid immediately upon exposure to heat, light, or oxygen.

The plastics we use also release other chemicals, because they all—even "biodegradable" plastics—contain numerous intentional and unintentional additives. Most additives are small molecular weight, relatively reactive compounds. Most are toxic; virtually all of them leach out of the plastic at an appreciable rate and they may equal as much as half the total weight of the finished product.

Intentional additives may include fibers, fillers, plasticizers, stabilizers, antioxidants, ultraviolet absorbers, biocides, antistatic agents, antifogging agents, flame retardants, colorants, fragrances, impact modifiers, and clarifying agents. Also added are processing aids, including lubricants and slip agents, plastisol viscosity depressants, mold releasing agents, and flow controls. The chemical reactions are controlled and directed with catalysts, coupling agents, emulsifiers, and inhibitors.

Unintentional additives, always present, include residues of solvents, as well as incompletely polymerized fractions, small molecular weight cyclic compounds, and side reaction products. Plastic foam contains residues of blowing agents, used to produce the foaming gas. Layered packaging also contains adhesives and primers, and may have solvent-based coatings, lacquers and inks.

Plasticizers—commonly phthalic acid esters—are liquids that make rigid plastics, such as polyvinyl chloride, flexible and soft. They usually make up 20 to 50 percent of the finished product. Stabilizers prevent polyvinyl chloride from losing hydrochloric acid, and are another 0.1 to 10 percent of the product. A "stabilizer" may itself be a conglomeration of 15 different types of chemicals.

Antistatic agents prevent plastics and synthetic fibers from having more "static cling" than they do, and make up .05 to 7 percent of the weight of the product. They are usually detergents, and work by causing an extremely thin, invisible film of water to form and continuously evaporate from the surface. Antifogging agents, another 0.5 to 4 percent of the product, prevent water condensation from fogging up food and agricultural packaging. Antioxidants are usually present in plastics in concentrations of up to 1 percent.

Impact modifiers make plastic bottles bounce instead of break, and they also make siding, window profiles, pipes, molded plastics, and plastic films absorb impacts without breaking.

Fragrances are added to mask the inherent odor of the plastic and all its additives and are present in food wrap films at a concentration of 10 parts per million. More conspicuous odors are added to some other plastics, such as the lemon scent added to garbage bags to hide the smell of garbage.

Biocides are often added to prevent mold and bacteria from feeding on all the other delicious chemical additives. Shower curtains, wire and cable, coated fabrics, wall coverings, automotive trim, marine upholstery, awning, pool and pit liners, and many other items are impregnated with them.

Title 21 of the United States Code of Federal Regulations lists thousands of chemicals and combinations of chemicals that are permitted to enter foods from plastic packaging.

The chemicals from which plastics are made are poisoning life all over the planet. Most of our river water, air, and drinking water contains measurable amounts of caprolactam, vinyl chloride, methyl methacrylate, and acrylonitrile. Five of the six chemicals whose production generates the most hazardous wastes are chemicals that are basic to the manufacture of plastics: propylene, phenol, ethylene, styrene, and benzene. These are the raw materials used to make polypropylene, phenolics, polyethylene and polystyrene. Benzene is the raw material for styrene and is also used as a solvent in making low density polyethylene and polyvinyl chloride.

Styrene is neurotoxic and mutagenic. Vinyl chloride causes multiple organ system damage and is carcinogenic, mutagenic, and teratogenic. Vinylidene chloride, the raw material for Saran wrap, is carcinogenic, mutagenic, neurotoxic, and causes liver and kidney damage. The synthetic rubber neoprene is made from a chemical that is embryotoxic, teratogenic, mutagenic, and damages the nervous system, cardiovascular system, liver, and kidneys. Dithiocarbamates, used in the rubber industry, are so poisonous that many of them are also used as insecticides. Caprolactam, the raw material for nylon, damages the nervous, genitourinary, and cardiovascular systems. Acrylic acid is embryotoxic and teratogenic. Polyurethane is made from toluene diisocyanate, which is derived from phosgene, a poison gas used in chemical warfare. Acrylonitrile, or vinyl cyanide, is widely used in making acrylic fibers, plastics, surface coatings, pharmaceuticals, and dyes. From 5 to 30 percent of the acrylonitrile may remain as unreacted monomer or be liberated with the aging of polymers. It has many of the toxic properties of the cyanide ion, and is a registered pesticide, used to fumigate tobacco, flour milling, and bakery food processing equipment.

The additives are just as bad. Heavy metals, including cadmium and lead, are used in pigments and in polyvinyl chloride heat stabilizers. Lead is universally used as a stabilizer in wire and cable because it is also a good insulator. Most flame retardants are halogenated compounds. Plasticizers, including phthalic acid esters, are some of the most widely dispersed synthetic chemicals in the environment. They are used in building and constructions, automobiles, home furnishings, toys, clothing, food coverings, and medical products. They evaporate from plastic upholstery, causing the "new car smell." They ooze from plastic bags storing whole blood or intravenous solutions and accumulate in human spleen and fatty tissue after blood transfusions; anyone who has been in a hospital with an IV line has some plasticizer in his or her body. Chemicals from the plastic bag or tubing can cause local inflammation at the IV site or systemic illness that is usually attributed to some other cause. Phthalic acid esters also enter our milk supply from the PVC tubing in milking equipment. They are biodegradable, but we use so much of them that some levels of phthalic acid esters are in all of our food, water, and air, and they have even been detected in the tissues of jellyfish 3,000 feet deep in the North Atlantic Ocean. These chemicals have caused birth defects, mutations, and liver cancer in laboratory animals, and are present in some bodies of water at levels high enough to interfere with the reproduction of water fleas.

Some of these compounds double as pesticides. Di-2-ethyl-hexyl phthalate, the most widely used plasticizer, is also used in orchards to kill mites. Replacement chemicals, such as dioctyl adipate, may be more toxic, but less studied and less regulated.

Most rubber manufactured today is either totally synthetic or a mixture of natural latex and synthetic materials. Scrap rubber is often disposed of by grinding it up and using it to "improve" soil, thus introducing synthetic chemicals into groundwater and therefore into our food supply.

The plastic pipes that carry water into our houses contaminate the water with small amounts of organic chemicals, but this is not their worst sin. Polyvinyl chloride, polybutylene, and polyethylene plastics are all capable of absorbing insecticides, herbicides, and solvents from outside, through the pipe and into the water supply. If toxic chemicals are used in the soil above a water line, this can be a very dangerous situation.

Plastic packaging of drugs has created a host of serious problems that few doctors or their patients are aware of. Because of the permeability of plastic, air can pass through it to oxidize the drug, change its pH or contaminate it with a foreign substance. Or ingredients in a drug can permeate the plastic, decreasing the potency of the medicine. Bacterial contamination may result when preservatives escape. Even solid materials can migrate through the container walls and reach the outer surface. It is well-known, for example, that nitroglycerin tablets will lose their potency if stored in plastic containers. Chemical ingredients in the plastic itself can leach into solutions stored in them and be administered to patients. Many drugs undergo chemical reactions with the constituents in the plastic, creating new compounds that discolor the container, but since most drug containers are already colored, this will rarely be noticed. Problems become evident only much later, when the container begins to crack or disintegrate.

I am still surprised by how many people are unaware that plastics are porous, and that a thin film of plastic offers little protection for food that is wrapped in it. You can discover this for yourself at home, or in a supermarket. Pick up any item that is packaged in plastic, such as cheese, or bread, or any bulk food such as a grain, or bean, or flour, that is in a plastic bag, and smell it. The odor comes right through the plastic, and you can easily detect whether the food is fresh or rancid. I know many people who re-use plastic containers, and even plastic bags, many times until they wear out, but I do not recommend it. If you store water in a plastic jug, you can taste the contaminant from the jug. But it is not even sanitary. Most plastic items not only can't be sterilized—at least not with home equipment (hospitals use gamma radiation)—but they contain a lot of juicy chemicals that bacteria love to eat. In the decade following 1950, about the time when plastic packaging for food was introduced, reported cases of food poisoning increased tenfold. We have been conditioned to believe that plastic is sanitary and protects the food, but this is basically a deception.

Any scientists whose experiments depend on the purity of chemicals are in a lot of trouble today unless they are still available in glass containers. I walked into a chemical supply house once looking for diatomaceous earth, and it was even worse than walking across the Goethals Bridge into Elizabeth, New Jersey on the peace walk. The air was permeated with a lethal-smelling mix of whatever thousands of chemicals in plastic bottles were on the shelves, each one contaminating and contaminated by all the other chemicals. I became nauseated within two minutes and walked out. I don't know how the very-unhealthy-looking man in the storeroom could survive in there, or if he did.

Whether or not some plastics are, or can be made to be, really biodegradable or recyclable is questionable. Many of the additives added to plastics—the antioxidants, the heat stabilizers, the ultraviolet absorbers and the biocides—are there to *prevent* biodegradation, and with good reason: in order to make the products last and not break down during use into their toxic components. Nevertheless, some manufacturers have attempted to produce biodegradable plastics by designing them to break down when exposed to ultraviolet light. This involves adding yet another toxic chemical to the mix, and is basically useless for plastics that will be buried underground in a landfill and exposed to no light whatsoever. Other plastics contain about 7 percent cornstarch and disintegrate in a matter of months into plastic debris or dust. This is not true biodegradation. Other plastics are on the market that are more truly degradable. But the problem of toxicity—of the plastic itself or of the degradation products—has not been addressed at all.

There is some recycling of plastic going on, especially where there is mandatory separation of garbage. But you can never melt down an old piece of plastic and get back the original item, not without a lot of extra chemistry and the input of new toxic materials.

The campaign to recycle more plastic and make it all biodegradable, so vigorously promoted by environmental organizations, may help us to feel less guilty about living wastefully and producing such a mountain of garbage, but it will not solve any of our actual problems. Only eliminating plastic altogether will prevent it from further accumulating in all of our air, water, and bodies.

CHAPTER 20

Detergents

Substitutes for soap were first manufactured in the 1920s, but until the end of World War II they saw only limited use, mostly in fine laundering, dish washing, and a few specialty shampoos. Heavy-duty detergents with condensed phosphates were introduced in 1947, and they dominated the clothes-washing market so quickly that only six years later more detergents were sold in the United States than soaps. Today few people even know the difference. Almost all products for washing clothes, almost all household cleansers, almost all shampoos and toothpastes, and a good many liquid and solid hand "soaps" are today detergent-based.

Although detergents have been marketed well, they are inferior in so many ways to the soaps they replaced, and cause immensely more pollution to our groundwater and our waterways. The superiority of soap has been forgotten, but in 1950 there were chemists who warned that we were going in the wrong direction:

> Built synthetic detergents often are so heavily built that they are really alkalies disguised as synthetic detergents and as such are often harsh and harmful to the skin and to textiles... Any synthetic detergent, in order to clean well, must be heavily built and this departs too much from the original mildness and neutrality... Soap, when applied under ideal conditions, still seems the best washing agent for laundry use... Why not build soap more extensively and, by so doing, increase its efficiency in hard water?" (H. C. Borghetty and C. A. Bergman, in *Journal of the American Oil Chemists' Society*)[151]

Soap is actually a much better cleanser than any detergent: the power of a pure soap to suspend soil and prevent its redeposition on clothes is ten to fifteen times greater than the power of an equal concentration of pure, unbuilt detergent.[152]

The problem with soap comes when it is used in hard water. Calcium and magnesium ions then combine with soap to make "lime soap"—a film that forms on your clothes and dulls your hair. The cure for this is to use a water softener, and many such have been included in soap formulations—polyphosphates, washing soda, borax, salts of EDTA, citric acid and others. The most efficient way to use a water softener, however, is to put it in the first rinse water.

Detergents to not deposit lime soap in hard water. But unlike soaps, detergents lose their cleaning power in hard water, and are always formulated with water softeners anyway. Laundry detergents, liquid or solid, usually contain substantial amounts of these and other chemicals:

- builders to soften the water, usually phosphates, silicates, carbonates or citrates (50% to 75% of the product)
- anti-redeposition agents, usually carboxymethylcellulose (1% to 3%)
- optical brighteners, which are really fluorescent dyes that absorb ultraviolet and give off blue light (small amounts)
- anticorrosion agents (small amounts)
- foam stabilizers (small amounts)
- hydrotopes to keep all the other chemicals in solution
- perfume
- color

Often added are:

- enzymes to remove stains
- sodium perborate, a non-chlorine bleach

All of their chemical ingredients make detergents a major pollution problem, because virtually all of them go down the drain and into our rivers, lakes, and groundwater. Phosphate builders, in particular, over-fertilize our waterways, causing lakes and slow-flowing streams to become overgrown with algae and depleted of oxygen. Other plants, fish, and aquatic animals eventually disappear. And so over the years we have become concerned about the phosphates in our detergents, and their use has been restricted by many governments and phased out by some manufacturers of detergents.

But detergents themselves kill fish by drowning them: gilled creatures depend on the surface tension of water for breathing, and all detergents reduce surface tension. So do soaps, but soaps in waterways are instantly inactivated by minerals and rapidly destroyed by bacteria. The precipitation of lime soap, which is a problem in washing, makes soap completely harmless to rivers and lakes. Detergents, on the other hand, biodegrade more slowly. During the 1950s and 1960s sewage treatment facilities were plagued with persistent foam, and streams and rivers and even well water in some places acquired a layer of suds. In 1965 manufacturers began to change over to formulations that broke down more quickly, and today all household detergents must pass specified tests for biodegradability. This is a vast improvement. However, no detergent breaks down as quickly as soap. Seventh Generation and other so-called "natural" laundry products are no more biodegradable than Tide or All. Their virtue is that they probably contain no other harmful chemicals besides the detergent.

We are polluting not only our environment but also ourselves. Soap is a simple substance that for thousands of years was made in barrels from wood ashes and animal fats. Detergents are usually petroleum products, but even those made from vegetable oils are the complex creations of a chemist's laboratory. They differ widely in their toxicity—those made from benzene are more poisonous than those made from vegetable oils—but they are all more toxic than soaps, and also more irritating to the skin. Even the mildest detergents are readily absorbed by skin and exert a tanning effect on it. Cationic detergents, one of the four main classes, are also effective antiseptics, and in large doses they can cause curare-like paralysis.

Exposure to these chemicals is continuous. People absorb them all day long from their clothing, all night long from their bed sheets and pillow, brush their teeth with them morning and night, add them to their hair when they bathe, and eat, drink, and cook with utensils, dishes, glassware, and pots that are coated with them.

Detergents are very difficult to rinse off. Soap residues leave spots on surfaces, but detergent residues are invisible. They can, however, be smelled. Washing machines do a very poor job of rinsing clothes, and indeed some of the chemicals, like optical brighteners, fabric softeners, and perfumes, are designed to stay *in* your clothes. Dishes may sparkle, but it's hard to get rid of the last traces of invisible detergent. Plastic dishes, which absorb chemicals, may even acquire a permanent odor after many washings with dishwashing liquid. Detergent shampoos always leave a coating on your hair; the chemicals that impart body and manageability are supposed to stay on your hair, not rinse off.

Practically all of the shampoos and toothpastes advertised as "natural," even the ones "made from coconut oil," are chemical detergents. They are not heavily built like laundry products because they do not need to clean as well, but they are irritants that leave residues on us. The detergent ingredient in many of these cosmetics is called sodium lauryl sulfate. I do not use products containing this or other synthetic ingredients. I usually wash my clothes by hand, and they get clean just fine with ordinary soap.

The natural world is abundant with possibilities for making things clean. The earth contains thousands of plants that have in their roots, leaves, bark, fruits, or berries, substances that foam up in water and make excellent cleansers for body, hair, or clothing. More than a hundred saponin-containing plants are known in North America alone, including:

agave – roots of some species
buckeye, *Aesculus pavia* – roots
Missouri gourd, *Cucurbita foetidissima* – fruits
wild lilac, *Caenothus* species – flowers and young seed pods
pigweed, *Chenopodium Californicum* – roots
red-root, *Caenothus Americanus* – young fruits
soap plant, *Chlorogalum pomeridianum* – bulbs
soapwort, *Saponaria officianalis* – roots
soapweed, *Yucca glauca* – roots

Plants called soapbark (*Quillaja saponaria*), soapberry (*Sapindus saponaria*) and soapnut (*Sapindus mukorossi*) come from the tropics; eighty years ago these could be bought in American shops for the washing of delicate fabrics.

Bentonite clay is a natural cleanser and water softener. Fresh egg yolks make a fine shampoo all by themselves. On camping trips, I have found that nothing scours pots and pans quite so well as plain dirt. The natural world is all around us to play in. Synthetic chemicals are a poor and lonely substitute.

CHAPTER 21

Biocides

These sprays, dusts, and aerosols are now applied almost universally to farms, gardens, forests, and homes—nonselective chemicals that have the power to kill every insect, the "good" and the "bad," to still the song of birds and the leaping of fish in the streams, to coat the leaves with a deadly film and to linger on in soil—all this though the intended target may be only a few weeds or insects. Can anyone believe it is possible to lay down such a barrage of poisons on the surface of the earth without making it unfit for all life? They should not be called "insecticides," but "biocides."

— Rachel Carson, in *Silent Spring*

Insecticides, herbicides, fungicides, acaricides, nematocides—they all have one purpose only—to kill life.

Biocides are everywhere. Our farmland is saturated with them, and therefore our groundwater, rivers, and lakes are contaminated with them. Our forests are routinely sprayed with herbicides because varieties of trees that we do not prefer have the audacity to grow in them. We poison our lakes and our reservoirs to get rid of plants we'd rather not have there, or to get rid of eels that prey on sport fish, or even just to replace existing fish with other kinds that we like better. Cotton is heavily sprayed, and so we have poisons in our clothing. Food crops are sprayed and food storehouses are fumigated, and so they are in our food. Roadsides, railroad corridors and power line rights-of-way are saturated with herbicides. Golf courses are treated with both herbicides and insecticides. Spongy moths in our oak forests, grasshoppers on our grain fields, fruit flies in our orchards, mosquitoes just about anywhere, any animals or plants that human beings decide bother us are dealt with in the same way: spray and poison. It has even become acceptable routine to spray clouds of insecticide over crowded metropolitan areas from helicopters. In my youth, the countryside was regularly dosed with DDT. Today farmland as well as entire cities are sprayed with malathion. DDT is still used in India, China, North Korea, and some countries in Africa.

Extermination is big business. Every grocery, every restaurant, every school, every airport, every office, every factory, every shopping mall is periodically fumigated or sprayed. There are few exceptions. There are few American homes without a can of bug spray somewhere in them. We treat our lawns with herbicides so no crabgrass shall grow. We poison our flowers for no better reason than cosmetics. We poison all our pets so no fleas shall bother them. We disinfect the air and all the indoor surfaces of our homes and offices as if our lives did not depend on the existence of bacteria. We chlorinate our swimming pools and our drinking water so nothing at all shall live in them. Our carpets and upholstery have biocides in them. So does the paint on our walls and the leather in our shoes. The wood in log homes is often permeated with them. The soil underneath most American homes is saturated with chlordane to kill any termites that might come around.

In the words of one salesman of swimming pool supplies: "We don't want anything alive in the pool but the kids."[153]

When the rats are gone, owls do not eat. When mosquitoes and flies disappear, bats starve. Insect-eating birds go hungry. Frogs have no dinner. When bees are gone, flowers are not pollinated. Fruit does not set.

Our biocide habit grows larger with each passing year. Insects and rats have short life spans. They eventually evolve and adapt to our poisons while their longer-lived predators die off, killed by the same chemicals. Owls, bats, and frogs become rare while rats and bugs survive. American farmers today use sixty times as much biocides as they used a century ago, but they lose 10 percent more of their crops to insects, fungi, and weeds.

Ducks, woodpeckers, bats, frogs, lizards, porcupines, skunks, shrews, and spiders once did a much better job on the insects of our fields and forests than the chemicals we now apply.

I cannot improve on Rachel Carson's *Silent Spring*. Her story is as valid now as the day it was published in 1962. She has not been heeded. Everything she warned us of has come to pass.

Alan Devoe wrote of our dependence on all other forms of life, insects included:

> We are one small ingredient in a whole of unimaginable vastness. More, we are a part of a general and embracing interdependence. We are here at all because there is a woven relationship, so intricately threaded that the whole pattern of the weave is beyond our understanding, between ourselves and birds, between birds and vegetation, between mammals and fish, between suns and seeds, between every ingredient of the totality and every other ingredient. We are supported by starfish. An owl props us. Earthworms minister to hold us upright . . . When we regard a toad or a snail, we are regarding a part of our own total selves. We are not a lone thing, huge and isolated. We are a very small thing, precarious with creaturely dependence. We may not despise, any more than we may despise a brother or an organ of our own body, even a being so small, so dim, so remote from us, as, say, this common grasshopper rustling in the grass here beside me on the hot sidehill under the June sun. (from *Our Animal Neighbors*)[154]

CHAPTER 22

Farewell to Silence

On January 26, 1991, a team of scientists from eight nations met near Heard Island in the southern Indian Ocean. They lowered five loudspeakers underwater to a depth of 820 feet, and for the next five days a tremendous blast of sound issued forth that was heard round the world. A signal was being sent, one hour on, two hours off. Other scientists waited with deep underwater microphones, some nearby in the Indian Ocean, some in the south Atlantic and middle Pacific Oceans, others near Bermuda, San Francisco, and Florida, still others along the coasts of Antarctica, South Africa, India, Australia, New Zealand, Brazil, and Canada. By precisely timing the arrival of the signal, scientists could measure exactly the speed of sound in the world's oceans.

Because the velocity of sound is dependent on water temperature, any warming of the oceans due to the greenhouse effect will speed up the journey of sound through thousands of miles of water. This was an environmental research project and was continued on a more limited scale in the North Pacific from 1996 through 2006. Said Dr. Walter Munk of the Scripps Institution of Oceanography, "There have been estimates that it would take one hundred years of measurements of the atmosphere to achieve 85% certainty that the greenhouse effect is, or is not, operating." These experiments were supposed to achieve the same degree of certainty in just ten years. "Surely, that is information the world needs," said Dr. Munk.

But at what price? These low-frequency sounds were louder than a jet airliner taking off. Fish and marine mammals for hundreds of miles around must have found this an annoying if not painful and deafening noise. For whales and Weddell seals, who use low-frequency sound to communicate with one another across whole oceans, these experiments were serious indeed. These animals already avoided harbors, oil-drilling operations, and major shipping lanes because of the terrible din.

The fact is that we human beings have become insensitive about noise, and we have ceased to care about how much of a racket we make in our own or anyone else's back yard.

Sound travels much farther in the ocean than in the air, as Leonardo da Vinci noted in 1490:

> If you cause your ship to stop, and place the head of a long tube in the water and place the outer extremity to your ear, you will hear ships at a great distance from you.[155]

These were, of course, sailing ships he was talking about. Today a good-size diesel ship can often be heard at a distance of 1,000 miles or more underwater.

The speed of sound in the ocean varies with temperature and pressure and therefore also with depth. Deep beneath the sea is a layer where sound's speed reaches a minimum; it increases upward because the temperature rises, and it increases downward because the pressure rises. At this depth there is in effect a sound channel which carries sound extremely far with only small losses. It lies at a depth of about 4,000 feet in mid-Atlantic, and 3,000 feet in mid-Pacific, but only about 820 feet near Heard Island in the southern Indian Ocean. Experiments by the United States Navy during the 1940s showed that the sound from an explosion of four pounds of TNT in the deep sound channel was quite capable of traveling 25,000 miles, or around the entire earth, before it diminished to background levels.

Before industrial times, most of the sounds heard beneath the great waters of our planet were made by wind, weather, sailing vessels gliding along the surface, and the conversations of fishes, lobsters, shrimps, crabs, seals, dolphins, and whales. Today, natural sounds still dominate the higher frequencies, but in the lower frequencies they are drowned out by modern shipping and other noises, because sound waves are used underwater for all the same purposes as radio waves in the atmosphere. The oceans, particularly near harbors and along the continental shores, are now filled with the sounds of:

- harbor noises, including the comings and goings of ships, dredging operations, and underwater construction;
- seismic exploration for minerals and oil, or for the scientific study of the earth's crust. Air gun or dynamite explosions can be heard several hundred miles away. 5 to 100 Hz.
- sonar, used in mapping the ocean floor; in mineral exploration; in bedrock studies for dredging, tunneling, and construction projects. 100 to 5,000 Hz.
- underwater telephones used by scuba divers, and homing tones that guide divers back to their boats. 8 to 11 kHz.
- depth sounders used by ships. 12 kHz.
- fish finders. All commercial fishing boats are equipped with sonar for locating schools of fish. 11 to 400 kHz.
- telemetry and remote control, used in drilling, and in operating undersea oil wells. 5 to 100 kHz.
- beepers that aid in locating lost objects; navigational beacons; survey markers for mining exploration. 15 to 100 kHz.
- short-range sonars for geological surveying; for dredging operations; for locating damaged or destroyed oil platforms or pipelines; and for various military uses. Up to a few hundred kHz.
- Doppler sonar for navigation or docking control. 300 and 600 kHz.

To this mix is now being added the Internet of Underwater Things, which is beginning to flood the oceans with sound in order to connect them to the Internet. And this sound is pulse-modulated with the same harmful frequencies as radio waves in order to carry the same data. And to communicate over large distances, some of the underwater acoustic modems that are being marketed are capable of producing sound as loud as 202 decibels. That is equivalent to 139 decibels in air. It is as loud as a jet engine at a distance of 100 feet and is above the threshold for pain in humans. These modems blast modulated sound at frequencies ranging from 7 kHz to 170 kHz, encompassing almost the entire hearing range of dolphins, which use sound for hunting and navigating.[156]

Sound is an important part of the environment for fish, crustaceans, and marine mammals, particularly so since daylight penetrates only about 1,500 feet into the clearest water, and vision is of even less use in muddy water and at night. In the lowest frequencies, most fish have better hearing than we do. Humpback whales sing melodious songs to one another, and deep fin whale voices converse across hundreds, perhaps even thousands of miles of ocean. Many whales and seals, and all dolphins use high-frequency sonar to locate fish and navigate through the dark seas.

Here on land also some creatures use their ears like we use eyes. The most famous are bats, those small winged mammals that roost in caves and hollow trees, and in barns, belfries and attics, those mysterious little mouse-like fliers who venture into the skies only under cover of night. Mystery, unfortunately, has bred fear and even hatred, but these persecuted dark-loving neighbors of ours are as colorful, in their own way, as the birds. They eat prodigious numbers of insects—at least most bats here in New Mexico do, the varieties of "cave bats" who hibernate during the winter, and the "forest bats" who fly south. Many bats in other parts of the world eat fruit only, and some Asian and South American varieties fish for a living. There are nectar-eating bats of tropical America, including a couple of species that spend summers here in New Mexico, that have long tongues with which they reach down into flowers that open up only at night. The famous vampire bats of South and Central America are unable to eat anything but blood. They make a sharp but superficial cut without waking a sleeping animal, and they lap rather than suck its blood.

All but the largest of these flying mammals make their way in the world, like dolphins, by hearing. Echoes of their ultrasonic clicks and chirps provide their picture of the world around them.

A few kinds of cave-dwelling birds also navigate by hearing, including the guacharo of northern South America and Trinidad, and the famous swiftlets of China, India, and Southeast Asia whose nests are used for soup. The American nighthawk probably uses sonar to hunt insects at night. Here on the ground, the tiny shrew—nocturnal, insect-eating, and small-eyed—uses its own form of sonar to get around.

Almost all blind people develop an ability to more or less successfully avoid obstacles, and it has been shown experimentally that they use sonar of a sort, and that their ability disappears completely when their ears are stopped up. This ability can also be learned by blindfolded sighted people.

Unfortunately, the acoustic environment in which we live is being polluted by ever louder and more prevalent noise. I am amazed at the ability of some people to tune out the constant roar of automobiles on roads and the passing of airplanes overhead, to tune it out so completely that they don't notice it anymore. I cannot do it. Few have been the days of my life that have passed wholly free of the jarring noise of motors. Some two-day adventures in summer camp in the late 1950s; several peaceful days in a canoe on the Raquette River in the late 1960s; a few one-week backpacking expeditions—in Yosemite National Park in 1970, in the Great Smoky Mountains in 1971, in the Gila Wilderness in 1973; occasional weekends camping in the Santa Catalina mountains, also in 1973; a week on a Mexican beach, later that year, listening to nothing but the sounds of the Caribbean Sea ebbing and flowing on the white sand; a few weekends at Sespe Hot Springs in the late 1970s. I have searched my memory to recall them. I so long for silence. Other people seem to drown out the din around them by seeking refuge in louder sound still. Rock concerts are really deafening, but it seems to me that all kinds of music concerts, and also music soundtracks are much louder than they used to be. I see people who are never without their smartphones piping music into their heads.

Perhaps the reader will think me sentimental, and perhaps that reader will be right. One must adjust to the times. But I think of my feathered, finned, and furry neighbors whose world is not electronic and who cannot possibly adjust to the times. I don't have to shout above the din to reach my friends—I can call them up on the telephone. But the animals are stuck with only the equipment they were born with.

There are far fewer buzzing insects and singing birds in my neighborhood than there once were, but there are some, and their voices are becoming lost in the surrounding cacophony. I can't hear what they are saying anymore. How can they?

CHAPTER 23

Transportation

A Boeing 747-200B jet airliner taking off consumes as much oxygen in a quarter of a second as a 120-year-old beech tree can produce in one sunny day. In half a minute it consumes a day's worth of oxygen from a one-acre forest. At cruising speed, it takes two minutes to use up one acre-day of oxygen.

An average passenger car burns up this much oxygen in twenty-four hours of driving.

One breathing human being uses up this much oxygen in about two and a half years.[157]

Airplanes, spewing water vapor into the air behind them, are responsible for a continuous increase in cloud cover over North America since the 1920s. Clouds covered, on average, 50 percent of the United States and about 55 percent of Canada in 1920. They cover over 60 percent of the United States and nearly 70 percent of Canada today.[158]

Jet planes suck a lot of birds into their turbines, just as automobiles flatten wild animals, and motorboats slice up fish. Some 200 manatees are killed every year in Florida's waterways, and many of their population bear scars left by boat propellers slicing across their backs. Many of the right whales in the North Atlantic—there are fewer than 350 left—bear scars from collisions with ships. I have never seen any statistics on air kills, but it was estimated in 1976 that the average road vehicle killed nearly four animals per year, which translated into five hundred million animals killed in the United States during that year, not counting insects. Today there are more roads, and more vehicles on them, and the only reason there are not so many roadkills anymore is there are not so many wild animals or insects anymore. Automobiles also kill 1.35 million people per year worldwide, up from 250,000 thirty years ago, and they injure up to 50,000,000 people.

What of the usefulness of our vehicles? Before the advent of public transportation in London, something like 50,000 people an hour used to cross London Bridge on foot on their way to work. Nowadays, railroads can transport about the same 50,000 people per hour along a single route, and buses can do about as well along a single highway lane. Private automobiles cannot accommodate more than about 6,000 people per hour, even if every car carries three people in it.[159]

The main purpose of both cars and airplanes is speed, and speed is an addiction we would do well to get over. In our haste to get from one place to another, we forget to enjoy the precious moments of our lives that pass while we are en route. The airplanes use words like "passenger-mile" as if we lived our lives in units of distance instead of time. We are each allotted a certain number of hours between birth and death. No matter how many miles we cram in, we have only those same number of hours to enjoy, no more, no less. If one is always going somewhere, one is never alive anywhere. The fuel efficiency and safety statistics for the airplane per passenger-mile look impressively good, until you remember that your life is still passing, even while you are on the airplane. The number of fatalities per passenger-*hour* is about the same in a plane as in a car. The fuel efficiency and pollution per passenger-*hour* are horrendously worse for the airplane.

No matter how fast we move around the world, our basic metabolism changes but little. Our blood flows at the same speed as our ancestors' blood a million years ago, and the earth keeps breathing at its own age-old rate. We biological beings cannot keep up with the pace set by airplanes and automobiles. Even if they were built of clay and ran on water and caused no air pollution, still automobiles would continue to smash up living bodies by the millions. Still we would continue to pave over the soft, porous earth with hard impermeable concrete and asphalt to accommodate them. Still, the roads on which they drove would continue to lay waste to forests, streams, wetlands, and prairies, fragmenting ecosystems wherever they are bulldozed through.

CHAPTER 24

Guns

In 2019, guns killed:

44 people in South Korea
101 people in Japan
112 people in Poland
163 people in the United Kingdom
239 people in Switzerland
241 people in Australia
303 people in Spain
440 people in Indonesia
37,040 people in the United States[160]

There are an estimated 400 million guns in circulation in the United States today. Eighty-one percent of American murders are committed with them.

Do guns kill people or do people kill people? The issue has been debated back and forth endlessly, but faced with such startling statistics, I can only conclude that America has such a high homicide rate because guns are so easily available here. I cannot believe that Americans are intrinsically tens or hundreds of times more murderous than other people. It is, however, tens or hundreds of times easier to kill someone with a gun than without one.

I bring this up because defenders of technology use arguments that are very similar to those of defenders of guns: they argue that technology is destructive only if it is not used wisely.

Is technology destructive or are people destructive? I do not suddenly become a wanton destroyer because you put a chain saw in my hand instead of an axe. Neither do I suddenly become wiser or have better self-control. But it is a lot more likely that I will ruin a neighboring forest with it than I would have with the axe.

CHAPTER 25

The Internet

Apart from the thousands of toxic chemicals, metals, and gases required to manufacture a computer (chapter 12); apart from the radiation emitted by computers, cell phones, and the global infrastructure that makes them work (chapter 18); apart from the enormous amount of electricity it consumes,[161] there is a much more fundamental problem with the worldwide web that has been woven in the last thirty years which allows any human being anywhere on earth to communicate at will, instantaneously and simultaneously, with any other person or persons anywhere in the world at any time of the day or night.

Ecosystems evolve because they are separated from one another by barriers of space and time—separated by a mountain, a river, a sea, an ocean, a continent. And on each side of the barrier, completely different communities of life evolve, each consisting of thousands and millions of unique species of insects, birds, amphibians, mammals, reptiles, fish, trees, plants, fungi, each with its own biological niche, all living together in harmony on their own side of the barrier, mixing only very slowly, if at all, with the species on the other side of the barrier. We have now punched holes in those barriers with bridges, and tunnels, and railroads, and highways, and airways. And to speed up travel through these holes, which by themselves would be relatively slow, there are telephone lines, and cables, and wireless connections to coordinate and direct the traffic. And to speed up travel even more, there are now an infinity of holes, not only between small ecosystems, not only between large ecosystems, but between every point on earth and every other point on earth, such a dense network of holes that it is called a "cloud."

To travel, or even send a message in one direction, ten thousand years ago, from what is now called England to what is now called the United States, took months. To travel from England to China took years. To travel from Uganda to Hawaii was impossible. You did not know Hawaii even existed, and if you did you could not get there. To travel just from the east coast to the west coast of what is now the United States took more than a year.

Holes between ecosystems did not begin to be punched until very recently. Less than two centuries ago, the first railroads were built, then the first telegraph systems. The punching of holes happened so quickly that by 1860, people were saying that time and space had been abolished. But even late in the nineteenth century, the idea of traveling around the world in just eighty days still sounded incredibly fast.

Today it sounds incredibly slow. Today anyone can travel from the east coast to the west coast of the United States, any time they wish, without any preparation, not in a year but in a few hours. Today, if someone in the United States wants spices from the Spice Islands, or ivory from Asian elephants, or paper pulp from the forests of Borneo, it does not take years to place the order, it takes a microsecond. Anyone can do it, at any time of the night or day, while sitting at their kitchen table, by moving their index finger a fraction of an inch on a handheld device. And it does not take years of laborious toil to bring such objects across the globe to you. Whatever you want, from anywhere in the world, can be destroyed, shot, harvested, leveled, captured, bulldozed, and brought to your doorstep, in just a few days.

Today's transportation and communication network is the equivalent of putting the Earth in a blender. The millions of species, which required billions of years of spatial and temporal separation to be created, are being destroyed in a couple of centuries, and the blender's motor is accelerating rapidly. Evolution has come to a sudden end.

This is not the fault of the human brain, or human culture, or human wishes, or evil people, it the result, purely and simply, of technology. It cannot be prevented. It is an inevitable result. It is required by the second law of thermodynamics, a law every student of physics learns in their first year in college.

The second law says, simply, that every closed system eventually runs down. Living things grow old. Rivers run to the sea. Apples fall. Fire burns. A tree can become sawdust, but sawdust cannot become a tree.

The Earth is not a closed system. Seeds sprout, new life is born, rivers are replenished by rain, oxygen is created, species proliferate, because the Earth receives energy from the Sun. It receives that energy at the same slow, constant rate at which it has always received it for billions of years. That energy turns disorder into order, creates biodiversity, maintains the fires of life, and drives evolution, but only at the same, steady rate. When the motor of the blender works faster than the motor of evolution, the end of life is in sight.

The engine that is driving the motor that is destroying the planet so quickly today is not the human mind. It is the transportation system that has reduced global travel time from years down to hours, and the communication system that has reduced global messaging time from years down to microseconds. It is air travel, and it is the Internet.

CHAPTER 26

Community

When I graduated from Cornell University in June of 1971, and the forty-odd people with whom I had lived so closely and cooperatively for four years went our own separate ways to all different parts of the country, I felt bewildered. I didn't know where I was going, and most of all I didn't know why we had left one another.

When the peace walk ended in November of 1986, and the six hundred-odd people with whom I had walked a couple of thousand miles, whom I had cooked meals for, who had shared with me a way of life more intimate with nature than I had ever known—when we left the District of Columbia to go our separate ways to all different corners of the world, I again felt bewildered. And the contrast between conventional American society and the life I had just been leading was much greater than it had been when I left Cornell. America had meanwhile become more electrified, more speedy, more automated, more isolating, more pretentious, and more mercenary; we had just been leading a slow-paced tribal existence in the open air, dependent largely on the good will of those we met, with few possessions, and little use for timepieces or money.

Approximately once a year after the peace walk, I journeyed westward seeking community—not a temporary community of friends such as I had known, come together for some other purpose, planning one day soon to disband, but a settlement of men, women, and children come together for the purpose of living—living, not in association with machines, but in association with plant and animal neighbors on the green earth under the blue sky.

In vain I sought the spirit of cooperative simplicity in North America. Only among the Amish did I see it, blossoming among a culture and a religion not my own. The communal movement in America was almost entirely automobile-based. All of the hippie communes begun in the 1960s had either failed or mechanized. Communes had all either plugged into the electric power grid, or else they were using solar and other alternative sources of energy to run the appliances and the computers they had acquired.

Wendell Berry, in response to my inquiry, wrote to me from Kentucky, "I don't believe any back-to-the-landers are left in my part of the country. My impression is that that population is shrinking." That was also my experience. The back-to-the-land way of life is not intrinsically so difficult or burdensome, but it is hard to keep at it in the face of the overwhelmingly mechanized society that surrounds us. The cultural support that the Amish have is just not there for most other communities and most other people to live that way. Wendell Berry farmed with a team of horses, and he wrote with pen and ink and not a computer. But he said, "I use more modern technology than I wish I did, and I don't know how to escape it. The modern world, for instance, was made by and for the automobile."

In July of 1989 I left New York once more, still under the illusion that if only I went far *enough* away from civilization I would find a corner of the earth that was safe from the ravages of our technology, where I could at least live my own small life in peace while the rest of humanity had their way with their portion of the world. I headed for British Columbia, journeying by train, by bus, by ship, by foot, and occasionally by canoe.

On the way west I noticed that the pure air over the Great Prairies was not so crystal clear nor so fresh-smelling as I remembered it being just three years before. I looked forward to the clean mountain air, the unbroken forests, and the pure ocean breezes I would find in British Columbia. But civilization, in the form of chain saws and pulp mills, motorboats and airplanes, had beaten me to the Canadian Northwest. Even in the remote Queen Charlotte Islands, summer breezes on a clear sunny day carried a faint sulfurous odor from the pulp mill in Prince Rupert 40 miles to the northeast.

My final disillusionment came in Yukon Territory. Here in the far northern forests and tundra, among the largest expanses of unbroken wilderness and the greatest density of wildlife on the continent, were evidences of a slaughter that had ended long before in the south but was still going on here. Lining the walls of saloons and the shelves of tourist shops in Whitehorse were the pelts of foxes, wolves, white bears, and other Arctic mammals. In the summer, residents of British Columbia drove hundreds of miles north to hunt moose and elk.

Gold and asbestos mines operated here in the north, and oil companies were exploring the Beaufort Sea. Airplanes were ever-present. In winter, snowmobiles had replaced dog sleds. The Yukon forests of short, thin trees were intact, and Yukon air was clear as a bell. But I hungered in vain for the refreshing, exhilarating air of my youth, the air of Ithaca of the 1960s, the air of Mendocino of the early 1980s, the air of the American cornfields of the mid-1980s—the life-giving air of a healthy breathing earth, the electric, sparkling, vibrating air that hardly exists anymore on the planet, even over mid-ocean or high in the tallest mountain ranges.

I spent the month of November in deep despair.

I shared my despair with Grant Smith, a house parent in the youth hostel in Edmonton, Alberta. Together we wrestled with the unanswered questions: How did the earth come to be in such a situation? Who are we human beings that we could do such a thing? Why are we overpopulating the planet? Where does the desire for such power and such wealth come from? What are we to make of war? What sense can be made of human history? and most of all: Is there any way out of this mess, and do any of the supposed experts really have a handle on what the environmental crisis is all about?

I lived in the Edmonton youth hostel until the end of December. During the day I was usually in the public library, immersing myself in books about anthropology, ethology, evolution, population, ecology, forestry, history—wherever my thoughts led me. In the evening I joined the New Zealanders, Australians, and other visitors around the fireplace, well-stoked to warm us against the -20° or -30° temperatures outside, chatting, listening to music, or playing cards—sometimes all three. And every few days Grant and I would exchange our thoughts and explore the earth's problems together, sometimes for hours at a time, well into the night. It became clear to us that several major topics, somehow related to one another, needed to be explored much more deeply before the confusion that clouded environmental thinking could begin to dissipate. Overpopulation. Economics. Technology. War.

Let us consider them.

PART TWO

DIGGING BELOW
THE SURFACE

CHAPTER 27

Economics, and Human Diversity

While I was a student at Cornell, the university changed the name of its College of Home Economics to the "College of Human Ecology." Although most of us made fun of the name change at the time, there was some wisdom in it. "Economics," like "ecology," comes from Greek roots—"oikos," meaning "house," and "nomos" meaning "law" or "management." The management of the house. Ecology is the study of the house—the natural environment, all of the creatures and growing things that live in it, and the ways in which they interact with one another. Economics is the study of human beings and our relationships with one another and with our environment. In other worlds, human ecology. "Management" comes in because presumably we human beings exercise some amount of choice and control over how we live.

All too often, the emphasis in the study of economics is on the workings of money and the financial system—profits, wages, rents, investment, interest rates, foreign exchange, and so on—and little attention is paid to the study of real human beings, their real environments, and real interactions of which the financial system is only a reflection. Economists can disagree so wildly with one another because the money system, which is but a representation of relationships in the physical world, is mistaken for the real thing. Cooperation and competition in a complex society are expressed through money. Giving somebody money is one way of giving that person power. In the buying and selling of merchandise, real objects move around, not just the "value" that we assign to them. Money is mainly a way of keeping track. At best it is a tool, at worst it is a fantasy. And, as it says in the Bible, the love of it can lead to considerable evil.

Since I was small, I wondered where money came from, but neither my parents nor anybody else ever gave me a satisfying answer. Most economics books don't give you a straight answer either, but I finally discovered the amazing truth. Where does money come from? Banks print it up. When a person wants to go into business, that person goes to a banker to borrow money. If the banker trusts the person, the following transaction takes place: the banker prints up some money and hands it over, and the businessperson writes up an IOU promising to pay it back with interest. That's the origin of money. Money is so imaginary that even the IOU is a sort of money that is part of the banker's assets. And if the borrower is the government, that IOU is as valuable and exchangeable as bank notes.

Nowadays there are strict rules about who is allowed to print up bank notes—in the United States, only the Federal Reserve Banks are allowed to do it—but most of the money that circulates is still manufactured by local banks and other lending institutions. Instead of currency, they create deposits that borrowers can spend by writing checks. These deposits are newly created money. The Federal Reserve System functions as the central bank that lends to the American government and to the banks the public deals with.

Roughly speaking, the amount of money in circulation in a modern industrial society is equal to the total amount of debt. Most of the $1 trillion in bank notes in circulation in the United States today represents government debt, and before deficit spending began in the 1980s most of it was printed in times of war. The total amount of *money* in circulation—including coins, bank notes, traveler's checks, bank deposits, and short-term government securities and commercial loans—is about $22 trillion. The total amount of *debt* in the United States is about $93 trillion; the federal debt is about a third of that.

If everyone paid off their debts in full, most of the money in the country would immediately disappear. All that would be left is about $272 billion in Federal Reserve notes that are backed by the gold in Fort Knox—about $800 per citizen.

The financial structure of the United States is just one among a great many possible organizations of the capitalist system, which in turn is one of many possible ways to distribute wealth in an industrial economy. Before exploring the industrial variety more deeply, however, let us sample some of the other types of human economies on this diverse earth. They have all been eminently more successful than ours, yet rarely have any modern economists or ecologists considered them seriously.

Agricultural Societies

Agriculture, in its varied forms, has been engaged in by men and women on this earth for many thousands of years—probably a good deal more than ten thousand. Until well into the twentieth century, every traditional variety of agriculture was still being practiced in those environments most suited to it.

Farmers cultivated the ground, planting seeds and tubers with their digging sticks, in many regions of the tropics. They grew food not to sell, but primarily to feed their families. Most of the tropical forest tribes of South America, living in the basins of the Amazon and Orinoco Rivers, were nomadic. They would clear a small section of forest, plant yams and other root crops for one or more years, and then move on to a new site, leaving the forest to nourish and reclaim the abandoned fields. In varying degrees they supplemented the fruits of their fields by hunting, fishing, and gathering wild plants in the forest. Similar forms of shifting agriculture were practiced in most of Central and South America, throughout the islands of the Pacific Ocean, in much of Southeast Asia and India, and in most of tropical Africa south of the Sahara.

Other people lived in permanent villages and farmed the same land year after year, often with the help of small or large scale irrigation systems. Many of these people have also kept cattle, and the plow has often replaced the digging stick for the cultivation of the soil. Out of thousands of years of settled agriculture grew the civilizations of China, Burma, Siam, Cambodia, Indonesia, India, Tibet, Mesopotamia, Asia Minor, Egypt, Rhodesia, Ghana, Congo, Mexico, the Mississippi Valley, Amazonia, and the Andean highlands. Until very recently, the majority of people in most of these places lived after the fashion of their ancestors. Most families had land to cultivate which, together with farm animals, provided the bulk of their food needs.

In still other parts of the world roam nomadic herders, who graze domestic animals but do not cultivate the soil. Bedouins and other peoples herd camels and sheep in arid regions of North Africa, Arabia, Iran, and Pakistan. In the East African savannahs live many tribes of cattle herders—the Nilotes and the Acholi in Sudan; the Lango and Bahima in Uganda; the Maasai, Nandi, Suk, and Kipsigis in Kenya and Tanzania. Some Bantu in South Africa are both herders and shifting cultivators. Other cattle herders live in Madagascar. In parts of China and Mongolia are raised horses, sheep, goats, and Bactrian camels, and in the Himalayas are yak and sheep herders. People still herd reindeer in Scandinavia and Siberia, and some people have taken up the practice in Alaska.

Each type of agriculture has its place in this world. The soils of tropical forests are exhausted after a couple of years of farming, and so for thousands of years, shifting or "swidden" agriculture has been practiced in the jungle. Permanent cultivation is more suited to river valleys and flood plains in the subtropics. Plows are better used on arid land that is not regularly flooded. In southeast Asia—Bengal, Thailand, Myanmar, and Vietnam—two rice harvests per year are produced on large alluvial plains with the use of irrigation and manure, but without plows. Mechanized farming with iron plows is more suited to the heavy soils of the mid-latitudes—North America, Europe, Russia, South Africa, and Australia. Pastoralists herd their animals on land that is poorly suited for any kind of cultivation—camels in deserts and semi-deserts; sheep and goats on the fringes of deserts and savannahs; horses, camels, sheep, and goats in steppe country; reindeer on sub-Arctic tundra. On the dry savannahs of Central and South Africa pastoralists may either engage in limited cultivation, or they may be neighbors to more settled farmers. The sheep herders in the mountains and valleys of southwest Asia and the yak herders of the Himalayas share the land and the economy with settled cultivators.

"To every thing there is a season, and a time to every purpose under the heaven." (Ecclesiastes 3:1)

And for every type of agricultural economy under the heaven, human beings have devised various means of distributing the wealth among the citizens. Some of them involve forms of money; others do not. None of them are as complicated as that of industrialized America, first because there is so much less wealth to distribute, and second because men and women who are farmers are already directly connected to the soil which sustains us, and which is the real source of all riches.

Among shifting cultivators, populations are small, and land is abundant and usually held communally and cooperatively.

Among cattle herders, private ownership of property is more the rule. The size of the herd usually becomes the measure of a family's wealth.

Where land is permanently and intensively cultivated and therefore scarce, the land itself becomes a measure of wealth. Land distribution becomes the basis of economic organization, and its ownership becomes the source of economic and political power. The organization of a settled farming economy may take any number of different forms. Land may be owned individually, or it may be owned collectively. Ownership may sometimes be claimed by powerful persons other than the farmers. Landlords may claim a fixed rent from their tenants, or they may claim a portion of the crop, or they may claim person services. There are no sharp dividing lines between landlord/tenant systems, feudalism, and slavery. The one blends into the other—it is only a matter of degree and type of power that one class of people holds over another.

People who engage in different forms of agriculture often live as close neighbors and trade with one another. Settled farmers may live in fertile river valleys while neighboring forest dwellers practice shifting cultivation, and nomadic herders live in the midst of both. Grains, forest products and animal products then circulate throughout the region. Such is the case in the Himalayan principality of Bhutan: Drukpa farmers live in the main north-south valleys, growing rice and other crops and keeping some cattle; their relations in the northern altitudes herd yaks and plant some crops during a short summer growing season; in the forests of the east and northeast the Monpa practice shifting agriculture. Money was entirely unknown in Bhutan until 1968, but these three groups of people have traded with one another for perhaps a thousand years.

In the twentieth and twenty-first century every kind of non-industrial society in the world, nomadic and settled alike, has come under intense attack from Western European civilization. I would like to examine the case of Java, a large tropical island in Southeast Asia.

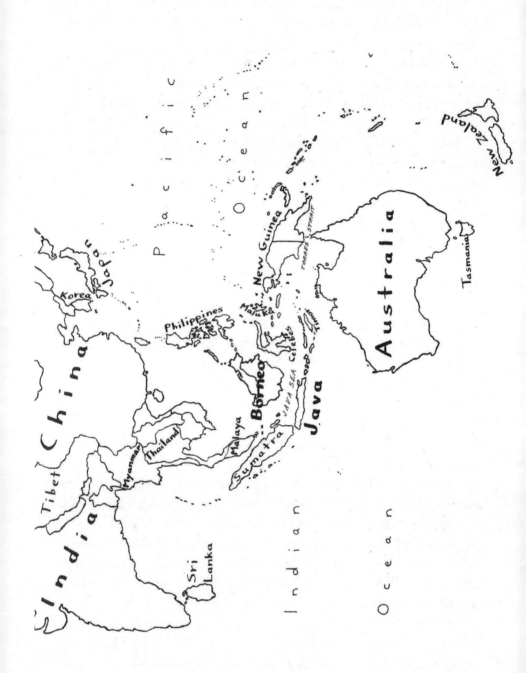

By any objective standards, Java ought to occupy an important place in the study of world history. One of the cradles of human beginnings, Java has been continuously inhabited for at least 1.6 million years. The oldest human skeletons outside of Africa have been unearthed there. The Javanese began to use bronze and iron at least as early as Europeans did, and at the dawn of recorded history they were growing rice in irrigated fields, keeping domestic animals, and navigating the seas in outrigger canoes.

Literature, dance, sculpture, theater, and music, the art of wayang, the gamelan orchestra, and the printing of batik cloth all have very long histories here. Today Java, about the same size as Ireland, is the world's second most densely populated land area. Its largest city, Jakarta, is the capital of Indonesia, the world's fourth most populous nation. More than half of Java's 152 million people make their living as subsistence farmers.

In Ice Age times, Java and neighboring Sumatra were often connected to the Asian continent by land. They are at the center of an ancient trade route linking China with India, Persia, Arabia, and Africa by sea, and are the gateway to the Spice Islands to the east. Javanese ships have always traveled far and wide, trading with distant kingdoms. Java's own kingdoms, and a few queendoms, built magnificent Hindu and Buddhist temples, some of which are 1,200 years old.

In the fifteenth century, before Columbus had set sail from Spain, before da Gama had left Portugal, and before the winds of Islam blew throughout the Indonesian archipelago to displace other faiths, Java was in its Golden Age. It had been united a century before under King Hayam Wuruk, and much of what is now Indonesia still sent tribute to the Javanese capital at Majapahit. But the life of the average Javanese villager was touched only little by these arrangements. The tax collector came periodically to make sure the royal court received its due. And another portion of the local rice crop was exported in exchange for iron, dishes, rattan, cotton, and spices from the eastern islands. But the villages, then as now, were largely self-sufficient both economically and socially. Each village allotted land to the individual families on the basis of what they were able to cultivate. The amount of this land was small and directly related to each family's needs—perhaps 2½ acres sufficed for a family of five. In addition to farmers—the majority of the population—the economy of the village included merchants, millers, tappers of sugar-palms, butchers, bleachers of textiles, dyers with indigo, noodle-makers, and lime burners. Production for export was limited. Most village work was done cooperatively, and the communal focus was more on social and religious affairs than economic ones. Building temples, hostels, and gamelan orchestras was at least as important as maintaining roads, bridges, and irrigation dikes. In the Javanese village—as in most agricultural villages of Asia—the individual feels herself or himself part of the whole and accepts traditions that have developed over centuries and millennia.

A traditional society such as this, in order to prosper, depends on a stable population, and fifteenth-century Java, prospering at the crossroads of the Spice Islands trade route, had had a more or less stable population for the previous two thousand years. Perhaps two million inhabitants cultivated only what land they needed—less than five percent of the total land area. Some of the low-lying land along the north coast was mangrove swamp, and most of the rest of the island was tropical rainforest, inhabited by tigers, elephants, orangutans, monkeys, buffalo, rhinoceros, crocodiles, wild oxen, and wild pigs. Thousands of species of plants, 400 species of birds, 100 species of snakes, and 500 species of butterflies inhabited these forests.

In the sixteenth century the Portuguese captured Goa in India, Malacca in Malaya, and Macau near China. Rival trading centers were established in Java and elsewhere in Indonesia, to which Muslim traders came to avoid the attempted Portuguese, and later the Dutch monopoly on the spice trade. But in most of Java, for a while longer at least, life went on as before. The arrival of Portuguese, British, and Dutch traders in this part of the world in sixteenth and seventeenth centuries signaled mainly the entry of one more element into an already rich and complex inter-island and intercontinental trade connecting such far-flung places as Italy, Egypt, Madagascar, Arabia, Persia, India, Burma, Siam, China, the Philippines, Borneo, and the Spice Islands —a trade that now expanded to encircle the globe, as first Christopher Columbus, then Ferdinand Magellan, and later Sir Francis Drake set sail searching for new routes to the Spice Islands.

During the seventeenth and eighteenth centuries the Dutch East India Company, by military might, won control of the seaways in the East Indies and established a monopoly on the world spice trade at its source.

The Dutch trading center in Batavia (now Jakarta) was becoming the center of a colonial empire that was interfering by degrees in the affairs of the local population. In Maluku—the Spice Islands—the Dutch destroyed the traditional way of life and decimated the population. To the west in Sumatra, they still had little impact except immediately near the trading posts. In Java, the Dutch penetrated the traditional economy and used it for their own ends. By 1799, when the Dutch government took over the East India Company, three-fifths of the island had been subjugated by military force. Western Java, near Batavia, was under direct rule, and the central and eastern Javanese kingdoms became protectorates, sending tribute to their new Dutch rules. Batavia became a thriving international city, host to merchants from all over the world who traded in spices, coffee, tea, opium, and tobacco, and in ivory, silk, copper, tin, and firearms. The city was surrounded by gardens and orchards, and fields of rice and sugar cane. In 1722, Batavia was being described by visitors as "one of the pleasantest cities in the world."[162] But by 1800, it had become known as one of the unhealthiest: the canals built by the Dutch in imitation of those at home were breeding grounds for mosquitoes, and malaria became epidemic where it had not been prevalent before.

Javanese village life was disturbed by a new system that forced farmers to devote as much as half their time to growing coffee, sugar, and tea instead of rice. They were required to deliver a portion of their rice crop to the Dutch conquerors in tribute, and under the "forced delivery" system they were also required to deliver export crops in amounts and at prices fixed by the Dutch. In 1710, there were already 130 sugar mills in Western Java owned by the Dutch, and others owned by Chinese in neighboring districts.

The Javanese, a proud and independent people who had a much older cultural tradition than the Dutch, were forced by greater military strength to the bottom of an economic pyramid in which the Dutch ruled and the Chinese served as middlemen. Javanese merchants were excluded from the coastal trading centers they had once dominated. Chinese merchants replaced them, and other Chinese, serving as moneylenders and tax collectors, became intermediaries between Dutch merchants and Javanese villagers.

During the nineteenth century the Netherlands gained control over all of Java and most of the rest of present-day Indonesia. The forced delivery system was replaced by a monetary land tax calculated at about two-fifths of the crop, and then by the "culture system," whereby a portion of the farmer's land and labor were at the government's disposal for the growing of export crops: coffee, sugar, indigo, tea, tobacco, pepper, cinnamon, cotton, cochineal, and silk. Large areas of Java became plantations. Wealth streamed out of the island to pay for the industrialization of the Netherlands. Many Javanese farmers were no longer self-sufficient, depending instead on income from working on the plantations. When the market failed, these farmers had no money to buy the rice they were no longer growing, and thus began a cycle of debt from which it was difficult to escape.

The first highways were built by the Dutch in the first decade of the 1800s. Dutch immigrants began to flow into Indonesia, and private capital along with them. Beginning in the 1860s, forced cultivation was abandoned and agriculture was turned over to private Dutch enterprise. Villagers soon became day laborers on European-owned plantations. New crops were introduced—kapok, copra, palm oil, cocoa, rubber. Railways were built. Europeans came searching for coal and iron and investing in the oilfields of nearby Sumatra and Kalimantan. The advent of steamships and the opening of the Suez Canal shortened travel time to Europe tremendously. Between 1870 and 1890, exports from Java doubled in volume and imports quadrupled.

The twentieth century saw Indonesia finally gain its independence from the Netherlands. During the First World War the Netherlands was distracted militarily and Dutch trade was temporarily cut off. The Indonesian nationalist movement gained in strength and American and Japanese merchants began to come in greater numbers. The Second World War involved Indonesia more directly. From the point of view of southeast Asia, this war was a conflict between the old colonial powers—England, the Netherlands, France and Belgium—and the more recent arrivals—Germany, Italy and Japan. At least in part, this was a war for control over colonial territories in Africa, Asia, and the Pacific, and the oil and mineral wealth that is the basis of the twentieth century version of the ancient spice trade.

In occupying Indonesia during World War II, Japan succeeded in finally destroying the old colonial system of the Netherlands. In defeating Japan, the Allies succeeded in destroying the new colonial empire of Japan, and in the power vacuum that followed, Indonesia threw off the yoke of colonialism altogether. In September of 1950, the Republic of Indonesia became the 60th member of the world community of the United Nations.

But a foreign economic system continues to industrialize Indonesia. Three-fifths of its people continue to be independent farmers or to work on plantations at low wages. The biggest exports are now petroleum, natural gas, minerals, palm oil, iron and steel, computers and electrical appliances, and wood products. Nominally the government of Indonesia "owns" these resources. But all of the high-paid "skilled" labor and capital to exploit them are supplied by foreign companies. Has colonialism really ended? Vast amounts of money still leave Indonesia, under a capitalist instead of a feudal system, to multinational corporations instead of powerful kings or colonial governments. What was formerly called tribute is now called interest on investment. Oil and other raw materials continue to flow out of Indonesia to Western countries. The power of the multinationals is backed by the collective military might of the Western nations.

The modern situation in Indonesia was analyzed by Dutch economist J. H. Boeke, in terms of what he called "dualistic economics," a state of affairs that results from the collision of Western European capitalism with native subsistence agriculture.

> Wherever Western capitalism in its triumphal march of world conquest collided with pre-capitalistic societies that could be neither destroyed nor assimilated, the dualistic economic problems loomed up and grew, until today in their totality they dominate not only the countries concerned but the whole world.[163]

Dr. Boeke concluded that the entry of capitalism into a subsistence society results mainly in gigantic problems of overpopulation, rural debt, and enormous suffering. Traditional crafts are destroyed because they cannot compete with cheap manufactured goods, and large numbers of skilled craftspeople are thrown out of work. The new dependence on manufactured goods creates the need for an ever-increasing supply of cash. Crop production for an external, undependable market controlled by large corporations and a long line of middlemen is far less secure than production for home consumption. The availability of work in urban factories only worsens the rural population problem because the village, not the city, remains the center of most people's lives, and city workers inevitably return to their families and villages after five or six years of factory work. All of the solutions to the poverty and suffering only make matters worse. More investment in industry leads to increasing mechanization of factories, putting more people out of work to join the swelling ranks of the unemployed. The so-called "green revolution," designed to increase agricultural efficiency, instead increases farmers' needs for cash to buy fertilizer, pesticides, and new seed varieties, and sinks them ever deeper into the spiral of debt.

These very serious problems are not, however, the result of capitalism specifically. They are rather more basically the result of an ever-increasing volume of foreign-controlled commerce, under *whatever* political or economic system. The population of Java began to grow with the building of the first Dutch cities in the seventeenth and eighteenth centuries, and it continued to grow under the forced delivery system, the forced culture system, and later the free enterprise system. It did not matter whether commerce was controlled by the Dutch East India Company, the Dutch government, or private investors, nor whether Indonesia was a colony or an independent nation. And the modern villagers' debts are in principle no different from the tribute once paid to the Dutch government through its representatives; money still flows from the villages to the cities and thence out of the country.

It becomes apparent that growth in trade and commerce, demanding as it does an ever-increasing number of workers, consumers, and merchants, is a tremendous stimulus for population growth. Although Java is the most spectacular example, this pattern has been repeated almost everywhere in the world during the past few centuries, as we will see in the chapter on population.

I would like next, however, to look at the food-gathering economies and how they participate in our colorful and diverse human world. Let us leave Java for now and cross the Java Sea to the island of Borneo.

Foragers

On the island of Borneo live many communities of food-gathering people known as "Punan" or "Penan." Today their traditional way of life is being destroyed just as utterly and rapidly as the rainforests they are part of. Only a few small communities have not yet been forced to take up farming, and many have died in the war of extermination against their world.

The foraging way of life is one of relative luxury and leisure. Luxury because wild nature is so prolific and so diverse. Leisure because the abundance of the wilderness need only be gathered, requiring no human labor to produce. Wherever there has been wilderness, there human beings have gathered, fished, and hunted, and wherever there is enough wilderness today, there human beings still today gather, fish, and hunt. They are no anachronism, they are part of the modern world. We kill them as we kill our world.

In the forests of Central Africa, on the Argentine savannahs, in the hills of India and Sri Lanka, in the Kalahari, in the Australian outback, on isolated islands in the Indian Ocean—rarely have any of the world's hundreds of nomadic tribes of foragers reached a population density of as much as one person per square mile; only some fishing societies, such as those of the northwest coast of North America, lived in densely populated villages in territories where there was no end to the salmon that returned yearly from the ocean to spawn in the rivers, streams, and lakes. Rarely, in any food-gathering societies, did gathering, hunting, and fishing take more than four or five hours of an average person's day. Rarely in all the world have any people had more leisure time to spend in visiting and socializing with one another, in dancing, making music, and relaxing. Food producers are almost always much busier than food gatherers and struggle much harder to survive.

Hunting and gathering cultures are as diverse in their languages and customs as any other human societies, but by their nature they share certain features. They always have a fairly stable population, which must be kept well within the limits of their environment to provide. They are almost always rather egalitarian societies, sharing most food and possessions communally. They are most often monogamous, and often men and women have equal status. Hunting is usually men's work, and gathering is usually women's work, but there are a good many societies in which women do some hunting and men do some of the gathering. By and large, food-gathering people are much less warlike than farmers, their religion is less authoritative, and they expect independence rather than obedience from their children. By no means are all hunters and gatherers peace-loving—they are real human beings like you and me. But most of the world's peaceful, sexually egalitarian *societies* have been foraging ones; agricultural people virtually all know regular warfare.

Hunting, fishing, and gathering has been an extraordinarily successful way of life, enduring through at least two million years of human history. In the beginning of the twentieth century there were more than a hundred societies of foragers scattered throughout the world still living in their traditional ways—wherever sizable enough wilderness areas remained:

In Africa: the Vezo, fishers of Madagascar
the San of the Kalahari (several groups)
the Koroca, the Kwepe, and the Kwise of southwestern Angola
the Aka, the Efe, the Mbuti, and the Binga of the forests of Central Africa. In 2023 there are at least 250,000 hunters and gatherers living in at least a dozen groups in what is perhaps the most extensive true wilderness left in the world.
the Dorobo and the Hadza of East Africa
the Fuga and half a dozen other groups in central Ethiopia

In Asia:	fishers and hunters of Siberia: the Samagires, the Goldes, the Gilyaks, the Nymylan, the Kamchadal, the Udege, the Nivkhi, the Nanai, the Ket, the Yukaghir, and the Chukchee
	the Ainu of Sakhalin and Hokkaido Islands
	hunters and gatherers of India: the Chenchu, the Korwa, the Birhor, the Irula, the Kurumba, the Kadar, the Malapandaram, the Paliyan, the Yanadi, the Allar, the Aranadan, the Eravallar, the Mala Vedan, and the Vettuvan
	the Vedda and Wanniya hunters and the Karava fishers of Sri Lanka
	the Kubu of Sumatra
	the Penan of Borneo
	the Ruc of central Vietnam
	the Yumbri, the Mrabria, and the Penang of northern Thailand
	the Mincopies, the Onge, the Jarawa, the Shompen, and the Sentenelese of the Andaman and Nicobar Islands
	the Semang and the Jakun of the Malay Peninsula
	the Agta of the Philippines
	native fishers of Taiwan
	native hunters of central Sulawesi
	the Raute of Nepal
In Oceania:	the Manus, fishers of Papua New Guinea
	the Asmat of West Papua
	the Bajau, fishers of the sea islands from Myanmar to New Guinea
	numerous tribes of Australia
In North America:	the Unangan of the Aleutian Islands

the northwestern Athapaskans of Canada and interior Alaska
the northeastern Algonkians of eastern Canada
the Inuit of the Arctic regions
the Seri of Sonora, Mexico

In South America: the Warrau of the Orinoco River delta
the Guahibo, the Piaroa, the Sanema, the Shirianá, the Waica, and the Macú of southern Venezuela and adjacent regions of Colombia and Brazil
the Aché of eastern Paraguay
the Ayoreodo of northern Paraguay and eastern Bolivia
the Selk'nam, the Haush, the Kaweshkar, the Yamana, and the Aonikenk of southernmost South America
the Krahó, the Canela, and the Apinagé of northeastern Brazil
the Cuiva of the eastern Orinoco plains

The tribes called "Penan" live in many different regions of coastal and inland Borneo. Borneo—the home, like Java, of ancient seagoing kingdoms—is today divided between Malaysia; Indonesia; and little independent Brunei. Penan groups are struggling for survival on both sides of the Malaysia/Indonesia border.

Each nomadic Penan group lives near the headwaters and tributaries of a river that is occupied in its lower reaches by a specific Kayan or Dayak or Kenyah agricultural people with whom they trade. If this affiliation lasts long enough, the two groups may intermarry, speak a similar language, and share customs. The villagers practice shifting cultivation. The forest people hunt using spears, blowpipes with poisoned darts, and hunting dogs. They fish with nets, and they collect honey, fruits, sago, and a wide variety of other wild plants.

The Penan and Kenyah have traded regularly with one another for millennia. It is simply a myth that there is any such thing as the "Wild Man of

Borneo" who lived isolated in the jungle, probably cannibalistic and unaware of the outside world until we modern Europeans discovered him and brought him up out of savagery. Nothing could be further from the truth.

First of all, no people in Borneo, New Guinea, Africa, Amazonia, or anywhere else on earth have ever been observed to eat human flesh under any but unusual starvation conditions. The "cannibal" epithet has been applied to virtually every group of people in the world by those that hated or feared them. The ancient Greeks supposed the Irish to be cannibals. The early Christians supposed the pagan Scots and Picts to be cannibals. The Romans accused the early Christians of drinking human blood in secret rites, and Christians have accused the Jews of the same thing until fairly recent times. The word "cannibal" itself is an accident of the fact that Columbus landed among the Arawaks in 1492, and the Arawaks told him stories about their enemies, the Caribs, who ate human beings. It is *also* recorded that the Arawaks thought Columbus and his men ate human beings. Neither Hernán Cortés nor any of the men with him ever observed anyone being eaten by the supposedly cannibalistic Aztecs. But the genocide that followed the Spanish conquest, reducing the population of Central Mexico from 28 million to 1 million in less than a century, could be justified by calling them cannibals and sodomists. Apparently the Aztecs also suspected the Spanish of cannibalism. A Nobel Prize in medicine was even awarded for the discovery that Kuru, a disease once prevalent among the Fore people of New Guinea, was caused by eating human brains—despite the fact that no one has ever observed such a practice.[164]

Secondly, the Penan—and most other hunters and gatherers in the world—have never been isolated. In fact, a thriving trade exists between the forest dwellers and their agricultural neighbors. From the Kenyah villagers, the Penan obtain machetes, metal knives, salt, tobacco, and cloth, and in exchange they provide the Kenyah with handicrafts and forest products: baskets, mats and blowpipes; rattan, dammar, fragrant aloes wood, illipe nuts, camphor, gutta-percha, beeswax, rhinoceros horn, edible birds' nests, and gold panned from rivers.

Such a trade was going on in the fifth century B.C. at the time of the kingdom of Kutai in eastern Borneo, and such a trade still goes on today. And the trade does not stop with the Kenyah, who themselves have no use for many of the forest products they obtain. The Kenyah, in turn, supply Chinese merchants with forest products, and the Chinese supply the Kenyah, and through them the Penan, with metal, salt, tobacco, silk, cotton cloth, and other goods. The system of trade goes something like this:

Penan people periodically make the several days' journey downriver to their Kenyah village bringing quantities of jungle products, and receive in exchange salt, tobacco, cloth, machetes, and knives. When trading and exchange of news and gossip have finished, the Penan return to the forest. Later the Kenyah paddle further downriver and exchange the forest products for much larger quantities of salt, tobacco, cloth, and metal, and such household utensils as woks and pots, which they receive from a Chinese or Malay trader running a small-scale operation from a town located on the lower stretches of the area's main river. The trader, in turn, brings these forest goods to sell to one or more large export firms operated by Chinese in the nearest city on the coast. There he stocks up on salt, tobacco, cloth and clothing, metal goods, and other items from various places in the world that are of use to the townspeople. The large export firms supply birds' nests to their customers in China, aloes wood for incense to their customers in India and Arabia, and rattan to their customers in Europe and America, who will make very expensive furniture out of it.

Down to the present day, the people at either end of this trade route have had no awareness of one another's existence. Even the urban Chinese exporters and the Penan in most cases do not know about each other. Yet, for centuries unnumbered, jungle products have found their way from forests to coasts to consumers throughout Europe, Asia, and Africa, and trade goods from the world round to the deepest recesses of the forests.

Some of the products which have flowed out of Borneo and other of the world's rainforests for thousands of years are to be found *only* in large tracts of primary forest. Gutta percha, a rubbery latex tapped mainly in the forests of the East Indies and Brazil, has long been used as an adhesive, as waterproof caulking for ships, and in medicines. In modern times it has been used for insulating electric cables, in golf ball covers, in dentistry, and in chewing gum. Illipe nuts provide an oil used for cooking and making tallow and wax, and in recent times for lubricating machinery, and in cosmetics. Rattan vines climb also in the forests of Burma, Malaysia,

Sri Lanka, and Sumatra, but the finest rattan in the world comes from Borneo, harvested in large part by nomadic hunters and gatherers.

At least this was the traditional arrangement. Swidden farmers have since ancient times been middlemen between forest gatherers and coastal merchants. Everyone, all the many links in the trade route—hunter-gatherer, farmer, trader, manufacturer, and urban consumer—was in some way dependent on everyone else, affected by and having an effect on unknown parties far away on the other side of the world.

Today both forest and forest-dwellers are being obliterated. Loggers are destroying the forests of Sarawak, on the Malaysian side of the island, at one of the fastest rates in the world. What jungle products are left are being harvested by outsiders who have no stake in preserving the forest. Since 1982, three hydroelectric dams have flooded Penan and other indigenous villages, a fourth is under construction, and eight more are planned. There are about 16,000 Penan today, of whom only about 200 still live as hunter-gatherers.

When the jungles are all gone, there will be no more gutta-percha, no more illipe oil, no more rattan, no more aloes wood, no more sandalwood, no more mahogany, no more teak, no more ebony. Already rattan and tropical woods are becoming scarce. Gutta percha and other natural resins are rapidly being replaced by synthetics. Camphor is being grown on plantations in Florida and California, and it too is being replaced by synthetic products.

Foragers in all parts of the world, with few exceptions, have always been active participants in international commerce. The Yumbri of the jungles of northeastern Thailand lived near specific villages of Kamuk and Meau farmers, trading beeswax and honey, elk meat, and ivory for iron, steel, and their scant clothing. The Bushmen, or San, of southern Africa, traded extensively among themselves and with their Bantu, Khoekhoe, and later European neighbors, using ostrich eggshell beads and tobacco as money. In the Nilgiri Hills in South India, the food-gathering Irula and Kurumba coexisted with the pastoral Toda, the agricultural Badag, and the artisan Kota; for untold centuries they all lived side by side in close economic symbiosis.

The Efe archers of the Ituri forest of Central Africa have had a symbiotic relationship with their Lese farming neighbors for centuries, trading meat, fish, honey, and forest produce for cloth, pots, metal, and garden produce. The Efe also come out of the forest to work on Lese farms during the dry season.

Even the Australians traded with the outside world on trade routes established long before any Europeans ever set foot on that continent. The islands of the Torres Strait were distribution centers for a thriving trade between Australia and New Guinea, and intermarriage took place among the Australians, the islands and the Papuans. From further away, people from Celebes and Timor used to make yearly sailing voyages down through the East Indies islands to northern Australia. There they traded dugout canoes, rice, cloth, knives, tomahawks, and pipes for pearls, tortoise shell, sandalwood, spears, and spear-throwers. Many Australians used to go back to Malay country with the returning fleets and stay until the following season. A few stayed permanently and married Malay women. Very likely the pearls, tortoise shell, and sandalwood of Australia entered the world-wide spice trade in which the East Indies people played such a central role, among them the sailors of Celebes and Timor. The pearl beds of northern Australia and the Torres Strait would have been especially prized, as they are the largest in the world and produce the most silvery-white mother-of-pearl.

It is important to realize that hunters and gatherers have always lived in the same world as farmers, herders, artisans, merchants, and city-dwellers, and that rigid categories of "Stone Age," "Bronze Age," and Iron Age" simply do not apply to real people in this small, interconnected world. When the first people began to plant crops, they did not leave the Paleolithic, or Old Stone Age behind them, nor did the first metal smelters leave the Neolithic, or New Stone Age behind them. What they did do is introduce new choices onto this interesting planet and give us all a lot more possible ways to live. And to each way there are places on this earth that are more or less suited. And people were hunters and gatherers and fishers, or they were slash-and-burn farmers, or they were settled farmers, or they were craftspeople, or they were merchants and city-dwellers, not because they didn't know any better, but because it pleased them to live that way in that place where they were born.

It is also important to realize that "hunter-gatherer," "herder," "fisher," "merchant," and "industrialist" are not fixed categories, and that human folk, liking our freedom as we do, move from time to time from one profession to another. There are Veddahs in Sri Lanka that have been farmers for a very long time, and some of their forest cousins probably have ancestors who once grew crops. In North America, the introduction of the horse gave many formerly agricultural tribes the opportunity to hunt buffalo, as the European invaders drove them from their prairie villages onto the Great Plains to the west. The Hidatsa and the Crow were originally the same people. The Hidatsa stayed behind on the prairies and remained settled farmers; the Crow became nomadic hunters on the plains. Throughout South America's rainforests, and on the Grán Chaco of Argentina, are many tribes who always hunted and gathered but who were so inconsistent in the extent of their planting that anthropologists do not agree among themselves as to whether they were farmers or not.

In the steppes of Central Asia, in Tibet, and in Sudan, there is a two-way street between settled farming and nomadic herding. Farmers often become herders, herders often become farmers, and family ties are thus maintained between wanderers and the settled folk they trade with. In areas where the soil and rainfall will support both herds of animals and cultivation, the distinction between nomad and farmer is only a matter of degree, and it changes with the weather and economic fortune.

In northern Siberia and Scandinavia dwell many interesting groups of people—Yukaghir, Tungu, Nentsy, Entsy, Nganasan, Sel'kup, Nymylan, Khanty, Ket, Oroki and Saami—who at some periods in their history and in some locations have been reindeer herders and at other times and in other places have been reindeer hunters. The recent introduction of the snowmobile has in modern times made the livelihood of some groups again more like hunting than herding. The distinction is not a sharp or an easy one.

The rubber tappers of Brazil are men and women of mixed European and native descent whose ancestors migrated into the rainforests, some during the rubber boom of the nineteenth century, others during the mini-rubber boom created by World War II. The first rubber tappers worked and died under conditions of virtual slavery. They did not know how to live off the abundance of the jungle, and they exterminated their Amazonian neighbors who did. But their descendants today are food gatherers as well as latex exporters, and they are joining with their indigenous compatriots in defense of their forest home and their chosen ways of life.

Even here in the modern United States I have known of a few people in the Pacific Northwest who were living nomadic lifestyles, traveling from place to place with their donkeys or their horses, living as much as they could by hunting, gathering, and fishing on the remaining public lands.

Hunting and gathering are activities that are so basic to human living that even modern corporate executives will not give them up—they want their money and their computers and cell phones and airplanes, but they must have their wilderness too. Millions of Americans hunt part time with rifles, and hundreds of thousands more use bows and arrows. No matter that the fish are wiped out from our rivers and lakes by all the hordes of human fisherman; our government breeds more fish and keeps our waterways well-stocked, maintaining the illusion that there is still wilderness out there instead of one big fish and game farm.

I myself have done my small share of fishing when I was younger, and I have gathered plenty of wild grapes, blueberries, huckleberries, mushrooms, sassafras, and other wild edible plants and herbs. I went to summer camp for seven years where I learned how to build a fire, make myself a shelter in the woods out of the plant life around me, and other wilderness skills that I value very much but have used so very little.

There is so little wilderness left in my country. The public lands are in such poor shape that they will no longer sustain even part-time hunting, gathering, and fishing as a way of life. The great tragedy of our time is the sudden elimination of choices in our nation and in our world. Two million years of human exploration and discovery, of ever-increasing diversity and the finding of so many ways to dwell with our family of fellow creatures and growing things in their myriad shapes and sizes and personalities, in all of the different climates of all of the forests and deserts and savannahs, in all of the mountains and river valleys and plateaus, large and small, on this magnificently colorful earth—all choice and diversity is being destroyed in a few short human lifetimes by a terrible insistence on the part of one culture that theirs is the only correct way to live upon this world; that a thousand other cultures and thousand other times are all wrong, and that we must have one government, one language, one religion, and above all one economy uniting all the earth. With the tremendous loss in human diversity comes also an enormous loss in the planet's biological diversity.

We no longer know how to adapt to our environment. Instead, we are forcing our environment to adapt to us. We are rapidly careening toward not just one government and one economy, but just one species left on a very boring and dying earth.

Europeans are not by any means the first people in the history of the world to think they are better than everyone else. But they are the first people that have ever had the technology to enforce their beliefs on a worldwide scale.

"Behold, my brother, the Spring has come; the earth has received the embraces of the sun and we shall soon see the results of that love!

"Every seed is awakened and so has all animal life. It is through this mysterious power that we too have our being and we therefore yield to our neighbors, even our animal neighbors, the same right as ourselves, to inhabit this land.

"Yet, hear me, people, we have no to deal with another race—small and feeble when our fathers first met them but now great and overbearing. Strangely enough they have a mind to till the soil and the love of possession is a disease with them. These people have made many rules that the rich may break but the poor may not. They take tithes from the poor and weak to support the rich and those who rule. They claim this mother of ours, the earth, for their own and fence their neighbors away; they deface her with their buildings and their refuse. That nation is like a spring freshet that overruns its banks and destroys all who are in its path.

"We cannot dwell side by side. Only seven years ago we made a treaty by which we were assured that the buffalo country should be left to us forever. Now they threaten to take that away from us. My brothers, shall we submit or shall we say to them: 'First kill me before you take possession of my Fatherland....'"[165]

— Sitting Bull, Sioux chief, at the
Powder River Council, 1877

"I am greatly astonished that the French have so little cleverness, as they seem to exhibit in the matter of which thou hast just told me on their behalf, in the effort to persuade us to convert our poles, our barks, and our wigwams into those houses of stone and of wood which are tall and lofty, according to their account, as these trees. Very well! But why now do men of five to six feet in height need houses which are sixty to eighty. For, in fact, as thou knowest very well thyself, Patriarch—do we not find in our own all the conveniences and the advantages that you have with yours, such as reposing, drinking, sleeping, eating, and amusing ourselves with our friends when we wish? This is not all, my brother, hast thou as much ingenuity and cleverness as the Indians, who carry their houses and their wigwams with them so that they may lodge wheresoever they please, independently of any seignior whatsoever? Thou art not as bold nor as stout as we, because when thou goest on a voyage thou canst not carry upon thy shoulders thy buildings and thy edifices.

"Now tell me this one little thing, if thou hast any sense: Which of these two is the wisest and happiest—he who labours without ceasing and only obtains, and that with great trouble, enough to live on, or he who rests in comfort and finds all that he needs in the pleasure of hunting and fishing? It is true that we have not always had the use of bread and of wine which your France produces; but, in fact, before the arrival of the French in these parts, did not the Gaspesians live much longer than now? And if we have not any longer among us any of those old men of a hundred and thirty to forty years, it is only because we are gradually adopting your manner of living, for experience is making it very plain that those of us live longest who, despising your bread, your wine, and your brandy, are content with their natural food of beaver, of moose, of waterfowl, and fish, in accord with the custom of our ancestors and of all the Gaspesian nation. Learn now, my brother, once for all, because I must open to thee my heart: there is no Indian who does not consider himself infinitely more happy and more powerful than the French."[166]

— a Gaspesian chief, quoted by father Chrétien Le Clercq, in *New Relation of Gaspesia*, 1691

"I heard that long ago there was a time when there were no people in this country except Indians. After that the people began to hear of men that had white skins; they had been seen far to the east. Before I was born they came out to our country and visited us. The man who came was from the Government. He wanted to make a treaty with us, and to give us presents, blankets and guns, and flint and steel and knives.

"The Head Chief told him that we needed none of these things. He said, 'We have our buffalo and our corn. These things the ruler gave to us, and they are all that we need. See this robe. This keeps me arm in winter. I need no blanket.'

"The white men had with them some cattle, and the Pawnee Chief said, 'Lead out a heifer here on the prairie.' They led her out, and the Chief, stepping up to her, shot her through behind the shoulder with the arrow, and she fell down and died. Then the chief said, "Will not my arrow kill? I do not need your guns.' Then he took his stone knife and skinned the heifer, and cut off a piece of fat meat. When he had done this he said, "Why should I take your knives? The Ruler has given me something to cut with.'

"Then taking the fire sticks, he kindled a fire to roast the meat, and while it was cooking, he spoke again and said, 'You see, my brother, the Ruler has given us all that we need; the buffalo for food and clothing; the corn to eat with our dried meat; bows, arrow, knives and hoes; all the implements which we need for killing meat, or for cultivating the ground. Now go back to the country from whence you came. We do not want your presents, and we do not want you to come into our country.'"[167]

— Curly Chief, Pawnee, quoted by George Bird Grinnell in *Pawnee Hero Stories and Folk Tales*, 1889

"You say, 'Why do not the Indians till the ground and live as we do?' May we not ask with equal propriety, 'Why do not the white people hunt and live as we do?'"[168]

— Chief Old Tassel, Cherokee, 1780s

"We do not want your civilization! We would live as our fathers did, and their fathers before them."[169]

— Crazy Horse, Oglala Sioux chief, about 1870

"Oh, we happy Greenlanders! Oh, dear native land! How well it is that you are covered with ice and snow; how well it is that if in your rocks there are gold and silver, for which the Christians are so greedy, it is covered with so much snow that they cannot get at it. Your unfruitfulness makes us happy and saves us from molestation! Pauia! We are indeed contented with our lot. Fish and flesh are our sole food; dainties seldom come in our way, but we are all the pleasanter when they do. Our drink is ice-cold water; it quenches thirst and does not steal away the understanding or the natural strength like that maddening drink of which your people are so fond. Our clothing is of unsightly thick-haired skins, but it is well suited to this country, both for the animals, while the skins are still theirs, and for us when we take them from them. Here, thank God, there is nothing to tempt anyone to come and kill us for its sake. We live without fear."[170]

— Paul Greenlander, in a letter to Paul Egede, 1756

Central Africa

The twelve groups of people called by outsiders "pygmies" are the most numerous and vigorous of the traditional hunters and gatherers living in the world today, numbering at least 250,000 men, women and children, including the 40,000 Mbuti of the Ituri forest of the Democratic Republic of Congo. That forest is under dire threat today from encroachment by outsiders. The widening of the one road that runs through the center of the forest has caused noticeable changes in rainfall along that belt. Where the farming population has grown in a scattered rather than linear fashion, the forest has been devastated.

Under the shade of the ancient forest the climate is even throughout the year. Variation in rainfall is minimal, and temperatures usually range from 70° to 80° F, with humidity about 95 percent. Forest water is pure and potable. The forest canopy arches high, and mosquitoes virtually never come below it except at river's edge. Flies are few, but abundant ants help dispose of refuse. Plant life is plentiful, and the people hunt fifty kinds of mammals for food. The forest is an eminently healthy place to live. And yet the Mbuti leave it once a year to toil as servants on their neighbors' farms, where they suffer from malnutrition, malaria, dysentery, yaws, measles, respiratory ailments, and parasites. The return they have always received is the continuance of their own way of life: by providing them with goods and labor, they keep their agricultural neighbors from expanding into the forest. The Mbuti, quite literally, are the protectors of the jungle. But both forest and forest-dwellers are threatened by roads, mining, and lumbering operations, and an ever-growing population of farmers seeking more land to cultivate. Today, the greatest threat to both the forest and its people is the insatiable demand for cobalt and tantalum, minerals that are contained in the batteries, capacitors, and wave filters in every cell phone in the world. The ground underneath the Democratic Republic of Congo supplies seventy percent of the world's cobalt and eighty percent of the world's tantalum, and supplying these metals for everyone's cell phones is driving the Mbuti toward extinction.

South America

In the Western Hemisphere, the lunacy that began in 1492 continues to this day. As long as there are still any wild "Indians" left, white Europeans continue to torture, rape, and kill them, and to sell them into slavery. It is a kind of lunacy that I think is at the heart of the disease that is destroying this earth. What does "wildness" mean to us, that we must want it and destroy it at the same time? It gives the white man status to claim Indian ancestry, but he cannot tolerate a living Indian. Cats and dogs have the run of his household, but wild cats and wild dogs must be exterminated. The word "wolf" conjures up so much more in our psyches than what a wolf actually is. Indian . . . wolf . . . bobcat . . . to us whom civilization has tamed, they represent our own wild selves that we must defeat in order to remain civilized. It will not do to see the Indian as just a man, a wolf as just a dog, a lynx as just a cat. They are symbols of our own wildness, and we need those symbols so that we can tame them. And when the last Indian, and the last wolf, and the last lion are finally gone, civilization will crumble, because its driving force will have vanished. Part of the thrill of warfare is to vanquish the wild enemy. When men emerge from the forest to settle down to farming and civilization, they begin to war upon one another. And when wild nature vanishes all around us, warfare becomes more insistent and more terrible. The enemy is the wild one. Saddam Hussein and the Iraqis. The nigger. The Jew. The Gypsy. They are the wild ones, and we must tame them. We in America have tamed our last frontiers. Alaska was still there, vast, wild, filled with Eskimos and Indians. But the excitement of the Oil Rush is over. The natives have settled for a billion dollars, and are now landowners like us. Then came the *Exxon Valdez*, which polluted the last American wilderness and took it away from us. We must have something more to tame. What will drive our civilization? Our youth want to know. They rampage through the streets of our cities, killing, raping, torturing, looking for the wild one to give their empty lives some meaning.

The stories that continue to come out of South America are sickening.

Bernard Arcand, in 1972, reported on the last 940 members of the Cuiva people in Colombia and Venezuela. 95 percent of their territory is grassy savannah with scattered palm trees and shrubs. But the Cuiva's home is within the cathedral forests that shade the banks of the many rivers and streams and provide both shelter and sustenance. For four centuries the Cuiva were pursued by European ranchers who often shot them on sight, and they retreated into ever more remote areas. Finally, having no place left to go, they built permanent houses and took to growing corn, manioc, and bananas to supplement hunting and gathering in their depleted woods. Yet again they were attacked by their rancher neighbors, and again they escaped into the forest. They built another village near the ranch of a friendly Swiss settler, and there they were living when Mr. Arcand met them. Almost all of their hunting territory was already gone, depleted of animals by ranchers hunting with guns, and by unending streams of poachers who flowed in from all directions. The remaining Cuiva were dying of malnutrition and disease.

Today the Cuiva are extinct.

From Eastern Paraguay in 1973, Mark Münzel reported on the desperate situation of the Aché. The first laws officially protecting them from being hunted or sold into slavery were passed in 1957 and 1958. But at the same time, the remote forests in which they lived were being opened to foreign investment, international roads, and commercial exploitation. The road through eastern Paraguay was completed in 1965. An additional road cut the forests of the northern Aché into two parts in 1968, and invasions by wood cutters, palmetto collectors, and cattle ranchers intensified. In 1972, Aché were still being shot on sight, and their children sold into slavery. Many were living captive in a concentration camp where they were cruelly mistreated and dying of malnutrition, disease, and neglect. As of April 1972, some 277 Aché were living on this "reservation." Perhaps 700 more were still in the forests being hunted down, and an unknown number of Aché slaves were living among Paraguayans in the outside world.

In the 1980s a five-mile-long dam was built on the Paraná River in Aché territory. Roads were cut through even the most isolated regions, and virtually all of the Aché's forests were sold to commercial farmers. Hundreds of thousands of acres were cleared of trees and planted in cotton and soybeans. Tens of thousands of Paraguayan settlers moved into newly built towns in the former wilderness. The displaced Aché were herded onto reservations administered by missionaries whose main goal was to civilize the Aché and convert them to Christianity. Today the Aché live in six reservations, including one inside the Mbaracayu Forest Reserve where the community make their living growing and selling yerba mate.

From northwestern Argentina in 1975 came a report by Nemesio Rodrigues on the Wichi (Mataco) and related tribes of the Grán Chaco, as follows: The Wichi have been fleeing the Spanish conquerors since the mid-sixteenth century. In 1975 they were living by food-gathering, fishing, hunting, limited agriculture, and wood-cutting for the timber companies. As the cheapest labor force in the country, they are paid not a salary but goods whose value is set by the timber company. The company raises the selling value of the goods 400 percent and undervalues the poles the natives cut, so that what is produced "does not cover" what is acquired from the store. It's an old story—"the company store." Debts are created which are impossible to pay. Native villages are considered by the government as "uninhabited territory," open for colonization by white farmers, stock breeders, and manufacturers. Deforestation and intensive cattle grazing are changing the Grán Chaco into a desert. Half of the Wichi villagers have tuberculosis, all are malnourished, and there is no medical care for them. Organized gangs kidnap native girls and sell them into slavery to be maid servants in Argentine cities.

Today all-weather roads extend hundreds of miles into the former wilderness where the Wichi lived. Loggers have penetrated the most remote areas, and five million hectares of forest have been logged in just the last twenty years. Cotton is growing on the most fertile land, and cattle are grazing elsewhere. The Wichi now live in shanty towns at the edge of northern Argentine settlements, with neither fields to cultivate nor forests to hunt in. Those who are not in debt-bondage to the timber lords have depended on seasonal work in sugar or cotton fields. A private association, independent of government or religious affiliation, has helped some Wichi begin their own businesses—commercial agriculture, ranching, lumbering, carpentry, brick-making—with the aid of loans from national and international development corporations. To pay off these loans the Wichi are forced to produce for national markets and to join the industrial Argentine economy. The traditional way of life is completely destroyed.

Slavery continues in more than one part of South America. Between 6,000 and 7,000 native boys presently work as slaves in the gold mines of eastern Perú. Most have been kidnaped from their homes in the Andes, and few ever return. They die of starvation, torture, or bullets, and end up buried in hidden cemeteries or floating down rivers.

Colombia, at least, has set aside almost one-third of its territory as reserves for its indigenous peoples. But the government still claims the rights to minerals and raw materials on all native land, and it is an old story in South America that official protection is not the same as actual protection. In a country partly ruled by paramilitary groups and drug cartels, what chance do the indigenous people have of receiving government protection? It is a story that will play itself out to the bitter end, as it played itself out to the bitter end in the United States over a century ago, leaving a long trail of broken treaties and promises behind. The white man will always find a way to encroach upon Indian lands when it becomes expedient for him to do so. Today those lands are remote and useless to him, tomorrow they will be the focus of a new gold rush. If not so many Indians are being murdered anymore, it is only because not so many Indians are alive anymore.

The story continues even here in the United States. The Navajo were left alone on their Southwestern reservation only so long as it was considered by white people worthless desert. With the discovery of coal and uranium came the mining companies demanding access to native land. And with the refusal of access came forced relocation of Navajo people, the confiscation of their livestock, the drilling of 500 uranium mines, and the contamination of much of the reservation with radioactivity. The Alaska Native Claims Settlement Act of 1971, granting native Alaskans $967 million in cash and 44 million acres of land—about one-tenth of Alaska—was passed not out of the goodness of the white man's heart, but to extinguish all native claims to Alaska, which could then be opened up to oil development. No reservations were set aside. Instead, the United States set up a multitude of land-owning corporations with native Alaskans as corporate executives and stockholders. This turned out to be the quickest way to develop Alaska. These now-wealthy corporations have logged off their own land, drilled for oil, built airstrips, and encouraged the industrial exploitation of their wilderness. Rich or poor, the native way of life does not survive, the minerals do not stay in the ground, and the trees do not stay on it.

Brazil is now finishing off the job it started 500 years ago. First came the Indian hunters of the sixteenth and seventeeth centuries, seeking slaves to work the vast sugar plantations of coastal Portuguese colonists. Then came the diamond hunters of the eighteenth century, and the rubber hunters of the nineteenth century, pushing ever deeper up the great rivers, ever farther into the jungle, rounding up the Indians and gunning them down, raping their women, burning their villages and their cities, pressing them into slavery. The great native civilizations on the banks of the lower Amazon, and on the banks of the Tapajos, and the Madeira—these were the first to go, obliterated so thoroughly from history that Old Amazonia is remembered only as the home of primitive hunters and gardeners. The rubber hunters ascended further up the great rivers, past the forbidding cataracts, ever deeper and wider into the wild jungle, until the last native tribes were no longer in the heart of Amazonia but on its periphery, protected only by the great waterfalls and rapids descending from the savannahs far above, into the lowlands far below. The rubber lords had their way in the lowlands, and the diamond and gold prospectors combed the plateau above, but wild men and women remained free in the Indian territories, the Mato Grosso and Rondônia. Until about 1940.

In the last eighty years the engines of progress have bulldozed and burned their way through the Mato Grosso and Rondônia, no longer Indian territories but states of an industrialized country struggling for a place of prominence in the world community of powerful nations. Where Indians and forests once were are now cities, roads, mines, cattle ranches, logging operations, and desert. Farmers typically clearcut the forest, harvest a few seasons' crops, and leave the fields to cattle ranchers. Herds then graze for a decade or so. Exhausted, the land is abandoned, once lush, transformed to desert in the space of a decade and a half. The forests were intact until 1973. The natives have been dying for much longer.

How many tribes have been completely wiped out in what is now the territory of Rondônia! Erased from the world, without even a few survivors. The list of tribes which no longer exist is striking: the Amniapés who used to live on the Rio Mequena; the Apiacas of the middle Madeira; the Aruas of the Rio Branco, a tributary of the Rio Guaporé; the Guarategajas of the Rio Guaporé; the Huaris of the Rio Corumbaria. The Ipotewats of the Rio Cacoal. The Iabutifeds of the Rio Rosinho. The Iabutis, also of the Rio Rosinho. The Caritianas, whose fief was close to the sources of the Rio Candeias. The Caxarabis of the Rio Abuna. The Querquircivats who once lived along the banks of the Rio Pimenta Bueno. The Macuraps, who used to haunt the rapids of the Rio Branco. The Mialats, formerly settled on the Rio Leitão. The Mondés, who roamed the upper part of the Rio Pimenta Bueno. The Palmeras, men of the interior of the jungle. The Rama-Ramas, who rode along the Rios Anaris and Machadinho. The Sanamaicas, who used to dominate the streams which flowed into the Rio Pimenta Bueno. The Taquateps of the Rio Tamuripa. The Toras, the big tribe of the Rio Madeira. The Urumis of the murkey Rio Roosevelt. The Wayoros, who lived by the Rio Colorado." (Lucien Bodard, in *Green Hell* [1972]).[171]

Some of the survivors are now protected on reserves—officially but not actually. The Uru-eu-wau-wau of the Guaporé Valley live on a forest reserve which used to measure 18,000 square kilometers. But with just 350 Uru-eu-wau-wau surviving, the Brazilian government reduced its size in 1989. In 2002 the population was less than 200. In Mato Grosso a reserve of 85,000 square kilometers was declared off-limits to white colonization in 1952. Later named Xingu National Park, its area has shrunk to 26,420 square kilometers, and it is still under assault from poachers and would-be ranchers. In the 1970s the government built a highway across the northern third of the park, opening the way for further invasion. Today the park is occupied by a few members of sixteen different tribes, which collectively number about 5,000 people.

The first road ever to link Brazil with the Pacific coast, called the Interoceanic Highway, was completed in 2011. Financed by loans from the World Bank and Japan, it was bulldozed westward from the state of Acre, over the Andes to coastal ports in southwestern Perú. Japan was looking for easy access to a large supply of tropical woods. As a result, previously intact forest in both countries has been opened up to widespread logging, commercial agriculture, and mining. A second road connecting Brazil with the Pacific Ocean, still in the planning stage, is a bit further north, and would shorten the journey over the Andes. It would begin at Cruzeiro do Sol, a remote port on a tributary to the Amazon River, cut through traditional territories of the Ashéninka, and continue to the city of Pucallpa, Peru. A third road linking Brazil to the Pacific Ocean is being constructed further south through Paraguay and Argentina to Chile. 2,200 kilometers long, it will cross the regions of Mato Grosso do Sul in Brazil, Gran Chaco in Paraguay, the provinces of Salta and Jujuy in Argentina, and the regions of Antofagasta and Tarapacá in Chile. It is being called a "new Panama Canal." About 525 kilometers of this new highway will pass through Paraguay's Gran Chaco, one of the main environmental reserves in the country, populated by savannahs and wetlands. Home to the Ayoreo indigenous community, and to jaguars, pumas, anteaters, and thousands of plant species, the Gran Chaco is one of the most biodiverse places on earth. Or it used to be. 20 percent of the Gran Chaco forest has already been converted into land for cattle grazing and industrial agriculture.

The remaining forests and forest-dwellers of tropical South America will not withstand the onslaught.

> When that whiteman shot at us with his rifle it offended me more than I can say. What if his bullet had killed me? My daughters, my little ones, would be left without me. They would grieve as only children can. They would suffer without their father. Why does the whiteman presume to make orphans of my children? Why does the whiteman seek to kill us?
>
> Does he think Indians do not have families? Perhaps he thinks that Indians do not have children. As he is, so are we. We are men! Doesn't he know that? Our thoughts, our desires, our lives, they are as his own! What we carry within our bellies, what is in our hearts, is the same. Can't he see that? (Chief Atamai, Wauja of the upper Xingu, September 1989)[172]

Europe

In the far north of Europe, Asia, and America are people whose lives for thousands of years have been linked with the migratory reindeer. The Americans remained hunters, but in Norway there is rock art that proves that reindeer were already used as pack animals shortly after the ice receded, perhaps nine thousand years ago.

Again, it is difficult to get rid of the old notion, so strongly drilled into us in our schools of modernity, that the Saami people of Scandinavia (often called Lapps) are primitive relics of human beings who simply never "caught up" with the inexorable advance of beneficent civilization. The fact is that fishing and herding reindeer are particularly good ways of making a living in the snowy north. The Saami language is closely related to Finnish and Estonian; settling to the north of their Finnish relations, Saami traders were originally in a position to supply codfish and furs to an international market. They were forced to increase their dependence on reindeer herding in the 1500s and 1600s because of a depression in the codfish trade and the opening up of America and Siberia as new sources of furs for the European market.

For several hundred years breeding reindeer and migrating with the herds became the center of the Saami economy, but that economy is now threatened by the same forces of "progress" that are destroying Africa and South America. Dams have flooded migration routes, and fences are destroying them entirely. Power lines, highways, and railroads crisscross the territory, and miners, loggers, and military bases have invaded it. But the biggest revolution in Saami life was caused by the extraordinary proliferation of the snowmobile. Invented about 1960, Ski-Doos had completely replaced dog sleds and reindeer sleds throughout the Arctic within a decade. The effects have been catastrophic: the damage to lichens and other plant life; the pollution; the fire danger; the snowmobile junkyards all over the Far North; the damaged hearing of all Arctic peoples; the sharp increase in the need for cash in Inuit, Indian, and Saami communities; the dependence on outside fuel sources; the stratification of native societies.

Pertti Pelto reported on the Skolt Saami of northeastern Finland. Presnowmobile Skolt society was egalitarian. Young children received "first-tooth reindeer," "name-day reindeer," and other occasional gifts that assured them the acquisition of a small herd by the time they entered the responsibilities of adult life. All able-bodied men and some women were active herders in a local association, and all were of equal status receiving equal pay for their work. Gathering the herds in the fall, after a summer of letting them roam free, was a leisurely affair lasting several weeks, involving a close relationship between human and animal, and between human and nature. This has all been destroyed. Speeding snowmobiles made it all violent.

Contacts between people and animals became minimal since herders no longer spent weeks and months living with their reindeer; because motors cost a lot to operate, they rounded up as many animals as they could in just a few days. But the quickly gathered herds were no longer social systems, just bunches of frightened animals forced to run in the same direction, terrified by the noise and smell of the machines. The herders lost control of their herds to such an extent that it was no longer possible to bring home family herds for the winter, earmark the newborn calves in the spring, and harness reindeer as draught animals. The reindeer were reverting to a wild state.

During the succeeding decades a new system has been invented. In the meantime, the Skolt have acquired all-terrain vehicles in addition to their snowmobiles. The herders no longer live in the woods with their reindeer at all, they commute home every night in their vehicles. Roundups of more or less wild reindeer are on a smaller scale and take place repeatedly throughout the winter. The reindeer, including pregnant cows, are driven too hard and too often, and their health and numbers have suffered. The Skolt have built wire-mesh fences so they can again keep winter herds and earmark newborn calves in the spring. The docile, tractable family herds of old, driven slowly home and turned loose to graze together in the neighborhood, are no more. Now the herders capture a few of their female reindeer at each winter roundup and transport them bodily to the enclosures, where they have to be fed all winter with purchased lichen and hay. Contract herders are paid for each adult reindeer they manage to bring to the corral, instead of a fixed daily rate. Ski herders can no longer compete, and many families have dropped out of herding altogether due to the high cost of owning, operating, and repairing snowmobiles. Herders who best knew the ways of the reindeer from long years of intimate association with them have been replaced by people who know how to handle machinery. Today the addition of helicopters, drones, radio collars, and satellite tracking have estranged the Skolt even further from their animals and from nature.

What has emerged is a stratified society in which a few families own a large number of reindeer, and a majority have few or none. Jobs in stores and construction are scarce, and welfare and unemployment checks support many. Local schools teach the Skolt language in addition to Finnish, and traditional crafts, music, and other aspects of Skolt culture are encouraged—all except the economy it is based on. Even reindeer-herding is encouraged, but it is now a heavily subsidized industry, in which roundups are organized and managed by a small number of active herders, while for most other participants, who drop in for only a few hours, the roundup is reduced to an enjoyable social event—"an early spring picnic in the scenic backlands."

From the point of view of the national economic and social system of Finland, the roundup event can be considered part of an overall system that subsidizes the maintenance of a rural population in a quasi-traditional economic pattern and which keeps off *some* of the pressures of population shifts to urban areas. Without a variety of subsidies, overt and covert, the local association and individual herders would not be able to maintain the full system of roundups, plus the relatively costly additional activities now underway in the "re-domestication" of the reindeer.[173]

Violence has been done to the land, to the animals, to the Saami livelihood. It has been done largely by a single violent piece of technology.

Industrial Societies

It is time to return to the industrial economy to explore the forces that the rest of the world seems so powerless against. Let us go back once more to the question of what money is.

Trade has probably been conducted virtually everywhere on earth between human beings and their neighbors for as long as there have been human beings on the earth, either through barter or using money. In southern Africa, eggshell beads and tobacco were recently used as money. Clams and other shells circulated in North America, among both the native tribes and the early British colonists. Cattle served as money in Asia and Africa. Ivory, rice, furs, whiskey, and many other substances have served as media of exchange in various times and places—all more or less common objects with intrinsic value aside from their use as money.

Gold and silver differ from earlier forms of money in several important respects: they are virtually indestructible; the supply changes only very slowly, increasing but never decreasing; and people have very unequal access to them. For the first time, the flow of goods that deteriorate is represented by a medium of exchange that does not. And the control of the medium of exchange—i.e. the means of power—is taken out of the hands of the many and put into the hands of the few. Anyone can raise cattle, or grow tobacco, or gather seashells. Gold is not so easy to find and is therefore a more suitable currency for an authoritarian society with centralized power.

As so many environmentalists have pointed out, continuous economic growth must eventually exhaust a finite earth. Yet the present-day economy of the entire world is based on continuous economic growth. A non-growing capitalist economy fails and plunges into depression. How did such a situation come into being? I have spent many hours—days, weeks and months, in fact—thinking about this problem because it seems so central to the problem of environmental devastation we and our fellow creatures face. Bearing in mind that money is not so much a cause as a reflection of human relationships with each other and with the earth, let us look at the monetary system of our capitalistic industrial economy.

To begin with, we invented a medium of exchange—gold—that always increases in quantity. This puts a certain pressure on the world to continuously increase the total supply of wealth—not rapidly, but slowly over long periods of time. An increase in wealth means, basically, an increased extraction of raw materials from the earth and an increase in the number of human beings producing goods and services. In 1850, California alone produced as much gold as the whole world had in an average year of the preceding decade. Before the end of the 19th century, huge gold rushes in Australia, the Yukon, and South Africa added to the now rapidly expanding world supply. The 19th century was also a time of rapidly increasing exploitation of the earth's resources, and population increases the likes of which the world had never known before.

Gold is still a relatively rare commodity—the total world supply would certainly fit onto one modern supertanker—although since gold is so heavy, it would probably sink to the bottom of the ocean and the world would be vastly better off!

The use of credit on an enormous scale became necessary to keep pace with the extraordinary expansion of world commerce in the twentieth century—gold and silver simply would not do anymore. By the use of credit instead of precious metal as the medium of exchange, authority has been concentrated even more—or, rather, into different hands. In theory, any citizen can mine gold, or, easier, pan for it in the rivers of gold country. But only banks and governments are allowed to print paper money. So authority has been concentrated in the government and in the banks and in the corporations, which also issue a sort of money called commercial paper.

The result is our modern world in which merchants and bankers hold all of the power. The rules of the game are somewhat different under socialist or communist systems, but the essence of the transformation which imperils us all is this: the merchants have taken over the world from the farmers, the artisans, the pastoralists, the hunters, the fishers, and the gatherers. There is no balance anymore. The soils, and the oceans, and the forests, and the rivers, and the lakes, and all of the millions of kinds of plants and animals are still the source of all wealth, the means of all survival, and the objects of all work. We are living in a society which pretends that we must all become businesspeople, but in which the reality is that the majority of workers must remain farmers, and miners, and woodcutters, and fishers, and factory workers, and artisans, and pastoralists, and hunters, and gatherers, yet to the degree that we pursue any of those latter occupations—no matter how mechanized they may become—we are more or less impoverished, and only to the degree that we become merchants to we accumulate wealth and power.

Because people have always traded with one another, there have always been merchants who shared this world with the farmers and artisans and hunters and the people of all the other professions. Often the traders have been a distinct ethnic group. We have seen how in Borneo, Chinese merchants live among Kenyah farmers and Penan hunter-gatherers. In the Himalayas distinct cultural groups live as nomadic traders: the Sherpas; the Bhotias of eastern Nepal and Bhutan; the Thakalis of central Nepal; the traders of the Karnali Zone in far western Nepal. Some of these people engage only in local trade, making annual treks between the Tibetan and Indian frontiers, bartering salt for Nepalese rice. Others became famous as wealthy merchants who once traded over long distances from as far as Lhasa and Shigatse in Tibet to Calcutta and Hong Kong.

Merchants and bankers have not always been held in high esteem. Throughout the ancient Mediterranean world, lending money at interest was condemned as usury, and merchants were tolerated as a necessary evil in a world where it was desired to trade with foreigners. Such attitudes are expressed in the Bible, and in the writings of Plato, Xenophon, and Aristotle, and of Chinese and, later Japanese Confucianists. In the Middle Ages, Thomas Aquinas held that to lend money at interest was to take from another what one had not earned. Thomas Jefferson wrote to John Taylor in 1816 that banking establishments were more to be feared than standing armies. A century ago, the Armenians in Turkey suffered dearly from a reputation of being that nation's bankers. The Jews have carried a reputation as bankers and traders wherever they have gone, and it has been a reason for hating them.

It is a strange new world we live in today, where merchants rule. In our world, lending money at interest is so commonplace that people think they are being robbed if their money does not grow. No business succeeds without providing its investors with a return on their investment. Our economy fails unless it grows continuously.

Now money does not grow. Trees are cut, iron is mined, animals are killed, rivers are dammed, soil is tilled, oil wells are drilled, factories are built, shoes are shined, haircuts are given, concerts are performed, literature is written. More, and more, and more, faster, and faster, until the earth is exhausted and overpopulated with seven billion people too many.

I am not pretending to give a detailed economic analysis here. I am concerned mostly with the relationship of human beings to one another and to the earth, and how our monetary system reflects that. It is perfectly possible to design a theoretical world whose investors receive interest on their money, and where the economy nevertheless does not grow. Such a world would necessarily be more egalitarian than ours, and merchants would not rule.

In our society, all other professions are permitted to survive only at the suffrage of the merchant class, and hence our economy thrives or fails as our banks and corporations thrive or fail, not, as in all other times in human history, as our crops and our farmers thrive or fail, or as the grazing remains favorable or not, or as the salmon run is good or bad, or as the caribou return in numbers or not. We are so estranged from our biology that we live or die, not according to whether or not there is plenty in the world but according to whether or not bankers and traders are doing a good business.

Karl Marx, and socialists after him, have attempted to design more egalitarian societies where the merchant class, or bourgeoisie, would not rule. They have had varying degrees of success. In my opinion the Communists have been much less successful than the socialists of Scandinavia. Where, as in China and the former Soviet Union, the state controls the distribution of wealth instead of the banks and corporations, power has simply been removed from one authority and placed in another. The Communist Party becomes the new merchant class.

Socialists and capitalists alike have *all* failed to realize just where wealth comes from. Money is not the source of wealth, nor is labor, nor is intelligence. It is the earth who makes our wealth. Nature does by far most of the work; we humans do only a very tiny amount of the work that sustains us. The earth is the source of all wealth, and any economy that extracts it faster than it is made will eventually fail.

I began by calling this section "Industrial societies." "Merchant societies" might be as good a description, but it is actually a partnership between commerce and industry which rules us. Where merchants are in power, factories always displace handicrafts, because they produce goods so much faster and in greater quantity than hand labor; merchants make their maximum profits by trading in manufactured goods rather than traditional crafts. Conversely, industry cannot mechanize to any great extent without enormous funding. So it's very much a cooperative affair. What is more, the merchant class wields its power by virtue of its monopoly on high-power weaponry, which in turn requires the funding of the merchant class. It really is a "military-industrial complex." Once technology reaches a certain scale and a certain level of power, a partnership between commerce and industry comes into being which places farmers, artisans, pastoralists, and foragers at a permanent disadvantage. Production for a market replaces production for home use. All nomadic people are eventually forced to settle down. Farmers are forced to mechanize, and the factory farm replaces the family farm. Artisans are replaced by factories. And foragers can no longer survive at all.

The invention of guns and artillery in Western Europe early in the 14th century was a very momentous and tragic occasion for this earth we live on. Before this time Europeans, like everyone else, fought each other with bows and arrows and swords. With their new toys, European kings allied themselves with European merchants and bankers, and for the next six centuries they terrorize the world, leaving death, destruction, and chaos wherever they went. The feudal system collapsed, and for four hundred years Europeans died by the millions of plague, malaria, smallpox, typhus, and other diseases. Next, European merchants took their new weapons and their new diseases to Africa, to India, to the Spice Islands, to America, to Australia, to Asia, to the Pacific. They had a field day! The world was helpless before their new explosive toys, and they eventually conquered practically every country on earth. In the twentieth century this had the consequence that petty quarrels between small European nations became worldwide conflagrations. I have been able to discover only four countries in the world that have never been colonized by Europeans: Japan, Thailand, Ethiopia, and Turkey (also Korea, which however was a colony of Japan; and Yemen and Saudi Arabia, former colonies of Turkey).

The merchants of this world will not disarm, because it is only force of arms that keeps them in power. When the United States, for example, concludes a trade agreement with a foreign country, the first thing it does is sell that country weapons. The United States is merely supplying foreign merchants round the world with the means to gain and maintain power—particularly important in the case of what we call "Third World" countries which still have subsistence economies. In European and American eyes, subsistence economies are subversive to world commerce.

The overwhelming power of European weaponry has been matched by overwhelmingly powerful technology in all other areas of industry, made possible by the new alliance of industry with commerce. The earth itself is helpless before it. Neither a human body, nor an animal body, nor a plant body, nor even a river, or a mountain, is any match for an explosive bomb. No bird is any match for an airplane. No animal, not even an elephant, is any match for a train or a truck. As fifteenth-century Europe made the entire world its enemies instead of its trading partners, our industrial society has made the earth and all its creatures our enemies instead of our neighbors.

The Great Depression of the 1930s

In the aftermath of the First World War, American and European agriculture were dramatically transformed by new machines and techniques. Gasoline-powered tractors replaced horses. Large gang plows, pulverizing harrows, and grain drills enabled farmers to plow, harrow, and sow grain in one process. A combined header, thresher, cleaner, and bagger revolutionized harvesting. Synthetic fertilizers were applied to soils, and newly discovered insecticides, gentle by today's standards, killed insect pests. Milking machines, motorized trucks, and refrigerator cars transformed the dairy industry. New, more productive breeds of cattle and new varieties of crops were introduced. In 1900, it required three hours of man and horse time to produce a bushel of wheat in the United States. In 1930 the same man, with machines, could produce the same bushel of wheat in three minutes. From 1925 to 1929, world farm prices dropped 30 percent while food stockpiles increased 75 percent.

Industry was also being transformed, grandly, almost overnight, with billion-dollar American corporations leading the way. New industries, producing automobiles, petroleum products, chemicals, electricity, rubber, and machinery, were booming. Old industries, such as textiles, were transformed by mass production and automation. A shoemaker who once made six pairs of shoes a week by hand could now make thirty-three pairs a week using machines. In 1920, there were 9.2 million registered motor vehicles in the United States; in 1929, 26.5 million. The total quantity of American manufactures doubled or tripled in a decade.

The industrializing world was over-producing. There was a worldwide glut of wheat, rubber, coffee, sugar, rice, cotton, silver, zinc, and other commodities. Technological unemployment was a sudden glaring fact. For every shoemaker who could make five times as many shoes as before, four shoemakers were now unemployed. Hand spinners and weavers were replaced by machinery. Corporate farms replaced large numbers of family farms, and people migrated in droves from farms to cities looking for work. The urban population of the United States equaled the rural population for the first time.

The tremendous expansion and diversification of industry also provided a great many new jobs, and the economic boom continued as long as new employment opportunities kept pace with the numbers of people looking for work. But soon machines were throwing people out of work faster than industry was creating new jobs for them, and both industry and agriculture were producing huge surpluses that could not be consumed. The Great Depression of the 1930s brought the military-industrial complex face to face with its own contradictions more dramatically than ever before. Simply put: what passes for economic development puts people out of work. Efficiency in industry is the production of the maximum quantity of goods using the least amount of labor. The result is mass unemployment and starvation amidst fabulous surpluses of food and manufactured goods. How are we going to distribute the wealth?

Since the manufacture of food and essential goods and the providing of useful services could no longer employ more than a small fraction of our population, the solution evidently has been to employ the rest of us at manufacturing useless goods, and providing nonessential services. In particular, the United States has stayed out of depression since World War II by having a permanent wartime economy and by having American industry produce garbage and sell it to the American public. Planned obsolescence employs people. Plastic packing, to be used once and thrown away, employs people. Government bureaucracy employs people. The army, navy, air force, and marines, and all of the weapons manufacturers employ huge numbers of people. Pollution, waste, and war. They are ways to distribute the wealth, to prevent the masses of people in the machine age from sitting idly and collecting welfare while a small fraction of the people go to work to provide for them.

And we have sold ourselves a bill of goods that all this is good for us. But as long as the suggested solution to poverty and unemployment is "more development," neither they in Europe and Asia nor we in America are facing our true problems, the contradictions of industrialization that threaten us all, not just with economic depression, but with physical extinction.

Hippies

From my own point of view, the hippie movement and student rebellions of the late 1960s in my country were a last-ditch effort to stop the subjugation of nature, to prevent the transformation, then going on, of a world dominated by forests, mountains, and mystery, into a world dominated by cities, highways, and artificial landscapes. Ours was the last generation of Americans who still knew true wilderness, the last generation who experienced an earth most of whose surface was not accessible to machinery. We were the last generation for whom quiet was still a choice and to whom clean rivers and lakes still beckoned from throughout the land. We were the last generation who still had real hope for a human future. We were really fighting against the takeover of the machine, our erstwhile servants threatening to become our masters. We had not the strength by ourselves to become free of them, and so we turned to marijuana and LSD. Drugs were seen as our liberators, but they were in reality just different masters. And the battle was finally lost, the fight given up.

Those of us who, during and in the aftermath of the failure of our movement, returned to the land, gradually gave up even that remnant of hope for the good life. The back-to-the-landers succumbed to the tyranny of the machine, like the yuppies who bowed in obeisance to it and sang its praises. We students and young people who could not see the sense of entering the rat race in which we saw our parents trapped, ended up having to enter it after all, because in the meantime the natural world was being destroyed, and with it, any possibility of living the good life in harmony with nature. As this world, to which we longed to retreat to escape our parents' fate, disappeared from view, most of us blinked in confusion as if at a vanishing mirage and concluded that it was after all but a fantasy that had never been real to begin with. We concluded that it had all been a delusion and that we had better wake up and go about the business of supporting ourselves, since it appeared we were after all only getting a free ride off of somebody else.

But we were not deluded. "There's no such thing as a free lunch" is a lie. The natural world is free to all creatures. Human workers do not create what we use, we get it all from the bounty of nature. It begins to be costly when we interfere in natural processes and redesign our world until in fact it is a rat race. And until the late 1960s there was still a choice on this earth to enter the rat race or, for those of us who wished, not to. Until the late 1960s there were enormous wildernesses on every continent in the world, and there were tribes of food-gathering people living on every one of them but Antarctica. And since the 1960s, almost all of these tribes have ceased to live in their age-old ways, their traditional cultures destroyed, the last of the great wildernesses on earth fragmented and invaded by settlers, roads, airplanes, computers, and cell phones.

It is human population growth that bears much of the responsibility, and the blame, for these invasions. And to the question of human numbers we now turn our attention.

CHAPTER 28

Population, and a Tour of the World

Which way is forward? I asked at the beginning of this book. Civilization has clothed my body with fabric and my mind with literacy, has substituted electricity for sunshine, machinery for vitality. But these additions are not laws of nature, nor of history. Nudists remind us of the façade we carry, that time has no social direction but of our own choosing.

One ought take nothing for granted. Our overcrowded world grows daily more populous, and as far as we can remember has always done so, yet memory has its limits, and its distortions. Although widely accepted, it is a misconception that human numbers have grown steadily in all places and at all times.

We obey the same laws of biology as the rest of our earthly neighbors, and permanent population growth is not normal. The carrying capacity for human beings depends on our economy, that is, on the tools we use and the amount of trade we carry on with one another. Foragers are the least populous human beings. Swidden farmers are less numerous than those who farm intensively. Industrial techniques increase the carrying capacity still further. So does increased trade, by the sharing of resources of distant places. But for a given culture with a given economy, population density does not change all that much; it goes down about as often as it goes up.

The past few centuries of steady world population growth have been due to the domination of the world by Western European commercial interests, driving and driven by technology of ever-increasing power. The result is a world economy whose so-called health depends upon continuous growth. The value of goods and services produced must rise. More trees must be cut down this year than last, more iron and more uranium taken out of the ground, more computers and more cars made, more houses built, more pesticides sold, more cattle slaughtered, more food grown, more fish caught. There also must be more people to consume all these things, and to provide more services than last year. If production does not rise, investors will withdraw their investments and businesses will fail. In other words, it is in the interest of business in an industrial economy to have a continuously growing population of human beings on this earth. Even in today's overcrowded world, many governments make it a conscious policy.

> A society no longer capable of ensuring the replacement of generations is a condemned society.
> — Valéry Giscard d'Estaing, President of France, 1978[174]

> An increase in the birth rate, efforts to ensure adequate population growth, and the strengthening of the family must constitute priority objectives for the development of our socialist nation in order to ensure the economic and social progress of the country and preserve the vigor and youth of the entire people.
> — Resolution of the Central Committee of the Romanian Communist Party, March 3, 1984[175]

On April 12, 1984, the European Community, whose countries were already some of the most densely populated in the world, passed this resolution:

> The European Parliament,
> (A) aware that Europe's standing and influence in the world depend largely on the vitality of its population and on the confidence placed by parents in the future and well-being of their children and the prospect of giving them a proper upbringing and education in a balanced family environment,

(B) seriously disturbed by the recent statistics showing a rapid decline in the total fertility rate in the European Economic Community, which fell from 2.79 in 1964 to 1.68 in 1982,

(C) whereas, unless steps are taken to reverse this trend, the population of the Europe of Ten will account for only 4.5% of the total world population by the year 2000 and only 2.3% by 2025, as opposed to 8.8% in 1950,

(D) having regard to the disappointing outcome of the informal meeting of the Ministers for Social Affairs and Employment in Paris on 5 April 1984,

1. Considers that population trends in Europe will have a decisive effect on the development of Europe and will determine the significance of the role which Europe will play in the world in future decades;

2. Considers that measures to combat this marked trend towards population decline, which is common to all the Member States, could usefully be taken at the Community level and would be of both political and social significance;

3. Calls on the Council of Social Affairs Ministers of the European Economic Community to hold a further meeting to study the practical measures which could be taken, notably on the basis of suggestion to submit proposals on this subject.[176]

In 2023, contrary to predictions, the population of the countries that were the Europe of Ten has *increased* by 60 percent and is still over 8 percent of the total world population. Yet it is still Europe's policy to grow its population some more. These excerpts are from a resolution of the European Parliament dated May 20, 2021:

A. whereas population distribution at local, regional, national and EU level, as well as its stability or change, have very different dynamics across Member States and their regions, with unequal impacts on the depopulation phenomenon, and ultimately on the social, economic and territorial cohesion of the Union . . .

C. whereas demographic trends are also influenced by climate change and in particular by floods and heat waves related to this process; whereas a coordinated approach integrating principles of sustainability, greening and digitalisation across different EU policies could also contribute to reversing negative demographic trends;

D. whereas there is a high correlation between the provision of social services, physical and ICT connectivity, education and labour opportunities on the one hand, and the ability to retain and attract population to certain areas on the other . . .

F. whereas, although the EU population has seen substantial growth in previous decades, the growth rate is now falling and the population is expected to decrease significantly in the longer term . . .

Policy recommendations. . .

34. Urges Member States and regional authorities to implement an integrated approach to addressing demographic challenges through cohesion policy instruments, and encourages the promotion of smart villages and other incentive schemes to retain population and attract young people to rural and semi-urban areas . . .

37. Reiterates that demographic change is a fundamental challenge for the EU, and that addressing it should be prioritised in the design and implementation of programmes; . . . that particular support should be given to NUTS level 3 areas or clusters of local administrative units with a population density of below 12,5 inhabitants per square kilometer . . .

42. Calls on the local, regional and national authorities in regions at risk of depopulation to focus investments on ways of encouraging young families to settle in those regions . . .

It is often claimed that eight billion people are not too many to have on the earth at one time. About one person for every four acres of habitable land. Perhaps that's not too crowded. But consider that we are just one among ten million or more species on the planet. The average human being's metabolism burns about 2,350 kilocalories in a day. That's 6.9×10^{15} kilocalories per year for the human race. The total amount of energy made available to the planet by all photosynthesizing plants is something under 10^{18} kilocalories per year, and we human beings use up around 3/4 of 1 percent of it, or 1 part in 140, just in breathing and going about our daily activities. That doesn't include the vast amount of the earth's productivity that we destroy because of industry, human-set fires, pollution, deforestation, desertification, and pavement. It seems to me that we humans are crowding out our ten million neighbors.

People have been arguing about the causes of overpopulation, and proposing remedies for it, for such a long time, with such little effect, that it is difficult to arrive at the truth of the matter. It is even difficult to arrive at the truth about the history of human numbers.

Thomas Malthus is famous for touching off the debate in modern times. In his *Essay* of 1798, he wrote that human numbers always tend to increase faster than the food supply, and that the only checks on population are misery and vice.[177] And to this day the debate rages on between those who say our increasing population is a calamity and those who believe the earth can accommodate our growing numbers indefinitely.

In the 1750s, there was another debate going on in Europe concerning historical trends. David Hume believed human numbers had always been increasing, and that in the times of classical Greece and Rome, the world had been less populous.[178] Montesquieu and Robert Wallace, disagreeing, thought ancient nations had been much *more* populous, and Montesquieu even claimed that Europe's population had declined to one-tenth of what it was in the days of the Roman Empire.[179] And the censuses and other records that have come down to us from ancient times show that they were right. They indicate without exception that the ancient world was a very crowded place. Those who believe that human numbers always grow and never get fewer simply reject all the ancient documents. But as T. H. Hollingsworth said, "To assume that the only figures we have are exaggerated because they make out the past to have been more populous than the present prevents research from ever bearing fruit."[180]

Our civilization is not the first to worry about overpopulation. Tertullian, the founder of Latin Christianity, wrote, in about 200 A.D.,

> Surely it is obvious enough, if one looks at the whole world, that it is becoming daily more civilized and more fully peopled than in

ancient times. All places are now accessible, all are well known, all open to commerce; most pleasant farms have replaced uninhabited wastes, cultivated fields have subdued forests, flocks and herds have expelled wild beasts; sandy deserts are sown, rocks planted, marshes drained; there are now great cities where formerly no houses stood. No longer are islands dreaded, nor their rocky shores feared; everywhere are houses, everywhere people, everywhere government, everywhere life.

There is abundant testimony to human populousness: we are burdensome to the world, the elements can hardly supply us, and our needs grow more keen, and complaints are heard everywhere, while now nature does not sustain us. In fact, disease and famine and wars and earthquakes have come to be regarded as remedies for nations, as if pruning the luxurious growth of the human race.[181]

Han Fei Tzu wrote in China, in about 240 B.C.,

In ancient times, people were few but wealthy and without strife.... People at present think that five sons are not too many, and each son has five sons also and before the death of the grandfather there are already twenty-five descendants. Therefore people are more and wealth is less; they work hard and receive little.[182]

I invite the reader to accompany me now on a journey around the world, inquiring into the economic and population history of each area, as much as it can be discovered. Perhaps a pattern will emerge, and the truth about population trends, and their causes, will become clearer. It is also a journey of self-discovery, of elucidating who the human animal really is, and what is our relationship with the earth. I hope the reader will join me in discovering that human beings are the same everywhere, and that human life, and human society, no matter where, and no matter when, under whatever type of economy, have changed only in very superficial ways.

Let us follow the ancient trade routes.

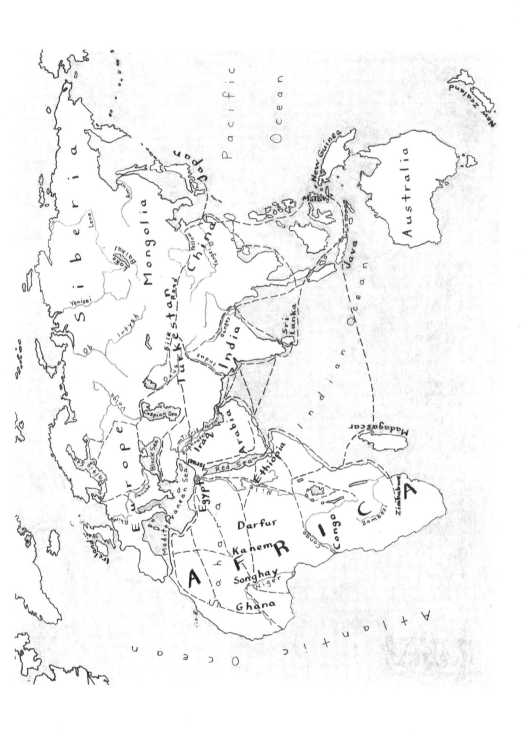

Java

We already began our exploration when we visited the island of Java in the last chapter. Old Java was the rice bowl of the East Indies, supplying especially the Spice Islands. A large merchant class sailed the world for thousands of years in outrigger canoes, trading in Australian pearls, East Indies spices, Chinese porcelain and silk, Indian cotton, African ivory, Mediterranean bronze ware and jewelry. Foreign ships, from Arabia, Persia, India, and China had safe passage in Javanese waters, and coastal trading centers did a booming international business. Madagascar was an Indonesian colony as early as the second millennium B.C.

This thriving tropical civilization cultivated its soil intensively and reached a stable population of probably a couple of million people—about forty or fifty people to the square mile.

In the sixteenth, seventeenth, and eighteenth centuries came first Portuguese and then British and Dutch traders. With their guns they were able to force an enormous expansion of trade, fueled by the labor of the Javanese people, while simultaneously destroying the livelihood of the Spice Islanders. Maluku became rapidly depopulated, and Java grew enormously more populous, until today 3,000 people crowd every square mile of Javanese land.

China

From Java we sail north across the sea until we reach a land famous for wealth and luxury in ancient times, a land whose merchants called at seaports from Southeast Asia to western India, doing business with Persians and Arabians and, once, with Greeks, and before them with Phoenicians.

Millennia ago, a well-worn trade route connecting China to a large part of the world became known as the Silk Road, the traversing of which, in Roman times, was a journey of more than six months. It wound its way through the vastness of Central Asia to the Caspian, the Black, and the Mediterranean Sea. Different parts of the trip were made on horseback, with donkeys, by camel-wagon, by ox-wagon, and by boat. Prosperous cities guarded these roads: Kashgar, Balkh, Samarkand, Herat, Merv, Otrar, Tashkent, to whose international markets merchants from east and west brought their goods. From Balkh a major trade route went southeast toward India, to the valleys of the Indus and Ganges Rivers and the Bay of Bengal. From Merv, another route went southwest through Herat toward the Persian Gulf and Arabia. Himalayan nomads carried the direct trade from China to India over the Tibetan plateau and the mountain passes of Nepal, Sikkim, and Bhutan. Another route went from China through Burma to the Bay of Bengal, and by sea to India.

The world's longest and oldest artificial waterway, the Grand Canal, dating to the sixth century B.C., flows 1,000 miles within China from Tianjin, the port for Beijing, southwards to Hangzhou, enabling ships to carry produce and trade goods between China's two largest rivers, the Yangzi and the Huang Ho.

Since very ancient times, in the marketplaces of Central Asia, in the port cities of coastal China, India, Persia, and Arabia, in the metropolises of Europe and Asia, a lively commerce has flourished in goods from all over the world: Chinese silks and porcelains, Indian muslins, Kashmir wool, Arabian alabaster, Persian tapestries, Italian wines, Baltic amber, spices and gems from every country, teak, ebony, skins, furs, clothing, tools, utensils, dyestuffs, foodstuffs, beasts of burden. The ancient world traded in parrots, lizards, and monkeys, and in lions, elephants, and giraffes. Empires came and went, and goods continued to flow eastward, westward, northward, southward, in torrents or in trickles, depending on political and economic fortune. When the seas became dangerous, overland trade picked up the slack. When one route was cut off, merchants found another. The world was interdependent, and a revolution in Mongolia had repercussions from the Atlantic to the Pacific, and from the Arctic to the Indian Ocean.

The first reliable census figures for all of China indicate a population of about 60 million in 2 A.D., and a reported population of between 50 and 60 million was usual for China all the way up through 1626. The numbers declined during times of war, floods, or other disasters, and rose again when peace and prosperity returned. The Sung dynasty (960 to 1279 A.D.) built extensive irrigation works, and settled new areas in the south. Census counts rose to about 125 million, and remained at about that number during Mongol rule. When Marco Polo visited Kublai Khan in the late fourteenth century, the city of Hangzhou, according to his detailed description, must have had at least 5 million inhabitants, four times as many as it had four centuries later. But the decline and fall of the Mongol Empire and the epidemic of plague in the 14th century reduced the Chinese population once again. All returns from the Ming dynasty, including the years 1361 through 1626, are in the vicinity of 50 to 60 million.

In addition to these records there are some for earlier dynasties dating back to 2200 B.C. indicating a similar density of people in northern China even at that early date. So there are good indications that for some 3,500 years China as a whole had a more or less stable population of between thirty and sixty people per square mile.

The Ming rulers maintained stability in a changing world by outlawing international trade and forbidding private citizens from traveling abroad. By 1500 the great Chinese shipyards were all closed down, and an edict of 1525 authorized the destruction of all sea-going vessels with more than two masts.

But this situation was not long tolerated by the newly mobile and powerful European fleets that were sailing the world's seas. Portugal, and then Spain, won trading concessions from China. In 1684, the new Manchu rulers finally lifted the general ban on maritime trade, and foreign ships once again flocked to Canton in great numbers—the British, the French, the Japanese; the Portuguese from their colony at Macau; the Spaniards from Manila; the Dutch from Indonesia; the Russians who by this time occupied large areas of Siberia and had ports on the Pacific. The Manchus, for their part, conquered Taiwan, and expanded in Central Asia, annexing Tibet and Xinjiang. From 1719 to 1833, the tonnage of foreign ships trading at Canton increased more than thirteen-fold. Tea exports increased more than twenty-eight-fold. Silk and porcelain flowed out of China, and silver flowed in. By 1749 the population of China had already tripled, to 177 million.

The nineteenth century saw the Manchus lose almost complete control of their country to foreign interests. In a series of costly wars, China lost Hong Kong to Great Britain, Taiwan to Japan, part of eastern Manchuria to Russia, and was forced to recognize Japan's conquest of Korea and

France's conquest of Vietnam. Peace treaties opened up numerous Chinese ports to foreign trade and residence, and granted foreigners exemption from local laws and the right to travel throughout the land. England, Germany, Russia, France and Japan began carving China into spheres of influence, and were well on the way to making them out-and-out colonies when the World War intervened.

Two and a half centuries of ever-expanding foreign trade then came to a halt. The Revolution of 1911–1912 was followed by years of civil war, war against Japan, and more civil war, until the Communists, led by Mao Zedong, finally reunified the country in 1949. They took China's first census in over a hundred years. In 1851, 432 million people had been counted; in 1953, 583 million. Seventy years of industrialization have now increased that number to 1.4 billion inhabitants, and a density of 670 people to the square mile for China proper (excluding the autonomous regions of Tibet, Xinjian, and Inner Mongolia).

India

A long and difficult summertime trek over Himalayan mountain passes brings us from China and Tibet southwestward into the richly diverse subcontinent that is India. Modern India's vast civilization holds together a multitude of ancient tribes and kingdoms and more than two hundred spoken languages.

No complete censuses have come down to us from the whole of ancient India, but we do have records of land use, army size, taxes, and population counts for parts of the country. Estimates of between 125 and 200 million have been made for the India of the third and fourth centuries B.C., higher than the commonly accepted figure of 110 million for 1605 A.D. In the intervening centuries empires prospered and declined, and many foreign invaders left their marks, among them the Greeks, who under Alexander the Great invaded the Punjab from the west in 326 B.C., and who returned to establish a kingdom in northwest India in the second century B.C.; the Scythians, or Sakas, of Central Asia, who ruled over much the same area from the first century B.C. to the second century A.D.; the Huns, who ruled parts of northern and central India in the sixth century;

the Turks, who conquered and ruled most of India in the thirteenth and fourteenth centuries, and who returned to found the Moghul Empire in the sixteenth century. During all of these 2000 years the population of India was apparently fairly stable.

The Europeans era in India began in 1498 with arrival of Vasco da Gama by sea from Portugal. The Portuguese built a colony in Goa, and for a century they controlled international trade in the Indian Ocean. Dutch ships arrived in 1595, and after them British, French, and Danish. The seventeenth century and half of the eighteenth was a period of relatively peaceful competition among the various European trading companies for the rich commerce in Indian cotton goods, indigo, silk, sugar, saltpeter, spices, and opium.

Violence erupted after 1740, and when the French were finally defeated in 1761, the British East India Company had secured a monopoly of the Indian trade. By 1818 the East India Company controlled almost the entire subcontinent, and the British government, with powerful weaponry, established direct rule over India in 1858.

British rule saw increasing agricultural production, rapidly expanding trade, early industrial development, and recurrent severe famine. The pattern we saw in Java is repeated here. Millions of farmers began to grow commercial crops instead of grain, and when world markets collapsed, they would starve. An expanding network of railroads brought cheap manufactured goods, shipped from England, to villages throughout India, and native handicraft industries were thereby destroyed, forcing more and more people to depend on agriculture for survival. The railroads, which had played a big part in causing this disaster, now were vital for transporting grain to starving people in times of famine—famine, while vast areas of farmland were being used for growing cotton, tea, indigo, and coffee for export.

A powerful nationalist movement, which grew after World War I under the leadership of Mahatma Gandhi and others, finally won independence in 1947. The former British colony was partitioned between a largely Hindu India and a Muslim Pakistan, and war between the two broke out immediately. In 1971, after bloody civil war, the state of Bangladesh was created out of East Pakistan, so that India's ancient kingdoms are now united into three modern countries.

It is estimated that the population of the Indian subcontinent rose from 110 million to 190 million during the time of peaceful trade from 1605 to 1750. During the years of British conquest the population rose hardly at all, to 195 million in 1800. The next two centuries tell a story similar to that of China and Java:

1850	233 million
1900	285 million
1950	434 million

Today 1.4 billion people live in India, 241 million in Pakistan, and 173 million in Bangladesh. The Indian subcontinent, which had a population density of about seventy per square mile in the sixteenth century, now has over 1,100 people to the square mile. Bangladesh, with about 3,000 people to the square mile, is as crowded today as Java.

Sri Lanka

Such great densities of people are not, however, unique to modern times. For thousands of years a large pearl-shaped island southeast of India was a main meeting place of merchants in the rich Indian Ocean trade, a place where Persian, Arabian, and Ethiopian sailors mingled with Chinese and Javanese. They returned home with tales of a fabulous kingdom perfumed with spices and haunted by demons and dragons. Perhaps some of them encountered Sri Lanka's real dragons—giant lizards called kabaragoyas, who share the island's lush jungles with iguanas and crocodiles, and with elephants, monkeys, and four hundred kinds of birds. Along with their stories, visitors also carried away native rubies, sapphires and pearls, and silks, spices and trade goods from around the world. The island became very wealthy indeed, and it has been estimated that in Sri Lanka's heyday in the twelfth century its fertile soils may have nourished 20 million people, more than 1,000 to the square mile. The people irrigated their rice fields with water from an extensive network of canals, reservoirs, and great water storage tanks.

After 1200 A.D. the Sinhalese, whose civilization had flourished there for seventeen centuries, were driven from the north by Tamil invaders from India, and by the fouteenth century most of northern Sri Lanka was a depopulated country. The great water systems fell into disrepair, tanks and reservoirs became muddy swamps, breeding grounds for malarial mosquitoes, and the jungle grew over the ruins of once-populous villages and cities. The people now became concentrated in the southwestern and central highlands, where agriculture was dependent not on irrigation systems but on the annual monsoon rains. Sinhalese sailors disappeared from the Indian Ocean, and trade, which still played a vital role in the economy, was carried to and from Sri Lanka largely by Arab merchants.

The first Portuguese fleet landed in Sri Lanka in 1505, and in 1518 the Portuguese built a fort at Colombo and obtained trading rights. During the next 120 years the foreigners' efforts to extend their control over the island resulted in continual warfare. They destroyed Hindu and Buddhist temples and sent Roman Catholic missionaries throughout the country. They pressed the salagama caste into service to gather cinnamon in the forest in a system that amounted to slavery. Cinnamon, elephants, pepper, areca nuts, and pearls flowed out of the country, but the embattled Portuguese did not show a profit.

In the seventeenth century the King of Kandy in the central highlands enlisted the aid of the Dutch to oust the Portuguese, but the result of another twenty years of warfare was the exchange of one oppressive ruler for another, still demanding cinnamon, elephants, and servitude on coffee plantations. The Dutch attempt to monopolize foreign trade caused shortages of food and clothing when Indian merchants stopped calling.

When the British conquered Sri Lanka in 1796 it had a total population of only about 900,000. The British, in contrast to their predecessors, paid all salaries in cash and did away with forced labor. They eliminated state monopolies and other restrictions on trade and encouraged European investors to purchase agricultural land. They also tamed the wilderness by cutting roads and railways through the mountains and clearing large areas of forest to plant coffee, cinnamon, pepper, coconuts, and later rubber. In the 1870s the coffee crop was destroyed by a leaf disease, and tea was planted in its place—so successfully that Sri Lanka became one of the world's foremost producers of tea, which it still is to this day. The island's population soared to 2.4 million in 1871, 3 million in 1891, 4.1 million in 1911.

The forces of nationalism that transformed so many European colonies in the decades after World War I also took root here, and in 1947 Sri Lanka won its independence in a peaceful and democratic process. Its population has continued to grow. At the time of independence it was 6.6 million. Today it is about 22 million, more or less what it was 900 years ago.

The Fertile Crescent

The land between the Nile and the Tigris Rivers is the crossroads of continents. The old Silk Road from China passes through Iraq; from here roads branch off westward to Egypt and northwestward to Asia Minor and Europe. The old Incense Road heads southward from Jordan into Arabia. The sea routes from Sri Lanka, India and Indonesia come either to Iraq on the Persian Gulf or to the Sinai Peninsula on the Red Sea. Any traffic continuing on from the Red Sea to the Mediterranean must pass through either Israel or Egypt. Most traffic between the Mediterranean and the Persian Gulf passes through Syria and Iraq, along the valleys of the Euphrates and Tigris Rivers. For two million years human beings have been living in Indonesia as well as Africa, and for two million years we have had to pass through this small and fertile area on our way from one to the other. It is no wonder that great religions have originated here, and that many wars have been fought for control of this small amount of territory.

Ancient Babylonia was a very wealthy and populous place. A sophisticated system of pools, dams, dikes, and canals supported intensive agriculture there five thousand years ago, and according to records of the Baghdad caliphates, Babylonia once had as many as 20 million inhabitants.

Baghdad was a wealthy city during the time of the caliphs. In the eighth century A.D., it was the capital of a Muslim empire that stretched from the Atlantic Ocean to the Indus Valley. Before its destruction by the Mongols in 1258 A.D., Baghdad occupied at least thirty square miles and had close to two million people.

For ancient Israel we have records from the Bible. Exodus 38:26 reports a census of 603,550 men over twenty years of age, implying a total population of about 2.5 million in the time of Joshua, about 1491 B.C. Numbers 26 reports a similar number, broken down in detail for each tribe. 2 Samuel 24:9 reports a census of the Israelites during the time of David, about 1017 B.C. In Israel there were 800,000 men of fighting age, and in Judah 500,000, implying a total population of about 5 million. 2 Chronicles 17 gives 1,160,000 men in the army of Judah at the time of Jehosephat, implying close to 5 million inhabitants of Judah alone in about 885 B.C., and perhaps 10 million Israelites in all. Present-day Israel does not have as many.

The Great Pyramid at Gizeh in Egypt was built in about 2680 B.C. by an already wealthy and populous civilization. In about 1300 B.C. the Egyptians dug a canal through from the Nile River to the Red Sea, for the same reason the French dug the Suez Canal in modern times: to accommodate the enormous amount of trade that passes in both directions between the Indian Ocean and the Mediterranean Sea. The population of Egypt during the greatest days of its ancient empire is not known, but it appears that at the time of the Persian Conquest in 525 B.C., Egypt had 20 to 25 million people. Some other estimates for Egypt are as follows:

525 B.C.	20–25 million
50	12–13 million
75 A.D.	7½ million
541	30 million
641	25 million
1798	2½ million
1907	11 million
1950	20 million
1990	52 million
2023	112 million

It is apparent that the Egyptian population has decreased as often as it has increased. Foreign invaders—the Persians in 525 B.C., the Macedonians in 332 B.C., the Romans in 50 B.C., the Arabs in 651, the Turks in 1517—usually killed a great many, as did epidemics of plague in the sixth and seventh centuries, and intermittently from 1010 to 1735. Egypt once prospered in the Roman Empire. Two recent centuries of French and British domination saw Egyptian numbers grow once more, nor has independence, achieved in 1937, slowed the rise.

Japan

The land of the rising sun is connected to the rest of the world by maritime trade routes—to the west, through Korea to China; to the south, through the Ryukyu Islands to the East Indies; to the north, through Sakhalin Island to southern Siberia; and to the northeast, through the Kuril Islands to the Kamchatka Peninsula and northern Siberia. Therefore, during most periods of its history Japanese ships have sailed the high seas in large numbers.

There have been some exceptions. After a military defeat in 663 by a combined Korean and Chinese fleet, Japan withdrew from world commerce for two centuries. The Japanese were also then at war with the Ainu, who at that time occupied northern Honshu. History shows that Japan's population was then declining.

The following population estimates are based on the recorded numbers of allotted rice fields:

610 A.D.	5,000,000
823	3,694,000
900	3,762,000
1050	4,417,000
1300	9,750,000

The population growth of the thirteenth and fourteenth centuries coincided with the flourishing of world trade fostered by the Mongol Empire. By the end of the fifteenth century Japan had broken up into several hundred semi-independent domains, intermittently at war with one another and only nominally ruled by the emperor. This state of affairs lasted until Lord Tokugawa Ieyasu reunited the country militarily in 1600, making himself Shogun. Economic and social life, meanwhile, were little affected. Buddhist monks maintained five universities, the smallest of them larger than the Oxford or Cambridge of the time. Agriculture flourished. Craftsmen prospered, notably papermakers, silk-spinners, weavers, swordsmiths, and later gunsmiths. And foreign trade continued unabated, while the population of Japan increased to about 18 million.

Direct contact with Europeans came in 1543 when a Portuguese ship was driven into local waters by a typhoon. Soon the Portuguese were sending regular fleets carrying both traders and Jesuit missionaries.

The new Shogun at first encouraged both foreign commerce and Japan's own merchant marine. The Japanese established trading settlements in Thailand, Annan, Cochin-China, Java, and the Philippines. During the first fifteen years of the seventeenth century Portuguese, Spanish, English, Dutch, and Chinese continued to travel and live in Japan. But Ieyasu's contacts with the Spanish in the Philippines and the Dutch in Java and Taiwan soon convinced him that missionary activity and European trade would lead to European conquest. Therefore, beginning in 1615, Japanese rulers made some momentous decisions which made Japan's future different from that of all her neighbors, with profound effects on the course of world history: they first banned Christianity; they deported all Spaniards in 1624; they expelled the Portuguese in 1638; they decreed that no Japanese could travel oversea, and that no Japanese who were overseas could return home. Limited contact remained with the outside world: Dutch and Chinese ships were permitted to come to Nagasaki, and traders continued to be received from Korea and the Ryukyus. At first all foreign books were banned, but in 1720 the restriction was lifted against all but religious works. A small group of scholars then took up the study of western medicine and science, but for two hundred years outside influence was extremely limited. Handicrafts continued to thrive, and agriculture remained the main occupation of most Japanese.

One other unprecedented and, for us, extremely significant decision was made by the Japanese people: to abandon gun-making, at which they had already become adept, and to which they probably owed some of their initial success in keeping away outsiders. Again, this brings one to question the inevitability of what is nowadays called progress. In the seventeenth century the Japanese actually abandoned guns and artillery in favor of bows and arrows and swords and spears, and they did not take up firearms again until forced to defend themselves against colonial powers two hundred years later. No law was ever passed forbidding the manufacture or possession of guns, nevertheless the people disliked them, and the demand for them virtually disappeared.

The peace and isolation of the Tokugawa period also brought population stability. The first census of the Tokugawa shogunate was taken in 1726, and counts were taken periodically thereafter until 1852.

1726   26.5 million
1852   27.2 million

All population counts during this time were between 24.9 and 27.2 million.

During the early 1800s ships from Russia, England, France, and the United States made repeated unsuccessful attempts to reestablish trade with Japan. The Japanese neither needed their goods nor trusted their motives, especially as these were the very powers that were already infringing on the sovereignty of neighboring China. Finally, a squadron of America warships commanded by Matthew C. Perry arrived in July 1853, and the shogunate, without modern weapons, could not protect Japan against them. Trade relations were forced open, as a series of treaties were signed granting American and European merchants most favorable customs rates and privileges of extraterritoriality (exemption from local laws)—privileges they would not consider dropping until Japanese legal institutions were in line with those of Europe.

The shogunate, which had been unable to protect Japan against the foreigners, was overthrown in favor of the Emperor Meiji. Meiji was a fifteen-year-old boy who was heir to a throne that had not exercised any real power for seven centuries. In his person now lay Japan's hopes for defending its integrity against fleets of iron steamships carrying 64-pound cannon. Isolation was no longer possible. Japan, if it was to keep its independence, would have to industrialize.

Meiji's government made haste to abolish the feudal domains and institute private land ownership. Internal and external commerce increased very rapidly, while printed money came into circulation throughout the country. Agriculture and industry adopted new foreign technologies, as Japan funneled its resources toward building a strong, modern military. An 1889 constitution gave Japan a European type of government, after which new foreign treaties abolished extraterritoriality. In 1894 Japan went to war against China, acquiring Formosa, the Pescadores Islands, and all the rights within China enjoyed by the European powers. The Ryukyus had already been annexed in 1879. In 1904 Japan went to war against Russia, acquiring Korea and the southern half of Sakhalin Island. Japan established a presence in Manchuria in 1904, and formally annexed it in 1932. Micronesia was occupied in 1914.

Japan was beating the Europeans at their own game—but at a price. Changes in agriculture made many farm workers unemployed. Household industries were destroyed. Household production of thread and textiles vanished with the importation of cheap factory-made cloth. Imported kerosene replaced locally produced wax-tree and rapeseed oil. Paper made in factories replaced handmade paper. As the populace moved gradually from farms to factories, cities grew in number and size. A new system of public education was opened to both men and women of all social groups. Japan whole-heartedly and rapidly transformed itself from a feudal agricultural society into a capitalistic industrial society with a growing empire. Its population soared:

1875	35.5 million
1895	42 million
1915	53.5 million
1940	73.1 million

Japan succeeded in this world by becoming colonizer rather than colonized and, like the European powers, was able to industrialize on a scale large enough, with enough local control, to provide employment for all those who were thrown out of work in the process. The Japanese empire reached its greatest extent during World War II, with the occupation of most of Southeast Asia, the Philippines, and large parts of China.

The ultimate price for beating the Europeans and Americans at their own game was the atom bomb. Dropped by the United States on Hiroshima on August 6, and on Nagasaki on August 9, 1945, it destroyed the Japanese empire. But it did not destroy Japan as a strong commercial power. While the United States became the world's largest debtor nation, Japan became the world's largest creditor. Until recently, Japanese corporations usurped more of the world's resources than those of any other country. It is Japanese lumber companies, Japanese pulp and paper mills, and Japanese whalers who have been assailed by environmentalists all over the world. In recent years China has become even a larger creditor. But Japan and China are not to blame. They are the lenders to nations in the present world, and to a large extent it is Japan and China that keep the world economy going. There are no nations, and precious few environmentalists, who want the world economy to collapse.

Japan's population rose until 1990, then stabilized:

1950	83.2 million
1960	94.3 million
1970	103.7 million
1980	116.8 million
1990	123.9 million
2023	123.3 million; 840 people per square mile

Mongolia and Turkestan

Central Asia is a vast territory of grassy plains, high plateaus, tall mountains, beautiful blue lakes, forested hills, and fertile valleys, stretching over three thousand miles from the shores of the Caspian Sea to the borders of Manchuria. In ancient times, those who ruled these lands controlled all of the overland trade between China and the Mediterranean world, and between China and Russia and the Baltic Sea. The high plateau of the Pamirs is the place where caravans from China and caravans from the Roman Empire used to meet and do business.

Native to this part of the world are many related peoples whom outsiders have called Mongols, Turks, Tartars, Huns, and Scythians—skilled horsemen all, whose style of fighting inspired terror throughout Europe and Asia during thousands of years of intermittent warfare. The great irrigated civilizations of Central Asia depended on keeping open the trade routes, and the control of these roads was the object of a great many terrible wars.

In studying Central Asia one is confronted with one of the greatest myths of our culture, because this huge land divides—or seemingly so—the white race from the yellow. I suspect this is one of the reasons the history of this part of the world has been so little studied. Myths that are so deeply rooted do not appreciate the light of day. Scholars of Central Asia have wasted a lot of ink trying to prove that the ancient Scythians, for example, spoke an Indo-European tongue and were therefore white-skinned and unrelated to the Huns. This despite the fact that no records of the Scythian language exist and that contemporary descriptions of that people are quite varied.

Let me get right to the point: this world, as disturbing a notion as it may be, is and always has been a melting pot. That this cuts right to the heart of the environmental dilemma is not obvious, but will become clearer the deeper we dig into the human condition. What divides civilized humanity from nature is also what divides human beings from one another. I am not merely repeating the politically correct dogma that race does not matter; I am saying rather that it does not exist. Anthropologists are finally coming to realize this, but society at large still takes race for granted. Two experiences brought home the falseness of the idea to me most clearly. The first was contemplation of the fact that going back only fifteen generations (to about 1600 A.D.) each one of us has over 32,000 theoretical ancestors and knows anything at all about at most one or two of them.[183] The second was meeting black Jews whose noses and cheekbones resembled mine so closely that it was evident we were cousins. It is folly, I realized then, to let the differing color of our skin blind us to our common ancestry.

And in Central Asia, people from East and West have met and mingled for so long as to make a mockery of the terms "Caucasian" and "Mongoloid."

Two thousand years ago the nomadic Huns and the Chinese were vying for control of the Pamirs, while three other great empires—that of the Kushāns, the Parthians, and the Romans—ruled the more westerly segments of the Silk Road. The decline of both Rome and Han Dynasty China after the second century A.D. disrupted world trade and left the empires of Asia open to conquest. Transoxiana—the fertile valley of the Zeravshan River west of the Pamirs—passed from the Kushāns to the Sasanians and then, in the 6th century, to the Turks. They in turn were conquered by the Arabs in the eighth century. The native Sāmānids ruled here in the ninth and tenth centuries, but they became subjects of the Seljuk Turks, and then the Khwarezmian Turks, in the eleventh and twelfth centuries.

East of the Pamirs, the basin of the Tarim River, also known as Kashgaria, was fought over by Turks and Chinese for centuries, changing overlords many times. In the ninth, tenth, and eleventh centuries the Uighurs ruled here, until they were conquered by the Kara Khitai in the twelfth century.

In the early 1200s Genghis Khan, after uniting the Mongol tribes, led his mounted soldiers from their homeland near beautiful Lake Baikal on a campaign of conquest that would eventually subdue not only all of Central Asia but a very large part of the Eurasian continent. His armies swept through northern China and the Asian kingdoms of Hsi-Hsia, Kara Khitai and Khwarezm, razing cities and spreading death and destruction wherever they went. From there they advanced through Persia into Georgia, crossed the Caucasus and invaded the Crimea, the Ukraine, and Russia. Genghis' successors waged war in Korea, Anatolia (modern Turkey), Syria, and the steppes of western Siberia. Tibet was conquered in 1239. The Mongols destroyed Moscow in 1237, and conquered Kiev in 1240. Mongol armies swept through Lithuania, Poland, and Silesia. To the south, other armies conquered eastern Hungary and raided into northern Italy before the death of their khan, Ogatai, called them home. Ogatai's successors, Mangu and Kublai, conquered Mesopotamia and completed the conquest of China.

In 1280 the Mongol Empire included all of China and Mongolia north to the Sea of Okhotsk and Lake Baikal, Russia west to the Dnieper and north to Moscow, and all of Tibet, Turkestan, Persia and Mesopotamia. At this point Kublai Khan devoted himself to reestablishing law and order throughout his vast empire. It is a remarkable fact that during the years of Mongol rule, world trade prospered as it had not for the previous thousand years. Within their territories the Mongols kept the peace and maintained public safety to a degree the world had not known for a very long time. Mongol rule meant free passage of traders between east and west, and it meant the toleration of all religions. Marco Polo sent glowing reports back to his native Venice.

It occurs to me that the Mongols resembled no one so much as the United States of America today, with their emphasis on free trade throughout the world, freedom of religion, tolerance of all nationalities within their borders, and utter ruthlessness toward their enemies.

The Mongols ruled southwestern Asia until 1335; China until 1388; Russia until 1480. The last reigning descendent of Genghis, Shanin Girai, Khan of the Crimea, was deposed by the Russians in 1783.

Tamerlane reunited all of western Asia under a Turkish empire in the late fourteenth century. At the end of the fifteenth century a new Mongol confederation encompassed the whole of northern Asia between the Ural Mountains and eastern Mongolia. The Uzbeks ruled all of Turkestan in the sixteenth century. And in 1552 the Khalkhas, under Altan Khan, reunited most of Mongolia.

But soon a change in the technology of warfare brought by the Europeans ended the domination of Central Asia by its native peoples. Beginning in the 17th century, Tsarist Russia conquered the western and northern parts of historical Mongolia, while Manchu China asserted sovereignty over the rest of Mongolia. Today western Turkestan is divided among Afghanistan, Iran, and several former Soviet republics; eastern Turkestan (Xinjiang) and Inner Mongolia are autonomous regions within China; the area around Lake Baikal belongs to Siberia; and Outer Mongolia, after seventy years of Soviet domination, is again an independent state.

It is commonly assumed that the Mongols, who once ruled a large part of the world, were nevertheless not very numerous. But this is almost certainly not true; there is ample evidence of a great depopulation before modern times. Sand blankets much of southern Mongolia today, but its many salt lakes once held fresh water. Both Inner and Outer Mongolia are covered with ancient ruined towns—at least twenty-six in the State of Mongolia alone. Khara-Khoto, capital of the Tangut kingdom of Hsi-Hsia, conquered by Genghis Khan in 1227, now lies ruined under the parched sands of the Gobi Desert. That city once boasted Buddhist temples, marketplaces, and libraries. The surrounding country was sown with grain and traversed by irrigation canals.

Karakorum, once the capital of the Mongol Empire, lies ruined in Outer Mongolia. In the thirteenth century it was one of the most important cities in the world, receiving emissaries from Popes and Kings of Europe. It had an Arab quarter, where markets were held, and a Chinese quarter, home to renowned artisans. Within its walls stood two mosques, a Christian church, and twelve temples dedicated to the gods of various peoples. Grain grew in the region, and sheep, goats, oxen, and horses were sold in the markets by nomadic tribes.

Both eastern and western Turkestan are now well into the process of desertification. Their growing populations are crowded into oases and dry grasslands, isolated from the rest of the country by empty deserts and salt lakes.

The city of Samarkand, on the Zeravshan River in Uzbekistan, stands at the junction of the main trade routes from India, Persia, Russia, and China. It is one of the oldest cities in the world. Once the capital of Soghdiana, it was ancient Persia's chief granary. A thousand years ago Samarkand consisted of a walled inner town with a citadel and four gates, and an outer town of beautiful suburbs covering forty-four square miles enclosed by a wall of earth twenty-seven miles long. Its irrigation system consisted of eight main canals lined with lead, and 680 sluices. Water was brought into almost every home. Each house had a garden with fruit trees, and the streets and public squares were shaded by cypresses and elms that were famous throughout Asia. There were two thousand places in its stone-paved streets and squares where iced water could be obtained for free from fountains or from copper or earthenware vessels. The city was famous for its paper manufacture, and in its markets were sold silk and cotton fabrics, and metal goods, all made in the surrounding countryside. At least 100,000 families lived in Samarkand before the Mongol conquest. Still a major cotton and silk center, Samarkand today has about 600,000 people, roughly the same as in ancient times.

Among the other cities in pre-Mongol Turkestan, Bukhara was known for its mosques, colleges and castles, its carpet industry, and its meats and melons, as well as for its overcrowded conditions and its bad water. Tashkent was the center of the arms industry, and the hub of trade for furs and other products shipped down the Syr Darya from the steppes of Russia. Herat had 12,000 shops, 6,000 hot baths, 659 colleges, and a population of over 400,000. Nishapur, the birthplace of poet Omar Khayyam, was famous for its healthy climate, its rose gardens, and its manufactured goods.

Bukhara today has about 280,000 residents. The surrounding countryside is desert, under whose sands are traces of ancient canals that once watered productive farms. Herat is the economic center of western Afghanistan, and the focus of one of that country's most densely peopled and fertile agricultural areas. Its population is about 600,000, about twice what it had in antiquity. Nishapur, situated in a wide and fertile plain in northern Iran, has about 260,000 residents. Tashkent is the capital of Uzbekistan and a thriving cotton-growing and textile-manufacturing center. With 3 million people, it is the largest city and the main economic and cultural center of present-day Central Asia.

Siberia

Before the Russian conquest began in 1580, Siberia—a land as big and beautiful as the continental United States—was inhabited in its southern grasslands by nomadic horse and sheep herders; in its great northern forests were hunters and gatherers who dwelled in tipis and kept reindeer for pack animals and milk; on its northern tundra lived reindeer hunters, reindeer herders, and fishers; on the shores of the Arctic Ocean and the Sea of Okhotsk dwelled hunters of whales, walruses, and seals.

All of these people traded their local produce for that of their neighbors. Furs and marine products from Siberia were exchanged through Mongol intermediaries for products of the agricultural settlements of China and Central Asia.

The Russian armies that marched across the Asian continent from the Ural Mountains reached the eastern sea in only fifty-six years, missionaries and merchants following close on their heels. The Kamchatka Peninsula was annexed in 1699. Continuing eastward, Vitus Bering followed the Inuit trade route to Alaska in 1741. Bering and his men took sea otter furs back across the Strait now named for him, and soon a rich Russian fur trade from America replaced the native trade that had been going on for thousands of years, probably since the ice last retreated. The first Russian settlement in Alaska was established near present-day Kodiak in 1784. By 1867, when the United States bought Alaska from Russia, the sea otters there had been hunted nearly to extinction.

Present-day Siberia has about 37 million inhabitants, most of them Russian colonists or their descendants, many of them offspring of political prisoners or criminals, exiled to Siberia in a steady stream since the beginning of the seventeenth century. A large number of the native people died during the wars of conquest. In large part, as Piotr Kropotkin said, the Russian conquest was one long hunting expedition that destroyed the source of livelihood for all of the native tribes who depended on hunting, and made life very difficult for the rest.

The Trans-Siberian Railway, completed in 1905, brought in a flood of new colonists. Soon forced labor was being used to mine coal, to produce iron and steel, and to mine nonferrous and precious metals. During World War II, Siberia became the industrial base of the Soviet war effort. In the 1950s and 1960s large oil and gas fields were developed in western Siberia; a series of giant hydroelectric dams tamed the Angara, Yenisey and Ob Rivers; aluminum smelters and pulp mills were built. Most food is imported, as agriculture is neither appropriate nor possible anywhere except the milder areas of the southwest and the Far East.

Eastern Siberia, between the Yenisey River and the Pacific Coast Ranges, has about 8 million people (about five per square mile). About a quarter million are native Buryats, and another quarter million are Sakha. In the north are remnants of many different tribes, who now number about 150,000 altogether. The Russian Far East, hugging the Pacific coast, has 8 million people (about three per square mile), of whom only three percent are native.

Europe

Europe today is among the most crowded places on earth—no surprise here, because Europe has colonized the entire rest of the world and made all of the earth's resources its own. Italy has 518 people per square mile. The United Kingdom has 725. Germany has 619. Belgium has 1,000. The Netherlands has 1,353.

Population figures for ancient Europe are difficult to come by—so much so, that it is not easy to resolve the controversy raised in the eighteenth century by Montesquieu and Hume. Twenty-first-century Europe is certainly more crowded than ancient Europe was, but by how much?

Clues are sparse. Plato, early in the *Symposium*, remarks that more than 30,000 people attended the theater of Dionysus in Athens.[184] The amphitheater in Paris during the Roman Empire is known to have held 15,000 or more spectators. The walls of ancient Rome could have contained up to two million inhabitants if they lived densely enough, but the records do not tell us. The Roman census of 14 A.D. has been variously interpreted by historians to imply a population of Italy of anywhere between 6 and 14 million. We are told that 300 African cities once gave their allegiance to Carthage, and that in Roman times, Italy had 1,197 cities, Gaul 1,200 and Spain 360. But we don't know how big all those cities were.

The Roman Empire in about 50 A.D. counted 6,945,000 adult male citizens, but citizens were considerably outnumbered by free non-citizens and slaves. Edward Gibbon, adding in women and children, estimated that the total population of the Empire (all of Europe plus Turkey) was 120 million.[185]

The situation for medieval Europe is not much clearer. For the plague epidemics of the 14th century, there are mortality reports from all over Europe, and the figures are so large that many historians don't believe them. Boccaccio wrote that 100,000 people died of plague in Florence between March and July of 1348. The chronicler Agnolo reported that 52,000 people died in the banking center of Siena, thirty miles to the south of Florence. According to city archives, 57,343 people died in Norwich, England in 1349. 50,000 people were reported to have been buried in one graveyard alone in London. Enrollment at Oxford was said to have declined from 30,000 to 6,000 between 1348 and 1350. 124,434 Franciscan friars were said to have died in Germany. 50,000 people were reported to have died in Marseille in 1348.

Hume's view that ancient nations were not very populous prevailed over Montesquieu's opposite opinion, and most modern historians regard all those reports of plague deaths as wild exaggerations. Medieval London is supposed to have had perhaps 50,000 residents altogether, for example, and Norwich 10,000. I find it easier to believe, however, that the numbers reported by contemporary observers were more or less accurate, and that a sparsely populated Europe would not have been dying in enormous numbers of typhoid fever, dysentery, diphtheria, and starvation, even before they began dying of the plague.

In the time of Tertullian, about 200 A.D., the population of Europe was probably quite large. The political stability of Rome, China, and Central Asia meant that world trade had prospered in peace for 200 years. But with the decline of Rome, the fragmentation of its empire, and warfare among its subjects, the European population must have begun a long decline. Great irrigation systems and aqueducts fell into disrepair. Trade with the east slowed to a trickle until the Middle Ages. Then, with the rise of Venice and the other commercial centers of Italy, and Mongol rule in the east, commerce prospered once more and, undoubtedly, populations grew. By the fourteenth century Europe was again overcrowded, and suffering from widespread famine and disease. The epidemic of plague that apparently began during the 1330s in China spread throughout the world—even, some historians speculate, to North America. A large fraction of Europe's population perished from the "Black Death" and from further epidemics of plague during the centuries of worldwide turmoil that followed the disintegration of both the Mongol and Byzantine Empires.

Eventually, the European conquest of the world would bring riches into Europe on a scale hitherto unimagined. The growth of that continent's population since the 1700s is well-documented:

1750	140 million
1800	190 million
1850	265 million
1900	400 million
1950	530 million
2000	727 million
2023	742 million

Ireland

Europe does hold one special surprise for us: Catholic Ireland is the only country in the entire world that has had a stable population for the past one hundred and eighty years. In fact, the population of the Republic of Ireland today is about 60 percent of what it was in 1841.

Until the 1840s Ireland, like many other colonized countries, suffered from both poverty and a steadily increasing population. In the 1600s England, regarding Ireland as conquered territory, confiscated most Irish land and parceled it out among soldiers and creditors of the Commonwealth. By 1703 Irish Roman Catholics owned less than ten percent of the land in their own country. Ireland had become a country of large estates, often owned by absentee landlords and worked by tenant farmers. This situation continued until the great potato famine.

Between 1845 and 1849, while large amounts of grain and meat continued to be exported from Ireland, perhaps one million Irish died of starvation and another million or more emigrated to foreign lands. The census of 1841 counted 8,200,000 people; that of 1851, only 6,514,000. The following decades witnessed a large-scale shift from food crops to cattle, as landlords enlarged their holdings and evicted their tenants. Between 1855 and 1866 some one million Irish people were replaced on the land by one million head of livestock. The population decline, begun during the famine, continued.

At the same time, the Irish began to organize to throw off British rule and take back their land. An Irish Land Act, passed by Parliament in 1870, protected tenants against eviction. Between 1881 and 1903 a series of new acts enabled tenant farmers to gradually obtain possession of their farms.

The Wyndham Act of 1903 provided government loans at reduced rates to tenants who wanted to buy their farms, and bonuses to landlords willing to sell. By 1922, when Ireland became independent, Irish farmers owned two-thirds of the land, and the rest was confiscated by law in 1923 and given to them.

Ireland was in the unique position of being a once-colonized nation whose people now owned their land free and clear, for the most part without encumbering debts. The subsequent population history, by modern standards, is remarkable:

1921	3,096,000
1931	2,933,000
1941	2,993,000
1951	2,961,000
1961	2,818,000
1971	2,978,000
1981	3,443,000
1991	3,500,000
2001	3,829,000
2011	4,545,000
2021	4,987,000

Today 97 percent of Irish farmers own their land, and two-thirds of all farms are under 50 acres in size.

In Northern Ireland it has been otherwise. There were two different forms of colonization in the seventeenth century. In most of Ireland, confiscated land was rented back to the Irish who remained in residence as tenant farmers. In the northeast, however, confiscated lands were settled by large numbers of English and Scottish immigrants. In this part of Ireland linen, ship-building and a host of smaller industries grew and prospered. During the late 19th century, while the population of the rest of Ireland was declining, the six northeastern counties of Ulster continued to grow in numbers. Northern Ireland has remained under British rule, and is much more crowded, with 346 people per square mile, than the independent Irish Republic, with 190.

Africa

Let us rest a moment on our journey, to take stock of where we have been and what we have learned. We have seen that ancient China, India, the Middle East, and Europe were densely populated—and that Java, Japan, and Siberia were not. We have learned that Israel and Sri Lanka had about as many people, or more, than today. And we have seen that all of the world's countries are now rapidly expanding in population except, surprisingly, for Ireland.

Connections among trade, technology, and human numbers have become apparent. Prevailing theories of population growth that attempt to analyze it as a personal rather than an economic matter are so inadequate to explain the facts that, as we have seen, historians have usually distorted the facts to fit their notions.

In Africa, the origin of all of us, and in America, the place of my own birth, recent history has been so violent and disruptive that mythology has all but obliterated the truth. If the world can again be made whole, it must certainly reintegrate these alienated continents into the one human story. Nowhere else has as much to tell us about the meaning of wilderness, and the parts of ourselves that technological man and woman have exiled.

Let us continue our trip.

We have already visited the shores of eastern Africa aboard the ships of Javanese traders who sailed across the Indian Ocean to Madagascar and the vicinity of present-day Mozambique. The Swahili language and culture bear elements of the three great civilizations—African, Arab, and Indonesian—that mixed and traded in this region of the world for so many centuries.

From Mozambique traders used to travel by sea, south toward the Cape of Good Hope, and north toward the Red Sea. All along the coast of East Africa were prosperous island and port cities: Kilwa, Mafia, Zanzibar, Pemba, Mombasa, Brava, Berbera. Many were centers not just for coastal commerce, but for overland trade with inland peoples, who brought cattle, ivory, iron wares, spices, and forest products to market. Further inland are still traces of an ancient system of roads running north and south on the east side of the Great Lakes in Kenya and Tanzania, and in the highland regions are stone ruins and irrigation works that supported a large population in ancient times. At Engaruka near the Kenya/Tanzania border, about 300 miles from the coast, is a ruined city whose 6,800 houses once must have had over 40,000 residents. Still further inland, in Uganda, lie the ruins of some of the largest earthworks in the world.

From the southeastern coast of Africa, the Zambezi, the Save, and the Limpopo Rivers have long carried travelers inland by boat. These rivers, and the highway that still leads inland from the coastal city of Sofala, once entered the Kingdom of Monomotapa, a civilization of which only ruins are left. The entire area of Zimbabwe, Zambia, Botswana, western Mozambique, the northern Transvaal, and the southern fringe of the Democratic Republic of Congo are covered with thousands of ancient stone ruins, the remains of large irrigation works, and sixty or seventy thousand ancient mine workings. Gold, copper, iron, and tin were extracted here for thousands of years. The Great Zimbabwe, the largest of the ruined cities, is a group of stone structures and walls covering nearly four square miles. At their center is a huge elliptical building, possibly a temple, decorated with soapstone birds and other sculptures. According to some, King Solomon's mines were here, and the Biblical "gold of Ofir" was imported from this area. Possibly some of it was. But the "Gold Coast" and other parts of Africa have also been sources of gold for millennia. "Ofir" is likely the same word as "Afer," meaning simply "Africa."

On the northeastern coast of Africa, tradition has it that the kingdom of Ethiopia was founded in the tenth century B.C. by King Solomon's first son Menelik. Ethiopia has prospered through the millennia from the Red Sea trade which passes by its coast, and from the overland trade which passes from the Red Sea through Ethiopia to the River Nile.

Trade also continued westward from the coast far across the middle of Africa. East coast cowries were used as money in the Uele valley of the northern Congo basin. And trade continued 1500 miles downriver to the west coast, where the Congo empties into the Atlantic Ocean.

The Kingdom of Congo, ruling the plateau south of the lower Congo River, was the most powerful of the Central Africa coastal states when the Portuguese arrived. Portuguese traveler Duarte Lopez described its palatial capital city in 1591:

The Royal City of the Kingdom of Congo ... is called San Salvador, and was formerly known as Banza in the language of the country, which generally means Court, where the King or governor resides, and is situated 150 miles from the sea, on a large and high mountain, almost entirely of rocks, in which nevertheless is a seam of ironstone, of which large houses are built. This mountain has on its summit a plain, entirely cultivated, and furnished with hamlets and villages, extending for about 10 miles in circumference, where more than 100,000 persons are located. The land is fruitful, and the air healthy, and fresh, and pure, and there are springs of moderately good water, never injurious.

... It is impossible to determine the size of this city, the whole country beyond the two boundaries of the walls being covered with houses and palaces, each noble having his houses and lands enclosed like a town. The Portuguese occupy a circuit of nearly a mile, and other buildings, such as the royal houses, about the same extent. The walls are of great thickness, the gates are not shut at night, nor even are sentinels posted ...

The whole plain is fruitful and cultivated, having verdant meadows and large trees, and produces grain of various kinds ... the variety of trees is so great as to produce sufficient fruit to supply nearly the whole population with food.[186]

When David Livingstone journeyed through Central Africa in the mid-nineteenth century he commented on the peace and security that reigned over great expanses of the interior. It seems that life for the traveler in middle Africa was a good deal safer at that time than it generally was in Europe.

When Henry Stanley explored the Congo region in the 1870s, he found trade flourishing all along the coast and for over a thousand miles inland up the Congo River. Among themselves the Congolese exchanged fish, pottery, baskets and cowries for copper, iron, farm implements and weapons. They exported red wood, wax, tin, iron, rubber, ivory and ground nuts in exchange for linen, cloth, copper kettles, guns and other merchandise from Europe. Much of the ivory and forest products were procured by forest dwellers—hunters and gatherers who bartered with their agricultural and pastoral neighbors of the savannah.

Stanley estimated that the country known today as the Democratic Republic of Congo had a population of 43 million in the 1870s.

From the mouth of the Congo River, the Atlantic coastal trade goes south toward Angola and South Africa, and north toward the ancient kingdoms of Benin, Ghana, and Mauretania. Traveling further north, one reaches Spain, France, and the British Isles. Turning eastward through the Strait of Gibraltar, one enters the Mediterranean Sea and travels along the north coast of Africa toward Egypt, following a trade route that has been controlled by a succession of great sea-going peoples since very ancient times: the Phoenicians; the Carthaginians; the Romans; the Arabs; the Ottomans.

The Sahara and Sahel regions, once better watered than today, were crisscrossed by roads, along which caravans of camels moved in a steady stream between east and west, and between north and south. The nations of this region—Ghana, Mali, Songhay, Kanem, Darfu—prospered from international trade, in particular the trade in gold, whose source was the forest belt to the south.

Ghana lay in the west between the salt deposits of the north and the gold deposits to the south, its empire built on the exchange between the two. Its capital, in the 11th century, had two cities six miles apart, while the space between them was also covered with houses. One city was the king's residence, fortified and enclosed within a wall. The other, containing a dozen mosques, was a merchant city of the Muslims; what are probably its ruins, covering over a square mile, have been discovered buried underneath the sand and scrub of the Sahel north of the upper Niger.

The cities of Timbuktu and Djenne were famous throughout the medieval Islamic world for their commerce and learning. Shoes and textiles were made here; gold, iron, and copper smiths practiced their professions. This was part of the empire of Mali, which at its greatest extent in the fourteenth century ruled a territory as large as Western Europe. In 1400, annual caravans crossing the Sahara by way of the Hoggar Mountains—only one of half a dozen well-used routes—numbered 12,000 camels. And to the south, caravans of donkeys carried salt, cloth dyed with indigo, and copper goods into the forests, and returned laden with gold and kola nuts.

Mali's eastern neighbor Songhay came to prominence in the late fifteenth century. Timbuktu and Djenne came under its rule, and for the next century Songhay ruled an empire as large as Mali's had once been.

To the east of Songhay lay the kingdom of Kanem, and its successor Bornu, around the region of Lake Chad. To the east of Bornu was the kingdom of Darfur. Their ruins are today buried underneath sand and sparse grassland. Among the dry and empty plains and hills of Darfur lies Jebel Uri, one of the largest ruined cities in Africa. Twenty miles to the south are the ruins of a smaller city that are notable for containing the remains of a Christian monastery.

All of these civilizations of the savannah were built on urban trade and a pastoral-agricultural economy. Ibn Battuta, traveling in Mali in the 14th century, wrote of the complete security of the country, and the absence of robbery and violence. Mungo Park, visiting Segu on the Niger River four centuries later, wrote, "The view of this extensive city; the numerous canoes; the crowded population; and the cultivated state of the surrounding country, found altogether a prospect of civilization and magnificence which I little expected to find in the bosom of Africa."[187]

To the south of the savannah lay the forest belt that was the source of gold, kola nuts, and pepper. Benin, Biafra and other kingdoms and city-states were located along the coast in this belt between the Senegal and Congo River estuaries. Early Portuguese explorers and missionaries were well-received there and invariably reported finding metal-working civilizations that built cities and kept peace and order within their domains.

Old Africa no longer exists. It was so thoroughly obliterated by four centuries of the brutal European slave trade that it has been almost entirely forgotten.

Slavery was not new to Africa. It was an institution in many African kingdoms, and the Arabs had been taking slaves out of Africa since the seventh century. But the Europeans came with guns, and their demand for black slaves was insatiable. In 1444 the first shipment of Africans was taken from Senegal to Lisbon. By the early 1500s, in some parts of Portugal, the number of African slaves was said to be larger than the number of native Portuguese. Later the demand for slaves in the Caribbean, in

Brazil, and in North America was much larger. Nobody knows just how many Africans were taken from all the coasts of Africa to North and South America between the fifteenth and nineteenth centuries. The estimates I have seen vary from 15 to 50 million. The number who died en route and during the slave hunts may have been equally large.

The peace and order of old Africa, both east and west, were permanently destroyed. Prosperous cities fell into ruin. Irrigation systems were abandoned. Whole countries were depopulated, their social systems and their economies devastated. Those who remained often kept their freedom by selling their countrymen into slavery. Parents sold their children, or children their parents, to European merchants who branded them like cattle with a hot iron.

In the nineteenth century visitors to coastal Africa brought back stories of a land whose people had never progressed past the Stone Age, a sparsely populated, dangerous land whose primitive tribes had little regard for life and were permanently at war with one another. This was "Deep Dark Africa." And so was born the myth of Stone Age Man. In Africa, it was thought, could still be seen the primitive state out of which we had all arisen thousands of years ago. In Stone Age times, life was brutal and short, warfare, disease, and starvation common—it was one long struggle for survival from birth until death.

But it was never so. There are no primitive societies on the earth today. We have all been here for more than two million years—those of us who use iron and steel and get our food from factory farms, and those of us who use stone and wood and gather our food from wild trees and shrubs. We all have old, complex cultures, social organizations, and economies, that work more or less well, and all of us are cousins to one another who live in the same world, more or less dependent on one another for our mutual welfare. Starvation has been the exception in this world, never the rule. Warfare, throughout most of the human story, has been relatively mild. I think it is fair to say, looking at the earth as a whole, that there has rarely been more warfare, famine, and disease in the world than there is today.

The Africa of nineteenth-century European repute was a creation of four hundred years of European brutality. Even in the nineteenth century those parts of interior Africa that had escaped the slave trade continued in relative peace and prosperity, as explorers like Livingstone and Stanley reported. The interior of the Congo basin was such a region. A vast area of grassy plains, wooded savannahs, and rainforests, the Congo basin teemed with human and animal life. Twenty thousand kinds of mammals lived here; one thousand kinds of fish swam in the rivers; reptiles, snakes, amphibians, birds, and insects were everywhere. And rubber trees grew in the forests.

From November 15, 1884 to February 26, 1885, representatives of fourteen European nations plus the United States met in Berlin and agreed to give possession of the Congo basin to King Leopold of Belgium. Under his rule, the Congo was going to help supply a growing worldwide demand for rubber—for shoes, raincoats, bicycle tires and, soon, for automobile tires.

The horrendous story of the holocaust which followed is detailed by E. D. Morel in his book, *Red Rubber*, and by Philippa Schuyler in *Who Killed the Congo?* In 1891 and 1892 Leopold issued edicts declaring all of the produce of the land to be the property of the government, and forbidding the Congolese to collect ivory and rubber except for the state. What Africans produced no longer belonged to them, and they could not sell it. It was taken from them by taxation: each village was required to deliver to the government quantities of ivory, rubber, and copal, and also to feed the black armies of the Belgians who enforced this tribute. The Congo became a gigantic property pillaged by slave labor to satisfy mainly an insatiable world demand for rubber.

Tribes were forced to trespass in each others' domains in order to get enough rubber trees to meet the tax. The penalties for non-compliance ranged from imprisonment to seizure of women and children to violence, torture, and murder. Whole villages were burned. Men were captured and forced to work in chain gangs gathering rubber. Africans were required to provide their masters not only with food but with women. They were denied free access to the land that provided them with their food and necessities. Often, they were not allowed even time to cultivate their fields, repair their houses, hunt or fish, and their hunting weapons were confiscated. They starved, their houses crumbled to ruin, and day in and day out from sunrise to sunset they toiled to provide rubber for their masters. They were not even treated as well as slaves—just as an inexhaustible supply of expendable machines whose sole purpose was to gather rubber until they dropped dead.

To the west, the French copied Belgian methods, except that the French Congo was not ruled directly by the French government but by some forty concessionaire companies, who were granted ownership of all the products of native soil.

In twenty years, the Belgian Congo was utterly destroyed, its population reduced from 43 million to less than 9 million. The French Congo suffered similar devastation.

The numbers are so unbelievable that many modern historians refuse to accept that the Congo ever had so many people. But Henry Stanley's are the only written estimates that have come down to us, and it is apparent that the Congo basin was once so populous that Europeans treated Africans as an inexhaustible resource.

In 1930 the Belgian Congo still had less than 9 million inhabitants and had not yet begun to recover from the devastation. In 1990, it was still less populous than it had been 120 years previously.

And who is to blame for such a holocaust? I have asked myself, similarly, who is to blame for the holocaust that is engulfing the earth today? In 1909 Arthur Conan Doyle wrote that the events in the Congo

> ... cast a strange light upon the real value of those sonorous words Christianity and civilization. What are they really worth in practice when all the Christian and civilized nations of the earth can stand round, and either from petty jealousy or from absolute moral indifference can for many years on end see a helpless race, whose safety they have guaranteed, robbed, debauched, mutilated, and murdered, without raising a hand or in most cases even a voice to protect them? ...

The incredibly brutal treatment of the natives passed for some years because men could not bring themselves to believe that in this age of progress it was possible that such things could actually exist. All the cruelties of Alva in the Lowlands, all the tortures of the Inquisition, all the savagery of the Spanish to the Caribs are as child's play compared with the deeds of the Belgians in the Congo.[188]

One must be cautious, however, in assigning blame. Perhaps it is "this age of progress" itself that causes such things to exist. Perhaps the Belgians and the French were only its instruments. Fifteen nations got together and awarded the Congo to Belgium and France for the purpose of providing the world with rubber. The internal combustion engine was invented in 1885, the pneumatic rubber tire was perfected in 1888, and the automobile industry took off, fueled by the blood of African slaves. It is an industry which more than any other drove the industrial expansion of Western Europe and especially the United States. The demand for rubber tires was enormous and insatiable. Wild rubber trees could not possibly satisfy the demand. They could for a short while, but only if millions of human beings were pushed way beyond their limits, not just in the Congo basin, but in the Amazon as well. The holocaust in the Amazon basin has been documented in Lucien Bodard's *Green Hell*.

The pressure on the wild rubber trees, and on African and American slaves, did not let up until 1913. In that year, for the first time, the output of plantation rubber in Indonesia, Ceylon and Malaya surpassed the world output from wild rubber trees. After 1913, the expanding automobile industry no longer depended on an expanding supply of wild rubber, and the enslavement on two continents could finally stop.

But today a second holocaust is occurring in the Congo in service of another technology for which the demand is insatiable. This time it is to supply cobalt for all the batteries, and tantalum for all the capacitors and wave filters required by all the cell phones in the world. And unlike the situation with rubber, which was eventually planted in India and Southeast Asia, these minerals cannot be planted elsewhere in the world to relieve conditions in the Congo. This holocaust is even worse, because in addition to enslaving Congolese people, it is obliterating Congolese wildlife, polluting air and water, and turning wilderness into barren landscapes. In the last couple of decades millions of trees have been clearcut, lakes and rivers poisoned, 80 percent of the area's elephants killed, the okapi driven to endangered status, and the eastern lowland gorilla driven almost to extinction. The Mbuti and other hunter-gatherers are losing their land. One million men, women, and children are living in conditions of slavery, having lost all other means of livelihood except working in the mines for $1 to $2 a day. As many as 40,000 children, some as young as six, are digging cobalt and coltan by hand in large mines and small, guarded, and sometimes tortured, raped, and murdered, by soldiers or militias. This in a virtually lawless part of Africa, producing raw materials for mostly Chinese companies for incorporation in the cell phones that are demanded by a majority of the 8 billion people in every country in the world.[189]

Slaves carrying sacks of cobalt for the world's cell phones.

Who, after all, is to blame?

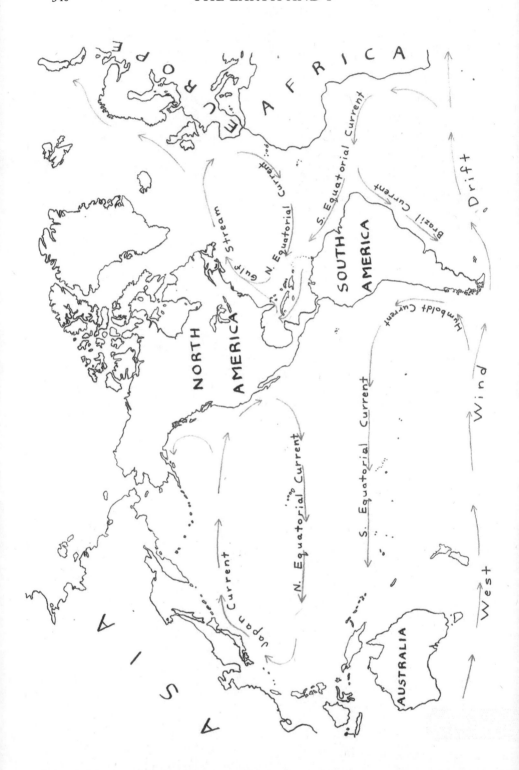

America

A river one hundred miles wide flows steadily from east to west across the Atlantic Ocean. Any boat or raft that enters this North Equatorial Current off the northwest coast of Africa will be carried by the current and blown by the trade winds, in two or three months' time, across the ocean to the Caribbean Sea. With sails the trip is reduced to a matter of weeks. It is a natural trade route, and it is the one Christopher Columbus followed in 1492. Another ocean river, the Gulf Stream, will carry a boat just as swiftly back from North America across the North Atlantic to Europe.

Farther north in the Atlantic is another route to America—from Norway to the Shetland Islands to the Faeroe Islands to Iceland to Greenland to Baffin Island to Newfoundland. On this, or a similar route from Scotland, one is never more than 150 miles from land, and one is never out of sight of land for more than a day or two. This is the route Leif Erikson took, about 1000 A.D.

In the southern Atlantic, the South Equatorial Current will carry a boat from southwest Africa swiftly to the coast of Brazil. Ferdinand Magellan sailed into this current in 1519.

Once arrived at South America, Magellan's ships sailed southward along the east coast, aided by the southward-flowing Brazil current. He then entered the Strait named for him at the southern end of the continent and sailed through to the Pacific Ocean. There his ships were carried to the northwest along the coast of Chile by the Humboldt Current, then westward across the Pacific by, again, the South Equatorial Current.

The South Equatorial Current flows westward from northern Chile and Perú, past all the numerous South Pacific Islands until it peters out by the time it gets to Australia and New Guinea. The North Equatorial Current flows westward unabated from Central America to the Philippines; the Hawaiian Islands are about midway on this trade route. The Japan Current will carry boats steadily and surely back from Japan across the North Pacific to the west coast of North America; or, voyagers from Asia to the Hawaiian Islands can ride the Japan Current to a point north of Hawai'i and then sail south.

Landlubbers wanting to sail from Japan to California without ever losing sight of land can just about do it, sailing northeast along the Kuril Islands, hugging the coastline of the Kamchatka Peninsula, and then island-hopping along the Aleutians to Alaska before turning southeast and sailing down along the coast of British Columbia, Washington, and Oregon.

Still further north, Asia and America are separated only by the fifty-five miles of the Bering Strait; this is the route Vitus Bering traveled in 1728.

A knowledge of the ocean currents has probably been learned and forgotten many times over the course of thousands of years of the rise and fall of the world's civilizations.

When Columbus reached the so-called "New World," he found black people already there. Probably they had come as traders; the distance from Africa to Brazil is less than half the distance Indonesians had been voyaging over open ocean to Madagascar for some three thousand years. The record of Columbus' third voyage to America, compiled from his own journal, says that he sailed south because he had heard from King John of Portugal "that there was continental land to the south." Furthermore it states "that canoes had been found which start from the coast of Guinea and navigate to the west with merchandise." The record of Columbus' voyage continues:

> He ordered the course laid to the way of the southwest, which is the route leading from these islands [the Cape Verdes] to the south, in the name, he says, of the Holy and Individual Trinity, because then he would be on a parallel with the lands of the Sierra Leona and Cape of Sancta Ana in Guinea, which is below the equinoctial line, where he says that below that line of the world are found more gold and things of value: and that after, he would navigate, the Lord pleasing, to the west, and from there would go to this Española, in which route he would prove the theory of King John aforesaid: and that he thought to investigate the report of the Indians of Española who said that there had come to Española from the south and southeast, a black people who have the tops of their spears made of a metal which they call

"guanine," of which he had sent samples to the Sovereigns to have them assayed, when it was found that of 32 parts, 18 were gold 6 of silver, and 8 of copper.[190]

Balboa and his men found black people in Panamá:

> The Spaniards found negro slaves in this province. They only live in a region one day's march from Quarequa, and they are fierce and cruel. It is thought that negro pirates of Ethiopia established themselves after the wreck of their ships in these mountains. The natives of Quarequa carry on incessant war with these negroes. Massacre or slavery is the alternative fortune of these two peoples.[191]

Early French explorers reported that among the natives of Labrador and Greenland were some who were "black like Ethiopians."[192]

Apparently there were white American natives as well:

> There came to see the captain four Indians, who approached us, and they were of such tall stature, that each one was a span taller than the tallest Christian, and they were very white and had very fine hair that reached their waists, and they were all decked out in gold and splendid clothing. (Friar Gaspar de Carvajal's account of the first European voyage down the Amazon, 1542).[193]

The conquering tribes of Europe do not want to believe it, but many had crossed the Atlantic Ocean before Columbus—not, however, with guns. In the thirteenth and fourteenth centuries numerous travelers, both African and Arab, brought back tales of trans-oceanic voyages to strange lands. Others left and never returned. Al-'Umarī, writing in about 1345 of the kingdom of Mali, tells that during the reign of Muhammad (1305–1310), an expedition of four hundred well-provisioned ships was sent on a voyage of exploration across the Atlantic Ocean. Only one ship returned, to tell of how the rest of the fleet had entered a swiftly flowing river in the open ocean and disappeared to the west. On hearing this, Sultan Muhammad led an even larger fleet of two thousand ships—"one thousand for him and his men who accompanied him and one thousand for water and provisions"—and embarked on the oceans, never to be seen again.[194]

The Scandinavians had a regular trade connection with Greenland and North America from the tenth through the fiteenth centuries, importing from that region sealskins, and narwhal and walrus tusks. Englishmen raided the Norse colonies on numerous occasions. And the Viking pioneers of the ninth and tenth centuries repeatedly said that the Irish were in both Iceland and Vineland (North America) before them.

Basque whalers, and English, Breton, and Portuguese fishermen were plying the waters of the Gulf of Saint Lawrence as early as the 1480s.

Medieval mapmakers placed a large island called "Antillia" in the ocean between Europe and Japan. It was supposed to have been colonized by seven refugee Portuguese bishops and their congregations escaping the Arab conquerors of Portugal in the eighth century. Antillia is found on most European maps drawn after 1424, sometimes placed in the actual location of the islands we still call the Antilles, the islands at which Columbus arrived in 1492. One map drawn by Andréa Biancho in London in 1448 depicts a stretch of coastline 1500 miles west of Africa, in the approximate true location of Brazil. Magellan, according to the journal of Antonio Pigafetta, a Portuguese explorer who accompanied him on his journey around the world, was looking for the passageway between the Atlantic and Pacific Oceans because had "seen it in a marine chart of the King of Portugal, which a great pilot and sailor named Martin of Bohemia had made."[195]

On their part, the natives in many different parts of America recorded the visits of strangers from across the ocean. Montezuma himself, Emperor of Mexico at the time of the Spanish conquest, was apparently of non-native origin. It is recorded that his ancestors came from across the Atlantic Ocean about six centuries previously, that they landed at the port of Panuco and subjected the Mexican people to their rule.[196]

In ancient times, Aristotle and Diodorus reported that the Carthaginians, "who were masters of the western ocean," settled a very large, populous, well-wooded fertile island many days' sailing distance west of Africa, with mountains, plains, navigable rivers, and every kind of wild animal. The description, if it is to be believed, could only fit America.[197]

Very likely many people had sailed across the oceans before even the Carthaginians. The oldest megalithic civilizations known to archaeologists—predating Egypt and Mesopotamia—were built by seagoing people and are located on islands—Cyprus, Crete, Malta, and Thera in the Mediterranean Sea; Bahrain in the Arabian Gulf.

Americans probably made it across the Atlantic at least occasionally. Pliny wrote that "Quintus Metellus Celer, pro-consul of Gaul, received from the King of the Swabians a present of some Indians, who on a trade voyage had been carried off their course by storm to Germany."[198] More likely they were Americans, carried across the North Atlantic by the Gulf Stream.

Crossing now to the west coast of South America, it is recorded that the Spanish who arrived there in the sixteenth century observed large fleets of ocean-going rafts sailing far to the north and south. From some of the old Inca mariners the Spanish were given sailing directions to some inhabited islands two months' voyage west into the Pacific. In 1572 Sarmiento de Gamboa recorded the story of a trans-Pacific voyage by Tupac Inca who, within living memory, had embarked westward with a fleet of ships carrying more than twenty thousand people, and returned after a year at sea with a great deal of treasure and many black prisoners.[199]

From the opposite side of the Pacific come tales of Chinese fleets that sailed across the ocean. In the Annals of the Chinese Empire is recorded the story of Hwui-Shin, a fifth-century Buddhist priest who visited the country of "Fu-sang," 7000 miles to the east. In the second century B.C. Tung Fang-soh, in the "Shin-I King," or "Record of Strange Wonders," wrote, "East of the Eastern Ocean is the country of Fu-sang. When one lands on its shores, if he continues to travel on land still further east 10,000 li [3,000 miles], he will again come to a blue sea, vast, immense, and boundless." The Shi Chi, the historical records of the Chin Dynasty, mentions an expedition of several thousand young men and women who set out across the Pacific Ocean in 219 B.C. and did not return. Kiuh Yuen, in about 300 B.C., placed Fu-sang across the Pacific Ocean in a poem called "Le Sao" or "Dissipation of Sorrows." The Shan Hai King, the world's oldest geography, is supposed to have been compiled by the Great Yu in 2205 B.C.; its fourteenth chapter purports to describe places "on the far side of the Eastern Sea, in the Great Beyond."[200]

Regardless of the historical accuracy of these and other Chinese tales, Polynesia has been inhabited for several thousand years by ocean-going peoples who traded with one another more or less regularly, even plying the waters of the Antarctic. And in the far northern Pacific, Aleuts and Inuit engaged in commerce long before any Europeans were in the area. Alaska and Siberia are so close to one another that related families live on both sides of the American-Russian border.

The physical evidence of all this trade is here to be seen. Many cultivated plants were already growing on both sides of the Atlantic and Pacific before Columbus' first voyage. Coconuts, sweet potatoes, yams, bananas, bottle gourds, and cotton all grew in America, Asia, Africa, and throughout the Pacific Islands. Kapok grew on both sides of the Atlantic. Taro grew on both sides of the Pacific and throughout Polynesia. Indigo and lotus grew in Asia and America. Grain amaranth and the common bean grew throughout the world. Chile peppers, tomatoes, cassava, arrowroot, tobacco, and totora reeds, all native to America, also grew on Easter Island, 2,300 miles to the west and the closest of the Polynesian islands. Pineapples, also native to America, grew on Easter Island, the Marquesas, and the Hawaiian Islands. Papayas grew in the Marquesas. There is some evidence that maize already grew in Asia before Columbus' time.

Silk was woven from the cocoons of silkworms in ancient China and in ancient Central America.

Chickens are native to Asia; a variety that lays green or blue eggs was being raised from Costa Rica to Chile before Columbus arrived.

The natives of Tierra del Fuego brought their hunting dogs with them on voyages to the Falkland Islands in the South Atlantic. They left no permanent settlements there, but the first European visitors to the Falklands found South American dogs still inhabiting the islands.

Travelers on long voyages spend a lot of time making music and playing games. Perhaps it is not surprising that identical pan pipes—identical even in pitch—have been collected by anthropologists in northwestern Brazil and the Solomon Islands. And that native variations on the game of backgammon exist throughout Europe, Asia, and the Americas. One of these games—India's parcheesi—is almost identical to the Mexican game of patolitzli, board, dice, and rules all being the same. The abacus, too, was once known throughout Europe, Asia, Africa, and the Americas, as a tool for computation; the game (and the name) of backgammon are derived from it.

Very similar sailing rafts with centerboards for steering were, and still are, used in Perú, Brazil, Formosa, and Southeast Asia. And an identical type of sail, carried on rafts and canoes, is found throughout the Pacific and in Brazil.

The evidence of long-established oceanic trade is so overwhelming that the Columbus myth survives only because the underpinnings of our technological way of life have come to depend on it. To admit that wilderness once coexisted successfully with populous civilizations would seriously undermine the foundations of our own. And yet today our way of life is being destroyed regardless, so there is really not much at risk anymore. Let us therefore continue to review the evidence, to help illuminate what has been hidden in darkness for so long.

Tools and techniques that crossed the oceans in antiquity continue to be discovered by those who dare to look for them. Archaeologists have found stone weapons, identical in blade, handle, and ornamentation, in western North and South America and in New Zealand. Conch shells were used as trumpets in old Europe, India, Japan, the East Indies, New Guinea, the South Sea Islands, Mexico, the Caribbean, Perú, and Brazil. The ancient Phoenicians and the ancient Mexicans extracted a brilliant purple dye from live sea snails; "Tyrian purple" disappeared from the Mediterranean millennia ago, but the industry still existed in Mexico in 1909.

Cultural evidence is also rich and abundant. The Aztec calendar, for example, was numbered in cycles of fifty-two years—one to thirteen in reeds, knives, houses, and rabbits—an arrangement exactly like our playing cards, which have their origin in astrology and divination. The rising and setting of the Pleiades in the sky marked the seasons for the Aztecs—as it did for the ancient Egyptians, Greeks, Israelites, Arabs, and Chinese.

The Spanish conquerors were astonished to find, in both Mexico and Perú, some very familiar religious practices: confession, communion, and infant baptism.

At least a dozen folk tales traveled across the oceans in ancient times. The turtle that holds up the earth belongs in North American mythology and also in Hindu mythology. Variations on the Orpheus and Euridice story have been collected from many peoples in Europe, Asia, and North America.

Many artistic themes from Indian and Indonesian temples—the tiger throne; the lotus staff; the lotus throne, on which a cross-legged god sits; fish eating lotus flowers; Atlas holding up the world—appear also in Central American temples. A figure very much like Vishnu appears in a temple at Alvarado, Mexico. Ganesha, the elephant-headed Hindu deity, seems to appear in Aztec and Mayan art as Tlaloc and Chac, respectively—a long-nosed god with tusk-like teeth. In the Aztec religion, the Creator of the Universe was a three-fold god who lived in a place called Zivenavichnepaniucha[201]—a name which to my ears recalls the Hindu Trinity—Shiva, Vishnu and Brahma. A wall panel at Piedras Negras in Guatemala remarkably resembles the "Life of Buddha" reliefs at Borobudur in Java.

Pre-Columbus contacts across the oceans would no doubt be better documented had not the Spanish been so thorough in their burning of American books—books whose paper was manufactured by techniques and tools that were identical to those of Southeast Asia, from the inner bark of trees that are closely related to Asian paper trees. So thorough was the destruction of American libraries that only three Mayan and fourteen Aztec books survive.

The destruction was not confined to books. The inhabitants of central Mexico before Cortés arrived numbered some 28,000,000. The subsequent population history is as follows:

1519	28,000,000
1532	16,870,000
1548	6,300,000
1568	2,650,000
1580	1,890,000
1595	1,370,000
1608	1,070,000

The holocaust on Hispaniola was even more horrible. When Columbus landed, he wrote in his journal that "the land is so fully cultivated that it looks like the countryside around Córdova." He estimated that Hispaniola was at least twice as populous as Portugal; the island must have had at least three million inhabitants.[202] Counts from later years gave the native population as follows:

1492	3,000,000
1508	60,000
1509	40,000
1510	46,000
1512	20,000
1514	27,800
1516–20	between 10,000 and 15,000
1535	200
1542	200
1548	500
1565	150
1571–74	50

The same thing happened to Cuba, Puerto Rico, and Jamaica. Thirty smaller islands in the vicinity of Puerto Rico were depopulated entirely.

Las Casas estimated that four to five million people died during the first fifteen years of the Spanish-licensed, German-led occupation of Venezuela; that two million people captured on the east coast of South America and shipped as slaves to Puerto Rico and Hispaniola died in the mines or other works; that over four million were killed in the Inca empire during the first decade of conquest there; that four to five million died in sixteen years in Guatemala, two million in eleven years in Honduras, eight hundred thousand in eight years in Panamá, five to six hundred thousand in fourteen years In Nicaragua; that half a million slaves were sold in ten years in Panamá and Perú, where they died.

When Europeans first explored the Amazon, paddling downriver from Ecuador's Napo River to the Atlantic Ocean, they discovered fertile floodplains nourishing very large communities. Friar Carvajal described one native nation

> ... which in the opinion of all extended more than 80 leagues [about 300 miles], for it was all of one tongue, these being all inhabited, for the distance from village to village was less than a crossbow shot, and never more than half a league, and there was one settlement where the houses stretched for 5 leagues [about 18 miles] without stopping, which was a marvelous thing to behold.

The Amazon cities had plazas and temples. The people wore fine cotton clothing, kept abundant stores of maize, oats, and dried fish, and traded over very large distances.

> From this village there were many roads, that entered the interior, very fine highways ... There was a villa in which there was much chinaware of various makes, both jars and pitchers, very large, of more than 100 gallons, and other smaller vessels such as plates and bowls and candelabra. This porcelain is the best that has been seen, because that of Málaga does not equal it, because it is all glazed and enameled with all colors, so vivid that they astonish, and more than this, the drawings and paintings that they make on them are so accurate that, with natural skill, they work and draw all these things like Roman articles.

And further downriver:

> Inland, at a distance of two leagues, more or less, there appeared very large cities that glistened white, and besides this the land is as good and as fertile and as normal in appearance as our Spain.[203]

The earliest pottery in America has been found by archaeologists here in the tropical lowlands—three thousand years older than any found in the Andes or coastal Perú. And similar cultures existed along the Ucayali River in eastern Perú, the Orinoco in Venezuela, the rivers of northern Bolivia, the tributaries of the Amazon in Brazil, and along the coastal plains of Guiana—wherever there were large expanses of fertile floodplains.

Within 150 years all of these civilizations had vanished so utterly that today it has been forgotten that native agriculture and civilization ever existed anywhere east of the Andes.

In North America the ancient trade routes are recorded in some detail. The Hopi, for example, were middlepeople in the American southwest, selling corn and blankets and trading in the goods of the nations that surrounded them. The Paiute brought forest products from the north; the Pueblos to the east brought jewelry, indigo, turquoise, and buffalo skins; the Apache to the south brought moccasins, bows and arrows, mescal, and goods from further south in Mexico; the Havasupai to the west brought trade goods from many tribes along the Colorado River. In Canada, the Lillooet in the west and the Montagnais-Naskapi in the east traveled well-worn routes of considerable length. Trade routes also connected the far northern American tribes to the nations of Siberia.

Of the population of old North America there is no record, and there are few clues. Most of the easternmost nations, from Newfoundland to Florida, were exterminated very early on by European settlers. The Russians decimated the Aleuts and the Alaskans through brutality and disease in the 18th century. Whalers on the Arctic coast brought whiskey and smallpox to the far north in the 19th century. The introduction of horses and guns to the North American interior made native warfare far more terrible. After four centuries of the European conquest, there are 11.5 million people in North America who identify as indigenous, and 368.5 million immigrants.

It is, however, sometimes difficult in the case of America to distinguish cultural genocide from purely physical genocide. When Indians did not actually disappear fast enough, they were often defined away through assimilation: those natives who no longer lived traditional lives weren't counted as Indians anymore. And, bearing in mind the 32,000 theoretical ancestors each one of us has from the time of the Pilgrims, and the half million unknown ancestors from the time of Columbus, anyone whose family has been in America for more than two or three generations without doubt has some native American blood flowing in his or her veins.

My best estimate is that there may have been 150 million inhabitants of North, Central, and South America in 1492. Today the American population—native, mixed, and foreign—is about 1 billion and rapidly increasing.

The Pacific Islands

We fought with spears, clubs, bows and arrows. The foreigners fought with cannons, guns, and bullets.
— F. Bugotu, Solomon Islands, 1968

The journey westward from America takes us through Polynesia to the islands of the western Pacific, past Australia and New Guinea, and finally back to Java, the beginning point of our round-the-world tour.

The people of the Pacific suffered immensely from European contact. The foreigners first came in large numbers early in the nineteenth century, looking for sandalwood to sell to customers in Asia, to supplement the more ancient Indian and Indonesian sources of this aromatic herb. When Europeans began exploring the Pacific Ocean, they found the trees growing on all of the islands. Sandalwood was discovered on Fiji in 1804, and by 1816 most of the trees were gone. It was discovered on the Marquesas in 1814, and that supply was depleted in just three years. From 1811 to 1818, Hawai'i's sandalwood trees were cut down. The exploiters turned next to Melanesia, depleting one island after another, until by 1865 there were not enough trees left in all of the Pacific Islands to support the sandalwood trade anymore.

Most of the labor for all of this and other enterprise was supplied by islanders. Sometimes they left home for months at a time to serve as crew on European ships, or to cut wood on other islands. Often they were kidnaped and kept as virtual slaves. After about 1860, and for the next fifty years, violence and exploitation were more the rule than the exception in the South Seas, as thousands of islanders were kidnaped and forced to work for years at a time as indentured servants on British cotton plantations in Australia and Fiji.

Europeans also brought to the Pacific terrible diseases, among them syphilis, gonorrhea, and tuberculosis. They brought grazing animals that devoured vegetation and changed fertile soil into desert. They brought missionaries. They brought tobacco. They brought guns. Disease, warfare, enslavement, and cultural and environmental destruction decimated the populations of all the islands.

Hawai'i probably had over a million inhabitants (about 500 per square mile) when Captain Cook visited there in 1778. By all reports the people were extraordinarily healthy and long-lived. A survey of ancient burial grounds reveals that infant mortality was between thirty-five and forty per thousand births—a rate that we in the United States did not improve on until 1950.

In 1832 only 130,000 natives were left. After reaching a low of 13,900 in 1876, Hawai'i's population then began a long steady increase to its present 1.4 million. About 300,000 are native Hawai'ians, less than a third of their original numbers.

Captain Cook estimated Tahiti's population at 204,000 (about 500 per square mile) when he visited there in 1769, and he was preceded by two other boatloads of sailors bringing venereal and other diseases. By 1797 the population had been reduced to less than 20,000. It is 308,000 today.

New Zealand had at least half a million Maoris in 1769. Only 85,000 were left in 1840, and 48,000 in the 1870s. Today they number about 850,000 of New Zealand's 5 million people.

The first English penal colony on Tasmania was established in 1803. Thirty years later only one hundred and eleven native Tasmanians were left alive. By 1877 they had vanished from the earth utterly.

Australians were more fortunate in possessing a whole continent, largely desert, in which to seek haven from their oppressors. They number some 745,000 today and are outnumbered by foreigners about 33 to 1.

Summary

1. It is unlikely that there have ever been any agricultural nations that have been sparsely populated.

2. Human numbers have decreased about as often as they have increased, until three or four centuries ago.

3. The European conquest, and modern European civilization, have been fueled by continuous economic growth, that is to say by population expansion.

4. The European conquest of the world was made possible by a revolution in weapons technology, in particular the invention of guns.

Behind modern technology there are certainly some noble motivations: we want to make life ever easier for ourselves, prolong its length, make it safer, cure diseases, reduce infant mortality, feed everybody, and so on. But by also making the world ever more crowded, we have in reality made life much more difficult for ourselves.

In fact, the effect of the European conquest on the world has been twofold: those who participated in the European economy multiplied out of control; those who did not were exterminated. Sometimes there was a period of transition, where the conquered people declined in number but ultimately were forced into the system and began increasing again; this happened to most of Africa and the Pacific Islands. In Ireland we have seen the reverse: an initial increase in numbers was followed by a decline and an eventual stabilization.

The factors to which most population theorists pay attention are not the driving economic forces behind population dynamics but rather the mechanisms of population change. There has been much confusion here. Sanitation and medicine are factors in controlling disease and prolonging life. But these are primarily improvements in the quality, not the quantity, of life. The population problem is usually thought of as resulting from a decreased death rate not accompanied by a decreased birth rate, and the solution is supposed to be education in birth control. But the fact is that the birth rate is determined mostly by powerful economic forces. Our present economy demands births, and Planned Parenthood is going to have no effect on the world's birth rate so long as its members participate in the world industrial economy.

Traditional methods of birth control, known to practically all of the nations and tribes of the world way before European civilization took over, were many:

Incest taboos often extended to all of the members of one's clan, limited the number of available sexual partners.

Initiation ceremonies had to occur prior to marriage.

A bride price or dowry often had to paid, delaying marriage even longer.

Babies were usually nursed for three or four years, inhibiting ovulation and pregnancy.

Very often there was a prohibition against intercourse for a period of time after the birth of a child.

Withdrawal was known and practiced in all cultures. Douching after intercourse was practiced in many. Abortion was widespread.

IUDs, barrier methods, and spermicides are all ancient and widespread.

In Indonesia, midwives used to induce a retroflexed uterus through abdominal manipulation after delivery. The object was the prevention of pregnancy.

Infanticide seems not to have been as prevalent or as culturally approved as has often been believed. Contraception, on the other hand, has been very widely known indeed.

Contraceptive herbs were known to virtually all peoples in all places. More than a hundred different contraceptive plants have been collected from tribes and nations throughout the world, and the effectiveness of some of them has been proven in modern scientific studies. Contraceptive plants include:

Plant	**Country**	**Formula**
yarrow	Northern Europe	hot infusion
pineapple	Malaya	unripe juice taken raw
dogbane	North America	roots boiled with water and liquid drunk once a week
Indian turnip	Hopi	decoction of powdered, dry root; larger dose confers permanent sterility
birthwort	Hungary	seed parts used to prevent pregnancy and menstrual pain
wild ginger	North America	root and rhizome boiled slowly for a long time, decoction drunk

milkweed	Navajo	infusion drunk after childbirth
asparagus	Southern Europe	decoction of berries
begonia	Central America	eaten to promote menstruation
thistle	Quinault (North America)	tea
castor beans	Morocco	eaten as a contraceptive
Indian paint brush	Hopi	decoction
squaw root	Chippewa	strong decoction of powdered root brings on menstruation, also helps childbirth
spotted cowbane	Cherokee	chew and swallow roots—confers permanent sterility
coconut	Pacific Isles	milk from green coconuts
antelope sage	Navajo	root boiled for 30 minutes and 1 cupful drunk—used by both sexes
barrenwort	ancient Greece and Rome	powdered leaves taken in wine after menstruation
willow	ancient Greece and Rome	potion of the leaves, used by both sexes
cotton root	Creoles (South America)	decoction for contraception; seeds to increase lactation
ivy	Mediterranean Europe	berries
Christmas rose		extract of root induces abortion

stoneseed	Shoshone	cold water infusion of roots taken daily for six months
marjoram	Germany	taken as a tea during menstruation
American mistletoe	Pomo	tea from leaves
rosemary	Opata (Central America)	tea
rue	Europe and America	hot decoction promotes menstruation; larger dose induces abortion
false Solomon's seal	Nevada	tea from leaves for contraception; root infusion regulates menstruation
Queen Anne's lace (wild carrot)	Appalachia, India, ancient Greece	1 tsp. seeds in glass of water, drunk after intercourse to prevent pregnancy

In recent experiments, the Shoshone herb stoneseed (*Lithospermum ruderale*), mixed with the normal diet, rapidly suppressed estrous cycles and lowered the birth rate in mice and rats. This was completely reversible and seemed to have no effect on any organs except the ovaries and uterus.

In Java and Bangladesh, the most crowded places in the world today, methods of birth control and abortion have been known and practiced for thousands of years; the major change is simply that modern methods have replaced ancient ones. It is abundantly apparent that the causes of overpopulation are not to be found in ignorance, and that the cure is to be found elsewhere than in contraceptive education.

CHAPTER 29

War

It would seem that ever since human beings became crowded enough, we have been a very aggressive sort of beast. We have made war. We have kept slaves.

But what is understood by "warfare" is not the same in all places and at all times. Centuries ago, among Africans, among Indians, among Americans, even among the knights of medieval Europe, warfare was often a ritualized affair in which a relatively few people died. The idea was to defeat one's enemies, not destroy them. It is eye-opening to survey the non-industrial world's erstwhile ideas of organized fighting. Observers from more "advanced" societies have sometimes seemed amused by the proceedings.

In southern India before the Portuguese came:

War had become with them a game governed by a series of elaborate rules, and to break one of these rules involved dishonor, which was worse than death. Their arms were lances, swords, and shields, and much taste was displayed in lacquering and polishing . . .

There was neither night fighting nor ambuscade. All fighting was in daytime when the sun had well risen. The opposing camps were pitched near each other and both sides slept securely. At sunrise the soldiers of both armies mingled at the tank, put on their armor, ate their rice and chewed their betel, gossiped and chatted together. At beat of drum either side drew apart and formed their ranks. It was creditable to be the first to beat the drum, but no attack was allowed until the other side had beaten theirs . . .

At times when the ranks on one side broke, the slaughter was very great, but after the drum sounded the two sides mingled together and there was no bad blood even when a man killed his own brother. (R. S. Whiteway, *The Rise of Portuguese Power in India*)[204]

In the Sudan:

Since the aim was to get the enemy to withdraw so that victory might be claimed with as little loss on your side as possible, you usually avoided complete encirclement for if the enemy was unable to withdraw they would, seeing that there was no hope, sell their lives as dearly as they could. You therefore left a gap in the rear. Moreover, there was a further convention, that fighting should begin about 4 P.M. so that those who were getting the worst of it could withdraw under cover of darkness. (E. E. Evans-Pritchard, "Zande Warfare")[205]

In West Africa:

Warfare among the Ibo was more a matter of affrays and raids than of organized campaigns. And it was waged not with the object of winning markets or annexing territory, but with the purely practical purpose of keeping one's neighbors in order. . . . There were certain rules which had the binding force of law for both sides. There was, for example, a recognized system of compensation at the close of the war, and there were rules prohibiting the use of lethal weapons in certain circumstances. (Charles K. Meek, *Law and Authority in a Nigerian Tribe*)[206]

Among the Tahitians:

It is much less disgraceful to run away from one's enemy with whole bones, than to fight and be wounded; for this, they say, would prove a man rather foolish than warlike. (William Wilson, A *missionary voyage to the Southern Pacific Ocean*)[207]

In Melanesia, before the coming of guns:

(The Admiralty Islands) Women are always left unmolested and are permitted to pass freely between the contending camps. (Camilla H. Wedgewood, "Some Aspects of Warfare in Melanesia")[208]

(The Trobriand Islands) They never fought without warning, nor would they fight at night . . .

Midway between the two capital villages a place was selected and a circular arena cleared, which would be the theatre of fighting. The opponents ranged themselves opposite each other, the warriors standing at a distance of some thirty to fifty meters apart and throwing their spears. Behind the warriors stood or sat the women, helping the men with water, coconuts, sugar-cane, as well as with verbal encouragement. The only weapons used in regular fighting were the spear for offense and the shield for defense . . .

Fighting lasted as long as both parties could resist the onrush of their opponents. When one party had to flee, the road to its village was open, and the enemy would rush on killing men, women, and children indiscriminately, burning the villages and destroying the trees. . . . As a rule, practically everybody, especially the defenseless ones, would succeed in escaping. (Bronislaw Malinowski, "War and Weapons Among the Natives of the Trobriand Islands")[209]

(Bougainville) On an appointed day the fighting men of both districts meet for battle. They range themselves on either side of the clearing and shoot at each other until their arrows are spent. Then they close in with clubs and thrusting spears and continue thus until one force has suffered a loss. After this they retire to allow the dead warrior to be cremated. The following day the armies meet again and the party of the dead man endeavors to avenge them. ... This type of combat may be repeated on two or three successive days, or it may be that both sides feel that they have had enough of fighting, and peace may be arranged. (Camilla H. Wedgewood)[210]

(Lifu) On the day appointed both parties meet on a clear spot of ground between the two tribes, and form in line abreast of each other about one hundred yards (or more) distant. The fight is then commenced by throwing spears from both armies and which they generally catch and throw back again. The two lines then make a charge, meet, exchange blows with their clubs in passing—and again halt, at about the same positions, they continue these manoevres, until some of either party is killed. (Dorothy Shineberg, *The Trading Voyages of Andrew Cheyne*)[211]

In New Guinea:

The largest military assemblage known to the Enga is that in which all the clans of one phratry combine to fight all the clans of another. Such great fights appear to have taken place only among the more densely settled Central Enga, and those rarely. Moreover, these contacts are very much in the nature of tourneys, bounded by conventions that minimize casualty rates in relation to the large number of warriors involved. ... Women, children, and the elderly belonging to each phratry may observe the fight from a safe distance (out of bowshot); they are not expected to participate, for instance by retrieving arrows, and they should be immune from attack. ... Should the weather hold, the fighting goes on until dusk, but usually it is halted by the late afternoon rain. ...

The participants are bound by generally recognized constraints ... that noncombatants are safe from attack, that equals should engage each other, that fighting should be confined to a delimited area. In addition, warriors should not burn down houses or damage gardens and valuable trees near the battlefield. Finally ...

even if one of the contending forces is so badly mauled that it must withdraw, the victors may profit only in terms of glory and the knowledge that they have weakened potential enemies. They have no right to turn the retreat into a rout and invade and occupy the losers' territory. And indeed I have not heard of this happening. (Mervyn Meggitt, *Blood is Their Argument*)[212]

Among the Maidu of California:

The women and children would gather on the knolls out of arrow range to watch the fun, while the young men, advancing unarmed, would skillfully dodge volleys of darts. For hours each side in turn would expose itself to the enemy bowmen, until in the end an accident would happen and someone would be hit—after which all the stricken men's comrades would flee, triumphantly chased by the victors, who caught and beat the slow of foot. But evidently they did not beat them too severely, for everyone would return to the battlefield, the women would pass the food, and all would feast together, while the winners would pay the losers an indemnity for the wounded men. (Stanton A. Coblenz, *From Atom to Atom Bomb*)[213]

Among the Papago of the American Southwest:

The warriors who killed Apache were required to undergo an ordeal of purification called lustration, during which they endured even greater hardships than on the warpath, thus expiating the crime of murder. (Frances Densmore, *Papago Music*)[214]

In Alaska's Yukon delta:

When a man on either side had relatives in the opposing party, and for this reason did not wish to take part in the battle, he would blacken his face with charcoal and remain a non-combatant, both sides respecting his neutrality. In this even a man with his face blackened had the privilege of going without danger among the people of either side during a truce. (Edward William Nelson, *The Eskimo About Bering Strait*)[215]

Among the Murngin of Australia:

The aggrieved group sent an invitation for the ceremony; the members of both clans arrived in war point and singing war songs; they stood a little more than spear distance apart, with a safe wall of mangrove jungle behind them in case it became necessary to temper valor with discretion; then the challengers danced over to their adversaries, stopped, and retreated; after which the other side repeated the performance. Spear-throwing now began according to a prescribed routine: two men ran across the field, accompanied by two close relations who were also kinsmen of the other side, and whose object was to prevent the spear from being thrown too seriously. Every member of the offended clan hurled at least one spear, sometimes with the result that the enemy was chased into the forest; and meanwhile the old men of both sides warned the throwers to be careful. But in any case the hostilities apparently did not last long; they were ended when one group danced up to the other and one of the aggrieved party pierced the thigh of a foe with a spear—if the wound was not too slight, this usually sufficed for the demands of vengeance. (Stanton A. Coblenz)[216]

Organized fighting in any form was unknown among the eastern Inuit, the Veddahs, the Andaman Islanders, the people of the Ituri forest, the Sakai, the Semang, the Kubu, the Penan, and many more cultures that practiced little or no agriculture. Violent deaths were rare among these people, and warfare was an incomprehensible thing for which no word existed in their languages.

What changed, to make our world so violent? What is the difference between the civilization that I live in, which keeps standing armies of professional fighters, where people grow up accustomed to brutality, and the hundreds of more peaceful civilizations that once existed, or that are now dying out? Has anything in fact changed? It is not an easy question to answer.

Violent warfare has been part of the scene in the Middle East for many thousands of years. The crossroads of continents, its civilizations have depended in large part on the enormous volume of trade that has flowed through from all directions, year in and year out, down through the ages. Greece, Rome, and Carthage were built on Mediterranean commerce, and they too knew organized killing. But the masters of killing in ancient times were the horsemen of Central Asia. They conducted warfare entirely on horseback, and it was a much more ferocious affair, as the speed and power of a horse were at each warrior's command. Not until the Europeans invented guns and cannon were the Mongols bested at the art of killing. All of the trade routes in the world then came under the control of the European nations; wherever they went, they changed the rules of warfare entirely, as the people and the animals of the earth looked on in amazement and terror. Blinded and benumbed by the might at their command, the men of Europe pillaged, raped, tortured, and killed their way around the world, as their civilization lost all of its restraint to the power of its own machinery. Guns and cannon escalated war to new heights—war on people and war on the planet.

And the rules keep on changing. The invention of bombs and airplanes made the war which began in 1914 a holocaust the likes of which the earth and its residents had never met before. The invention of the atom bomb obliterated any remaining rules of conduct with a vengeance. One man's little finger can now destroy a city, or an island, or a mountain, in just a few seconds. It is human to err, but we have saddled ourselves with technologies that allow for no error at all.

Even without atom bombs, unbelievable devastation was visited on the Middle East in 1991 in a war of just six weeks' duration. The Persian Gulf was massively polluted. Kuwait was on fire for the better part of a year, as crude oil rained out of the sky as far away as the Himalayas. Unprecedented levels of acid rain fell on southern Russia. Levels of soot five miles high in the atmosphere were from five to one hundred times normal over the entire northern hemisphere. Kuwait's countryside, and a good deal of Iraq's, was littered with craters. One hundred and fifty thousand people were bombed to death in this short war, and the country of Iraq was left a shambles, lacking even clean water for its people to drink, who died in very large numbers of water-borne diseases.

So what indeed has changed? The more self-sufficient people of the world—most of the old cultures of Africa, of North America, of Polynesia, of Australia, of southern Asia—have been relatively peaceable and gentle. Those who lived on the most prosperous trade routes and built large commercial empires—the Greeks, the Romans, the Arabs, the Persians, the Mongols, the Incas, the Aztecs—were more violent. When Europe was cut off from the mainstream of world trade during the Middle Ages, warfare there became ritualistic and confined to the upper classes. In the fifteenth century, Europe again began to command a larger share of world commerce; feudal class boundaries began to vanish; and European weaponry became very much more violent.

It is the military-industrial complex, the fatal alliance between trade and technology, that seems to be promoting violence in this world. What, then, has changed? Trade has expanded enormously, and technology has grown incredibly more powerful. Out of the partnership comes violence and brutality and destruction. The people of this world are engaged in commerce on such a tremendous scale that it is depleting all of the world's forests and oceans and minerals and wildlife in short order. The people of this world have such powerful technology literally at their fingertips that it is polluting all of the air, and all of the water, and all of the land on this beautiful planet beyond possibility of recovery.

Let us listen to the words of Saukamappee, an eighteenth-century Nahathaway (Cree) who described how technology alone changed the nature of native American warfare. The account begins with a battle against the Shoshone that took place in western Canada in about the year 1725:

> My father brought about twenty warriors with him. There were a few guns amongst us, but very little ammunition, and they were left to hunt for the families; our weapons was [sic] a lance, mostly pointed with iron, some few of stone, a bow and a quiver of arrows;

the bows were of larch, the length came to the chin; the quiver had about fifty arrows, of which ten had iron points, the others were headed with stone. He carried his knife on his breast and his axe in his belt. Such was my father's weapons, and those with him had much the same weapons. I had a Bow and arrows and a knife, of which I was very proud. We came to the Peeagans and their allies . . . we were about 350 warriors . . .

Both parties made a great show of their numbers, and I thought that they were more numerous than ourselves. After some singing and dancing, they sat down on the ground, and placed their large shields before them, which covered them: we did the same, but our shields were not so many, and some of our shields had to shelter two men. Theirs were all placed touching each other; their bows were not so long as ours, but of better wood, and the back covered with the sinews of the bisons which made them very elastic, and their arrows went a long way and whizzed about us as balls do from guns. They were all headed with a sharp, smooth black stone (flint) which broke when it struck anything. Our iron headed arrows did not go through their shields, but stuck in them; on both sides several were wounded, but none lay on the ground; and night put an end to the battle, without a scalp being taken on either side, and in those days such was the result, unless one party was more numerous than the other . . .

I grew to be a man, became a skillful and fortunate hunter, and my relations procured me a wife. She was young and handsome and we were fond of each other. We had passed a winter together, when messengers came from our allies to claim assistance.

By this time the affairs of both parties had much changed; we had more guns and iron headed arrows than before; but our enemies the Snake Indians and their allies had Big Dogs on which they rode, swift as the deer, on which they dashed at the Peeagans, and with their stone clubs knocked them on the head, and they lost several of their best men. This news we did not well comprehend and it alarmed us, for we had no idea of horses and could not make out what they were . . .

In the ensuing battle, the guns were decisive, the few defeating the many. The Piegan and their allies lost ten warriors; the Shoshone (Snake) lost more than fifty. Said Saukamappee,

> The terror of that battle and of our guns has prevented any more general battles, and our wars have since been carried by ambuscade and surprize, of small camps, in which we have generally the advantage, from the guns, arrow shods of iron, long knives, flat bayonets and axes from the Traders.[217]

Before the smallpox came like poisoned rain into the land; before the red deer and the bison faded away in front of the arriving waves of white settlers; long before the white people themselves came as conquerors into the land of the Nahathaway and the Piegan and the Shoshone, the white man's weapons came into the territory to sow violence like a noxious weed throughout the northern plains. And it found fertile soil, nor has it yet been uprooted.

Economics. Overpopulation. War. They are all aspects of the environmental problem, like interlinked parts of a single puzzle. It is time to go deeper still. At the beginning of this journey, I asked the question, "What is a human being?" and I would like to return to it. More particularly, what is human civilization? I want to turn next to the problem of slavery, which is so widespread among so many human cultures that it will have something important to teach us about our treatment of one another and about our treatment of our planet.

CHAPTER 30

Slavery

Slavery has been reported in the history of surprisingly many cultures:

- ancient Mesopotamia, Egypt, and Israel
- classical Greece and Rome
- the Byzantine empire
- the Ottoman empire
- all Islamic societies, including North Africa, Arabia, Iraq, Iran, Pakistan, and Indonesia
- EUROPE: England, France, Spain, Portugal, Italy, Germany, Poland, Lithuania, Russia, and Scandinavia
- ASIA: India, China, Nepal, Burma, Thailand, Laos, Cambodia, Malaya, Japan, Korea, the Philippines, Mongolia, and numerous nations in the Caucasus, Central Asia, and Siberia
- AFRICA: reported in many nations throughout the continent
- SOUTH AMERICA: the Incas, and many nations of Patagonia, the Grán Chaco, the Amazon basin, and the Guinea coast
- the ANTILLES: the Carib and Arawak
- CENTRAL AMERICA: the Aztec, Maya, and Cuna
- NORTH AMERICA: the Micmac, Iroquois, Comanche, Pawnee, Ojibway, Navajo, Kwakiutl, and many other nations throughout the continent
- the PACIFIC ISLANDS: Tahiti, Hawai'i, Rapa Nui, Lifu, New Zealand, New Guinea, and the Solomon Islands

Looking below the surface, however, it is apparent that neither anthropologists nor historians have been able to agree upon a definition of "slavery." Indeed, asking the question "What is slavery?" forces one to look into one's own soul and one's own culture in a way that is very pertinent to the questions of this book.

In my culture, "people get paid for their labor." What does this seemingly simple statement mean? It means that when one person works for another, money changes hands. But money is a relatively recent and rare item in the history of humankind. As we discovered in investigating economics, money is nothing but a way to keep track of human relationships with one another and with the physical world around us—relationships that existed long before the invention of money. When anthropologists look at a society that operates largely without money, they see people working for other people without getting paid, and most often they report it as "slavery." But there is a major problem here: people in my culture are under the illusion that the movement of money changes something. To get at the root of human relationships—with one another and with our planet—it will be necessary to take off our money-glasses.

What in former times were called "servant" and "lord" are now called "employee" and "boss." The fact that money changes hands alters nothing. Human beings have not changed, nor have human relationships. When you work for someone, you lose a piece of your freedom; someone else gets to tell you what to do with your life, someone else has power over you. When you lose all or most of your freedom to another human being, that is called slavery—but the difference is only one of degree, and the presence or absence of money is not at all a determining factor.

All of us belong to groups that we serve without pay, to whom we owe some loyalty: we serve our family, our school, our friends, our neighborhood, our town or city, our country, our culture, our church, our political party, and any other groups and organizations we belong to. These relationships are complex, and all of them involve the loss of some degree of our freedom.

As children we lose most of our freedom to our family and our school, some schools and some families wielding more authority than others. As adults, those of us who have jobs lose more or less freedom to our boss.

The degree of servitude may depend on the type of job; one's education; one's experience; whether one is an illegal alien, a legal resident, or a citizen; one's age; one's color; one's debts; and so on. Some adults lose all or most of their freedom for part or even all of their lives: convicts, prisoners of war, hospital patients, monks and nuns, soldiers, professional athletes.

Foreigners always have fewer rights and less freedom than the legal citizens of a country. There are always obstacles, both legal and social, to an outsider becoming an accepted and welcome resident of a community; it may take several generations for full integration to occur, if ever. In an increasingly mobile world, foreigners gain acceptance much more quickly than in times past, but even in the United States of today there are still traditional communities where one is not fully accepted on equal terms with everyone else unless one's grandparents were born there.

Slavery, then, arises out of the needs of human societies. It is:

1. an organized system by which human beings work for other human beings;
2. a method of integrating outsiders into a society as functional individuals; and, what is more pertinent to our present quest,
3. **an outgrowth of civilized people's need to tame wild nature—their own, each other's, and the planet's.**

Old Australia knew no such thing as slavery. Nor did most of the world's other hunting and gathering peoples. Foraging nations bring up their children to be free and independent, not to serve their parents obediently. Nor, as adults, do hunters and gatherers keep servants, in these mostly cooperative and egalitarian societies.

I am not suggesting that these folks are in any basic way different from you or me. Their economy and their technology are simply not geared toward domination. Human populations are small. Wild nature is abundant and is the source of everything people need.

Farmers, on the other hand, are in the business of replacing wild plants with cultivated crops, and wild animals with livestock; their survival depends on it. Except for a few fishing communities, the large populations made possible by farming cannot be supported by the wilderness.

Civilization as we know it has its basis in the taming of the wild. And this is reflected in the taming of each other and of ourselves. It is reflected in the keeping of servants and slaves. It is reflected in warfare and racism. And its tragic end result is the ultimate replacement of all wild nature with crops, livestock, and human artifacts; the tragic end result is the disappearance of our earth's life support systems.

The list of slave-holding societies at the beginning of this chapter should now be modified. Many reports of slavery came from early European explorers, missionaries, and conquerors—people who spoke of the natives as ignorant cannibalistic savages—hardly objective reporters. Many more reports of so-called "indigenous slavery" came from nations that had already suffered from centuries of European enslavement; and parts of Africa had exported slaves to Egypt, Greece, Rome, and Arabia for several thousand years. With this in mind, we can make a few generalizations:

1. Most hunters and gatherers keep neither servants nor slaves.
2. Most farmers and herders keep servants.
3. Slaves are found in those societies that engage in the most commerce and have the densest populations.
4. In modern times machines are replacing human beings as slaves.

Slavery as a concept is very widespread among human beings, but slavery as an important institution existed in antiquity only along the oldest and most well-traveled trade routes: in the civilizations of the Mediterranean region, the Middle East, southern and southeast Asia, and all along the northwest coast of North America from the Aleutian Islands to Oregon.

CHAPTER 31

Religion

The Judeo-Christian heritage in particular has been blamed by some for our environmental troubles. After all, it says in Chapter One of the Bible that God told man and woman to "Be fruitful and multiply, and replenish the earth, and subdue it, and have dominion over the fish of the sea, and over the fowl of the air, and over every living thing that moveth upon the earth."

But as we have seen, people have been subduing the earth since long before the Bible was written, and people of many different faiths have claimed dominion over their non-human neighbors. Let us once again look globally and ask what religion can teach us about ourselves.

It is a curious thing that so many of the world's scriptures that are in large part documents of warfare, nevertheless preach peace and brotherly love. To begin with, let us look again at the Bible.

> He maketh wars to cease unto the end of the earth; He breaketh the bow, and cutteth the spear in sunder; He burneth the chariot in the fire. (Psalms 46:9)

> If thine enemy be hungry, give him bread to eat; and if he be thirsty, give him water to drink. (Proverbs 25:21)

> And they shall beat their swords into plowshares, and their spears into pruninghooks: nation shall not lift up sword against nation, neither shall they learn war any more. (Isaiah 2:4)

And his name shall be called Wonderful, Counsellor, The mighty God, The everlasting Father, The Prince of Peace. (Isaiah 9:6)

Whosoever shall smite thee on thy right cheek, turn to him the other also. (Matthew 5:39)

All things whatsoever ye would that men should do to you, do ye even so to them. (Matthew 7:12)

But war begins in the first book of the Bible:

Blessed be the most high God, which hath delivered thine enemies into thy hand. (Genesis 14:20)

The Lord is a man of war. (Exodus 15:3)

The bloody conquest of Canaan is described in some detail in Numbers 21.

And the Lord harkened to the voice of Israel, and they delivered up the Canaanites; and they utterly destroyed them and their cities. (Numbers 21:3)

There is even a lost book of the wars of the Lord.

Wherefore it is said in the book of the wars of the Lord, what he did in the Red Sea, and in the brooks of Arnon, and at the stream of the brooks that goeth down to the dwelling of Ar, and lieth upon the border of Moab. (Numbers 21:14-15)

At Jericho, the Israelites "utterly destroyed all that was in the city, both men and women, young and old, and ox, and sheep, and ass, with the edge of the sword. (Joshua 6:21)

Said David to Goliath, "I come to thee in the name of the Lord of hosts, the God of the armies of Israel," and he slew Goliath with a slingshot, then cut off his head with a sword, and brought it to Jerusalem as a trophy.

The historical books of the Old Testament are all full of battles and conquest: Numbers; Deuteronomy; Joshua; Judges; I and II Samuel; I and II Kings; I and II Chronicles.

The New Testament, by contrast, contains no warfare, and seems to preach only peace. But in the name of Jesus a great many terrible wars have been fought: the Crusades; the Inquisition; all of the wars of conquest waged in recent centuries by Europeans in Africa, Asia, America, and the Pacific. In the two World Wars, God was supposed to be on everybody's side. Even the different sects of Christianity have shed a lot of blood among themselves. Protestants fought Catholics in Northern Ireland only recently. The truly pacifist sects—the Friends, the Brethren, the Hutterites, the Amish, the Mennonites—have always been few in number and usually persecuted for their pacifism.

What is going on here?

One has only to look at the first two Commandments for a clue:

1. I am the Lord they God.
2. Thou shalt have no other gods before me.

The world is clearly divided into "us" and "them." Above all else, religion is supposed to command one's loyalty, and it is our loyalties which generate conflict and warfare. The rules of peace are simply not meant to apply to foreigners. The business of missionaries is to convert the world so that one day there will *be* no foreigners. When all the foreigners are either killed or converted, evidently there will finally be peace in the world.

> Think not that I am come to send peace on earth: I came not send peace, but a sword . . . he that taketh not his cross, and followeth after me, is not worthy of me. (Matthew 10:34, 38)

> He that is not with me is against me. (Matthew 12:30)

> He that believeth on the Son hath everlasting life: and he that believeth not the Son shall not see life; but the wrath of God abideth on him. (John 3:36)

No man cometh unto the Father, but by me. (John 14:6)

At the very same time Moses gave the Israelites the Ten Commandments, which included "Thou shalt not kill," he also gave them laws of war, and this commandment:

> Of the cities of these people, which the Lord thy God doth give thee for an inheritance, thou shalt save alive nothing that breatheth; but thou shalt utterly destroy them: namely, the Hittites, and the Amorites, the Canaanites, and the Perizzites, the Hivites, and the Jebusites, as the Lord they God hath commanded thee." (Deuteronomy 20:16-17)

Our loyalties—to family, to friends, to clan, to tribe, to city, to nation, to culture, to church—necessarily bring us into conflict from time to time with other, different families, clans, tribes, cities, nations, cultures, and churches. Let us recall the two great forces of evolution, cooperation and competition. These are also the two great methods of resolving conflicts—between species and between human beings. The peaceful nations of the earth temper their disagreements with a large helping of cooperation. For hunters and gatherers it is a natural thing, because in order to survive they must cooperate in large measure with the forces of nature. Their competitive skills are channeled into the making of tools, but they do not subdue and dominate their entire neighborhood. The Amish and Mennonites who have remained pacifist while living amongst us have also rejected some of most of the violent modern technologies.

Religion is the bridge between our inner and outer worlds. Each of us experiences both, and each of us has some beliefs about how the two are connected—what one's own place is in the universe. Religion, therefore, is very deep-rooted inside of us, is liable to earn our strongest loyalties, and can lead to the most intense conflicts, between friends as between nations. The inner reflects the outer. Judaism and Christianity developed out of an agricultural and pastoral civilization at the crossroads of the world, whose livelihood depended on subduing and dominating nature, whose government was a kingdom, and who knew warfare as a regular part of life. Human nature is presumed inherently sinful as a result of the original fall from grace in the Garden of Eden. It is a civilization whose driving force is the taming of wild nature—the taming of the earth, of each other, of the foreign barbarians, and of our own wild selves. Technology. Warfare. Government. Religion. As the bridge to our souls, religion is perhaps the glue which holds civilizations together.

Let us look briefly at some of the other religions of our wide and varied human world.

Islam

The Qur'ân divides believers from non-believers from the very beginning. "This is the book," we read, "there is no doubt therein, a guide to the pious"; and for infidels, "on their eyes is dimness, and for them is grievous woe." (II, 1)

> Fight in God's way with those who fight with you, but transgress not; verily, God loves not those who do transgress. Kill them wherever you find them, and drive them out from where they drive you out; for sedition is worse than slaughter; but fight them not by the sacred Mosque until they fight you there; then kill them, for such is the recompense of those that misbelieve." (II, 186-7)

> Say to those who misbelieve, "You shall be overcome and driven together to hell." (III, 10)

> Kill the idolators wherever you may find them; and take them, and besiege them, and lie in wait for them in every place of observation; but if they repent, and are steadfast in prayer, and give alms, then let them go their way; verily, God is forgiving and merciful. (IX, 5)

Like Judaism and Christianity, Islam preaches peace among believers, and war against the heathen.

Hinduism

Ahimsa, or non-injury, appears as a virtue for yogis, and was championed by Mahatma Gandhi, one of the strongest voices for peace in modern times.

> In the presence of one firmly established in ahimsa, all hostilities cease. (The Yoga-Sutra of Patañjali, II.35)

> The killing of living beings is not conducive to heaven. (Manu Smriti 5, 48)

But there are four main Hindu castes—brahmins (priests), kshatriyas (warriors), vaishyas (craftspeople and farmers), and sudras (menial workers). Ahimsa is to be observed by priests, but what about warriors?

> There exists no greater good for a kshatriya than a battle enjoined by duty. (Bhagavad Gita 2, 31)

And indeed, the Mahabharata, a religious book of which the Gita is a part, is also the longest epic of war ever written. The Ramayana, too, is a document of warfare against the heathen. It is the story of the love between Rama and Sita. But it is also a tale of the hostilities between the gods of India and the demons of Sri Lanka.

Jainism

The Jains carry ahimsa to the extreme of not even harming insects. They neither hunt nor eat meat, and they do not engage in agriculture lest they harm living things by tilling the soil.

> One may not kill, nor ill-use, nor insult, nor torment, nor persecute any kind of living being, any kind of creature, any kind of thing having a soul, any kind of being. That is the pure, eternal, enduring commandment of religion which has been

proclaimed by the sages who comprehend the world. (Ayaranya Sutra I, 4, 11)

But Chapter 42 of the Adi-Purāna contains a treatise on the duties of the warrior. Evidently ahimsa and pacifism are not the same thing. From the second to the thirteenth centuries there were Jain kingdoms in the Indian state of Karnataka, defended by Jain armies and Jain generals.

Zoroastrianism

In this religion which originated in ancient Persia, life is seen as a perpetual battleground between the forces of good and evil. In the time of the Persian Empire, it was one of the most militaristic of the great religions. Of the thirty sins listed in its Scriptures, five have to do with straying from the one true faith: idol-worship; worship of other religions; heresy; apostasy; and demon-worship. (Dinai Mainogi Khirad 36, 11-19)

> It is requisite to abstain from the same cup as those of a different religion, and it is not desirable to drink the water of any goblet of theirs ... Because, when any one drinks with a stranger, it makes his heart inclined toward him, for it would be a sin; and, on account of the sin committed, he becomes bold, and his soul has an inclination toward wickedness. (Sad Dar 38, 1-4)

Sikhism

Those who belong to the Khalsa—the fellowship of the pure—carry the five K's with them at all times: Kash (uncut hair and beard), Kangha (wooden comb), Kach (shorts), Kara (steel bracelet) and Kirpan (steel dagger). The dagger is supposed to be an instrument of defense, a statement that Sikhs are slaves to no one. Like many other religions, Sikhism preaches love, service, and tolerance of one's fellow human beings. Today Sikhs are a persecuted minority in India and Pakistan, but at one time, before the British conquest, Sikh armies defended a small empire that included the Punjab and Kashmir.

In the Guru Granth we read:

Reflect ye on this, that without the Guru, no one is ever redeemed (Ramkali M.3, 22)

Save for the True Guru's, all other Word is false. (Ramkali M.3, 24)

And in India today, Hindus, Muslims, and Sikhs continue to kill one another in the name of God.

Buddhism

The Buddha, like Christ, preached only peace and non-violence, but many wars have nevertheless been fought in his name. Many Buddhist kings of India and Sri Lanka were great conquerors. Buddhist wars were fought in Burma and Thailand. Buddhist monks have fought in the armies of Korea, China, and Japan. And Buddhist missionaries have traveled the world over spreading their version of the Truth.

Baha'i

Founded in the nineteenth century, Baha'i preaches peace, love, and tolerance.

Let not a man glory in this, that he loves his country; let him rather glory in this, that he loves his kind. (Baha'u'llah)

Baha'is do not participate in government or political action of any kind. But the distinction between believers and heathens persists in Baha'i Scriptures:

Lauded be Thy name, O my God! Thou beholdest me in the clutches of my oppressors. Every time I turn to my right, I hear the voice of the lamentation of them that are dear to Thee, whom the infidels have made captive for having believed in Thee and in

Thy signs ... And when I turn to my left, I hear the clamor of the wicked doers who have disbelieved in Thee and in Thy signs. (*Prayers and Meditations* by Bahua'u'llah, page 7, number V)

Many religions have promised paradise to warriors who die in battle: Islam, Shinto, the faiths of the ancient Germans, Norse, Aztecs and others. All the Greek gods took sides in battle; in the Iliad Hera, Athena, Poseidon, Hermes, and Hephaestus fought on the side of the Achaeans, and Ares, Phoebus, Artemis, Leto, Xanthus, and Aphrodite fought on the side of the Trojans. The Norse gods fought a great war among themselves, the Aesir against the Vanir. The Aztecs hung the heads of their defeated enemies on skull racks in their temples.

Although China has known a great many wars, some Chinese religions have truly promoted peace and not war.

Confucianism

Tzu-Kung asked about government. The Master said, sufficient food, sufficient weapons, and the confluence of the common people. Tzu-Kung said, suppose you had no choice but to dispense with one of these three, which would you forego? The Master said, Weapons. (Analects, Book XII, No. 7)

Taoism

One who assists the ruler of men by means of the Way does not intimidate the empire by show of arms. (Tao Te Ching, Chapter 30)

There is no glory in victory, and to glorify it despite this is to exult in the killing of men. (Tao Te Ching, Chapter 31)

Among farmers, truly pacifist religions are rare. Among industrialized nations, even rarer. Mention has already been made of the Christian pacifists—the Friends, the Anabaptists (Amish, Mennonites and Hutterites) and the Brethren.

Foragers generally do not have religion in the same sense as farmers. Their magicians and shamans have a special relationship with the inner, or spirit world that connects us to plants, animals, and places, to the moon, to the winds, and to each other. But the idea of a divine ruler, of omnipotent gods or goddesses, of a government in the spirit world—such ideas are not meaningful to people who haven't got a government in the material world. Authoritarian religions have no place in cooperative societies.

CHAPTER 32

Sex

And the journey takes us still deeper into ourselves, into hidden, dangerous territory. The connection of human sexuality with human overpopulation is obvious enough. But what about economics, and war, and religion? What does environmental destruction, after all, have to do with sex? Why do we call it the rape of the earth? A curious figure of speech.

Sex certainly seems to be a motivating force for technological change. Guns, skyscrapers and rockets are so obviously phallic, and the ubiquitous push-button and the touch screen are perhaps as obviously clitoral—our machines, like perfectly designed lovers, are made to be more and more responsive to our slightest touch and caress.

Our religions are deeply involved with sex—they prohibit it altogether for our priests and regulate it severely for the rest of us.

And sex certainly is connected with violence, though at first sight it is not at all clear why. Conquering armies, since time immemorial, have pillaged and *raped*. Why? Why, too, does the most obscene and violent word in the English language mean "sexual intercourse"? I ask once again, what is going on here, here on this earth, here deep within our own male and female selves?

The connection between sex and violence is made even clearer by the violence we do to our own sexual organs. Circumcision and other mutilation of the male organs have been practiced in:

- AFRICA: almost everywhere, including Morocco, Algeria, Tunisia, Libya, Egypt, Ethiopia, Sudan, West Africa, Guinea, the

Congo basin, Zanzibar, Madagascar, Mozambique, South Africa, Botswana, Kenya, and Tanzania.
- AMERICA: Mexico, El Salvador, Nicaragua, some tribes of the Orinoco and Amazon basins, the Andean Highlands of Colombia, Perú, and Argentina, coastal Perú, eastern Bolivia, and among the Athapascans and McKenzie of western Canada
- ASIA: everywhere except Mongolia and Siberia
- the PACIFIC: Australia, New Guinea, and throughout Melanesia and Polynesia
- EUROPE: the Balkan Peninsula
- among all Jews and Muslims
- in ancient times, among the Phoenicians, Syrians, Ethiopians, Egyptians, Edomites, and Moabites
- in modern times, the United States, England, Canada, Australia, and New Zealand.

It would appear that the concept of original sin is not original to Jews and Christians. A great many tribes of us seem to want to make ourselves suffer for our pleasure, and the variety of ways we have found to do it is extraordinary. The Kikuyu and the Masai of Kenya and Tanzania are half-circumcised—the lower part of the foreskin is not cut away, but hangs atrophied instead. In parts of Indonesia and in the South Pacific Islands, the foreskin is split but not removed. Some Mexican peoples make only an incision to draw blood from the penis. The Wakamba of Kenya make an additional cut at the base of the circumcised glans. Many Australian and New Guinea tribes practice subincision in addition to circumcision—they split the underside of the penis from the base to the glans, the extent of the incision varying from tribe to tribe. The Yesidies of Yemen remove a strip of skin from navel to anus, including the skin of the penis and scrotum. In some parts of Mexico, young men formerly perforated their penis and drew a rope through it. On the Philippine island of Capul, a tin nail was drive through the glans after circumcision. In Sumatra, small stones are implanted into the skin of the penis. In parts of Borneo and North Celebes, men with a pierced glans place through it a small metal rod with a ball at each end before intercourse, for the pleasuring of the woman. Removal of the left testicle is reported among some tribes in Algeria, Egypt, Ethiopia, South Africa, the Tonga Islands, and the island of Panope in Micronesia.

Circumcision is almost always a rite of passage, usually done at puberty and a prerequisite for marriage. Among Jews, however, it is done on the eighth day after birth. The Jewish ritual consists of three parts: the chituch (cut); the periah (exposure of the glans and removal of the foreskin); and the mezizah (sucking of the wounded penis). Sucking the penis has usually been omitted since the nineteenth century, but Orthodox observance still requires that a drop of blood be drawn off.

Again, I ask, what on earth is really going on? In the modern United States, there is in general no religious or medical reason to circumcise male babies, yet it is still very widely done. Parents request it. Many hospitals still do it automatically. Why?

Mutilation of the female sex organs has also been, and still is, surprisingly common, although not as common as male mutilations. Female "circumcision" has been practiced in:

- AFRICA: almost all countries
- ASIA: Malaysia, Indonesia, parts of India, Pakistan, Bahrein, the United Arab Emirates, Oman, and South Yemen
- AMERICA: some tribes of the Amazon and Orinoco basins and eastern Mexico
- the PACIFIC: Australia and New Guinea

Among the Omagua of northern Perú, the female foreskin is cut away. The Chama and Cashibo of eastern Perú remove the clitoris and inner labia. Removal of the clitoris and often the inner labia is done in much of Africa, Indonesia, and in Islamic countries. Infibulation is done in Sudan, Somalia, Djibouti, Eritrea, South Egypt, northern Kenya, northern Nigeria, Mali, and the Central African Republic. This is the cutting off of the clitoris, the labia minora, and two-thirds of the labia majora which are then sewn together leaving only a small hole. Among these people sexual intercourse requires a new incision, and childbirth requires a larger one. Among those Australian tribes that subincise their men, the woman's hymen must be ruptured and the vaginal entrance widened to accommodate the exceptionally broadened penis.

Male infant circumcision was introduced in the United States in the early 1870s, not for health reasons, but to prevent masturbation and hypersexuality. By the early 20th century circumcision was common in the United States, England, Canada, Australia, and New Zealand. In the United States it is still routine hospital practice. Vasectomy and circumcision are the two most frequent surgical procedures performed on American males today.

Clitoridectomy was done to infant girls in Berlin in the 1820s, also as a cure for masturbation, and it was done in Vienna in the 1860s, and in England on a relatively large scale in the 1850s and 1860s. In the 1890s some French doctors were amputating the labia minora as well, before the practice was prohibited entirely by French law. Infibulation—the sewing up of the foreskin or labia—was done to infants of both sexes in the United States to stop masturbation. Clitoridectomy was very popular in the United States from the late 1860s until about 1910. It was gradually replaced by circumcision, the removal of the female foreskin, which was widely practiced until about 1940. Here in America, we no longer circumcise our young girls, but caesarean section and hysterectomy are the two most frequent surgical procedures performed on American women.

What is all this violence but the taming of our own wild sexual selves? It is a requirement for membership in a civilized society, a symbol of our enforced docility. It is a reminder that as we subdue and dominate the earth, so we also subdue and dominate our own wild and natural sexual beings.

And so we require our men to shave off their beards and cut off their hair, and our women to denude their armpits and their legs. And so we require men and women alike to mask their sexual odors with deodorant and—under threat of imprisonment—to hide their sexual organs with clothing. And today we are even sending our women to surgeons to remove the feminine curves from their hips and thighs with so-called "liposuction," curves that will not vanish with just dieting.

And our sexuality, which we suppress for our civilization, becomes blatantly expressed in the technology of that civilization. Seeking release, it explodes forth in the violence of rape.

Our dependence on increasingly powerful and violent technologies is an addiction that is well on its way to destroying us, and it is an addiction that I suspect we will not overcome without paying attention in some way, shape, or form to our own wild sexuality, finding some other ways to give it free expression besides the building of bigger and better guns, bombs, towers, and cars, and ever more perfect and responsive machinery.

The hippies were right in insisting upon sexual freedom and long hair. But they were unable to break their own addiction to technologies which thrive on sexual suppression. The feminist movement, which has grown so strong during the past fifty years, is again a recognition that our sexuality has taken a beating from our civilization, and that therein lie the roots of some of our problems. But the feminists, being part of the same society fifty years later, are that much more dependent than the hippies were on technologies that oppress them. It is a frustrating and impossible contradiction.

We are all of us equally involved in a world situation which had its beginnings thousands upon thousands of years ago—all of us, men and women alike. It will not help for us to go blaming one another for our troubles, because there is no one to blame, and no easy solutions. Cars are sexy because they are expressions of what we have suppressed inside of ourselves, all of us, men as well as women. Computers are popular because they make such ideal slaves—obedient and perfectly predictable, without a trace of wildness left in them. Wars continue to be fought because our economic systems will not run without them, and because there are so many of us crowding this little earth. It is high time for us all to stop trying to find someone to blame for this mess we're all in—for women and men to stop blaming one another, for Americans to stop blaming Chinese, for consumers to stop blaming corporations, for liberals to stop blaming conservatives, for socialists to stop blaming capitalists, and for us to stop blaming ourselves. It is high time that we began to rejoice in being alive, and to realize that we human beings are not such bad cats. Let us rejoice in our emotions, our spontaneity, our unpredictability, our imperfections, our curiosity. Our problem is that instead of designing our technologies to conform to living systems, we have insisted upon the impossible task of changing human nature—and indeed all of nature—to adapt to our technologies, and then we go about blaming each other because we haven't succeeded in becoming perfect machines. We have become so attached to our technological environment that we mistake it for an extension of our own selves. It is time for us to stop blaming each other, and to stop trying to save the earth, and to start making some crucial decisions about how we're going to live and what kinds of technologies we will allow in our lives. Until now very few of us have been making these kinds of decisions at all.

CHAPTER 33

Technological Trance

A baby's mother initially provides all of its needs, but as it grows up into a child, and later an adult, it learns to do more and more things for itself. This is the healthy development of human beings, and of all animals.

The mother supplies food through her breast, while the baby need do no more than suck. Later the breast is either withdrawn or rejected, and the child eats, on its own, food that is still provided by its parents. Eventually the adult learns to hunt and gather, or plant and harvest, or earn a paycheck and shop, on its own.

The baby, which was carried at all times, as a child first crawls, then walks on its own.

An adult feeds itself, clothes itself, makes its own shelter, and makes its own friends, as its own arms and legs come into their own power, doing the work of survival and continuing the species. This is biological, psychological, and sexual maturation. The work of separation and individuation is vital for personal, familial, societal, and ecological well-being.

The effect of modern technology is that separation and individuation, coming into one's own power, sexual maturation—it never happens anymore, not really.

Our legs are rendered useless by automatic transportation, so that we are carried not just as babies, but all of our lives. We never need raise our voices to be heard—machines will amplify our sound and carry it long distances. We need not use our arms to carry our own loads. We need not clean up after ourselves—machines wash our clothing and flush toilets take care of our bodily wastes. The television entertains us. Trucks bring us our food. Computers and cell phones are our companions. We need not lift our arms and legs and voices to survive and socialize, only our fingers. Our mother never really leaves us, indeed the mother that used to tire and sleep and yell at us is replaced by one that is always alert and efficient.

We never need to fly the nest, and even if we wish to, our machines are difficult to flee, for they pursue us wherever we go, begging to serve. The technological mother has become so powerful that to resist her requires a much bigger rebellion than the healthy adolescent need mount to assert his or her individuality and separate from a flesh and blood mother.

Our technology, more and more, is incompatible with sexual maturity, and though we don't know this consciously, still it is a biological law. Therefore, at a very young age we clothe our children and forbid them to expose their sexual organs in public—or to see ours. A modern technological society could not possibly function were all of its people to go naked. The whole game that maintains its structure would be over. For clothing today is not just protection and adornment: it is also bondage and, increasingly, violence.

Sexuality and personal power, hidden and suppressed, builds up tensions inside of us. The biological and psychological need for identity and making one's way in the world under one's own power is frustrated. Denied direct expression, our power is projected, transferred from self to technology. When the denial is not so complete, our sexual energies may be sublimated, as Freud recognized 120 years ago, to art, music, and other creative outlets. But we have passed the point where this is a healthy thing. Today not only technology but also art, music, films, and expression of all kinds are becoming more and more insistently and brutally sexual and violent. And the more personal power we give away to our machines, the less our own urges are satisfied, yet the frustrated forces inside us will not be denied. They explode into violence against one another, and drive an incessant demand for more and more powerful technology. It's like any other addiction: the more machines we have, the more we want. Our machines don't satisfy the need that gave rise to them, which is the need for *self*-expression.

When a man kills an animal with a gun, it is not the man's own strength and power that kills, but the power of an explosion not of his own making. It does not even require great skill to kill if the gun is machined well enough and carries a telescopic sight—still less if it is an automatic weapon. The greater the technology, the less effort, the less skill and the less power are required until it is no longer the man that kills, or overcomes, or succeeds, but the tool, almost independently. The ultimate tool, indeed, is the robot, and its day is upon us.

So many in the environmental movement would like to argue that technology is neutral, and can be used just as easily for good as for evil; that it is not even a proper topic for discussion. But without such discussion there is a certain futility to the efforts to clean up the earth, and all of the environmental books end up sounding pretty much the same. For technology today is so powerful that its mastery over and destruction of nature are not only inevitable but must continue to increase, because this has become a substitute for our own personal individual satisfying mastery over and coming to terms with our own environment, a mastery which is perpetually denied us, not by our culture or our upbringing, though that is secondarily true, but by our machines. As long as we remain enslaved to our mechanical mothers we must destroy anything and everything that reminds us that we also are wild, powerful, and sexual beings—lions, and tigers, and wolves, and bears, and insects, and snakes, and spiders, and rain, and germs—anything and everything that dares assert its own self against the world is intolerable, because we humans no longer dare to assert ourselves against the world. We have delegated our power to our machines, and unless our own wildness is directly and continuously attacked, it will emerge too obviously and boldly to tolerate.

Wilderness is a way of life. Those who, desiring to live more harmoniously with nature, would emulate the traditional native American way of life, would do well to look at the tools that produced it, and not assume that modern tools can substitute for them without consequences. The purpose of industrial technology, in essence, is to destroy wilderness. And I think most people, even in an industrial society such as mine, feel the need to preserve a balance of both in this world.

TOWARD THE FUTURE

CHAPTER 34

Slowing Down

The tractor in itself does not appear to be anything so bad. But when we look into the history of the Amish churches which have changed and permitted the use of the tractor for field work, we must conclude that it does have a significant influence on the life and thinking of a group. Only a few communities have been able to retain their Amish simplicity with the tractor. Dozens have been carried away from the Amish by the tractor.

— An Amishman from Iowa, 1985

Our life and our thinking are powerfully influenced by the tools that we use in ways that we can hardly imagine. It might be thought that this is not really a problem except for one thing: the earth is dying.

The earth is dying, and it is not because we are such evil people who live on it. If you take away our tools, we human beings are pretty much the same wherever you find us, and we cannot have changed all that much in the million or more years or more that we have shared this planet with our sister and brother animals and plants. Kenyans, Norwegians, Siberians, Australians, Peruvians, Jamaicans, Hawaiians, we are all the same, except perhaps for the shape of our ears or the texture of our hair or the shade of our skin, none of which have as much impact on any of our lives as the physical environment that we live in and the technology that our society makes available to us.

Growing up in Hawai'i is not at all the same experience as growing up in the Yukon. Hawai'i is a tropical island, surrounded by salt water, with tropical forests and animals and fishes and a strong sun and warm temperatures and plenty of rainfall year-round. The Yukon is an inland plateau that gets little rainfall and lots of snow, and is peopled by wolves, moose, and musk oxen in small-treed evergreen forests and tundra, with short warm summers and frigid dark long winters. To expect that the same agriculture and the same crops and the same livestock and the same industry and the same technology are appropriate to both Yukon and Hawai'i is to destroy both places. And to expect that one technology and one agriculture and one way of life are appropriate to the entire world is to destroy our world. Diversity—biological, cultural, economic, religious, technological—in short, total ecological diversity is the key to the perpetuation of life on planet earth.

The earth can take care of itself, only provided that it is diverse enough. Its prairies, and its many different kinds of forests, and its deserts, and its wetlands, and its rivers and lakes and oceans—its fishes, and its reptiles and amphibians and birds and mammals and insects and spiders and worms and trees and shrubs and weeds and cacti and grasses and mushrooms and humans—we are all one living system that will perpetuate itself forever, provided only that we are diverse enough.

WE CANNOT SAVE THE EARTH! Human beings cannot save the earth. We are not in charge here. All of our frantic activity on our planet's behalf can only make matters worse. All of our video conferencing with environmentalists on the other side of the planet, all of our flying around the earth in airplanes to environmental conferences and threatened rainforests, all of the weather balloons and satellites and automobiles and four-wheel-drive vehicles and snowmobiles and diesel ships and computers and radar and sonar that are at our service to study environmental problems, all of the hordes of scientists running around Antarctica and the Amazon, all of the animals running and crawling and swimming and flying around the wilderness with radio collars around their necks, all of the thousands upon thousands of scientific studies and manuscripts and theses and books and papers being published out of the bones of our fast-dwindling forests—all this does not help the earth! If my own book does not help reduce the pile, it too will have been a waste.

The only thing we can really do for the earth is to stop destroying it. Then the earth will take care of itself. Instead of trying to fix the whole planet, let us attend to our own simple lives.

I am optimistic about life, and about the essential goodness of humanity. But technology has made us spectators instead of participants in life. This in itself is not such an evil thing, but our technology has become so powerful that the spectators now dominate the scene, and life has become a curiosity dying in a zoo. We must quit being spectators and become again participants, breaking down the fences and regaining our freedom. The fences that confine us are not within us but out there in the real world—the airplanes, the computers, the televisions, the cell phones, the chain saws, the automobiles. When we have no choice but to operate within their confines, we are not free. And we are all starving for freedom. But we have tamed our own world to such an extent that we are now looking for wilderness in outer space, among the innumerable stars. It is not really other life or other intelligence that we seek out there—such are still abundant here on earth—but wild versions of ourselves.

As I close this book, as I end what was originally a four-year inquiry into the human-earth relationship, a four-year journey into the depths of my soul which has now become a journey of three decades, I want to make a short list of some of the things that must happen on the earth if it is to survive, and that we as individuals can do to make them happen. They are listed in the order of priority.

1. Our cell phones, upon which we rely to make us safer, and around which our social and professional lives have become structured, are destroying the earth more quickly and thoroughly than anything else (chapters 12, 14, 16, 18, 19, 27, and 28). With our cell phones we have become the terminals and control centers of the most violent technology every invented: the Internet (chapter 25). We have become the agents of planetary destruction. Our cell phones also emit powerful radiation, and they control, require, and create the infrastructure that makes them work: the cell towers, antennas and satellites that, together with our phones, are strafing the earth with bullets of radiation, day and night, year in and year out. From obesity to diabetes to heart disease to cancer to influenza

to birth defects; from forest die-off to insect armageddon to amphibian extinctions to birds falling dead out of the sky by the millions, the lethal effects of the radiation are everywhere. **We must throw all our cell phones away.** This is both the most important and easiest change to make and is under the control of every individual.

2. By changing over from natural to synthetic products, not only have we made our lives dependent on vast amounts of deadly poisons (chapters 10, 12, 16, 17, 19, and 21), but we have contaminated all of the air, water, and soil on earth, and our bodies and the bodies of all other living things with them. As a result, we have lost all our commitment to preserving our life support systems. **We must stop using plastics completely** (chapters 12 and 19). Plastics are choking and suffocating fish and wildlife. Microplastics are filling up the oceans, raining on us day and night from the air, displacing sand on all our beaches, poisoning us from the food and water we drink. This will be more difficult because almost everything that is manufactured today—cars, buildings, appliances, electronics, etc.—is made with large amounts of plastics, and foods and all kinds of goods are packaged in it. But we can start by never using plastic bags—not for groceries, not for produce, not for garbage, not for anything—and it will spread from there. We can also stop wearing polyester, acrylic and nylon, and return to cotton, linen, ramie and wool. Synthetic fabrics are the source of a huge percentage of the microplastics that are poisoning our planet.

3. **We must stop spraying our homes, yards, trees, pets, businesses, and farms with pesticides** (chapters 10 and 21), and buy and support only organic agriculture. This too is both easy and necessary.

4. The above is only a beginning, but they are things that individuals can easily do in large numbers. I have not ridden in an airplane since 1987. I have not watched television since 1967. I have no Internet at home. I bicycle most places, and rarely travel long distances. I have never used a cell phone or a pesticide. Saving the earth requires (a) not poisoning it, (b) slowing down, and (c) navigating the world under one's own power.

Computers and automobiles are incompatible with life, but abandoning them is not easy. It requires community. We must first deal with the emergencies. Cell phones. Pesticides. Plastics. The use of fossil fuels—coal, oil, and natural gas—is also an emergency (chapters 4, 10, 12, and

14). Their burning is rapidly heating up the earth, depleting our oxygen supply, and acidifying the oceans. But they are not under individual control, and the solutions that are being pushed are not solutions. Electrifying everything will kill us even faster than global warming. From the mining of cobalt for lithium-ion batteries (chapters 27 and 28) to the thousands of chemicals in the manufacture and composition of solar cells (chapter 10) to electromagnetic armageddon (chapter 18), there will be no people or animals left to enjoy a cooler earth. It is not yet clear that there *is* a solution, but if we can regain our power as individuals by giving up cell phones, plastics, and biocides, we can learn to cooperate with one another again to make intelligent decisions as communities.

There are already a multitude of well-written books available on living simply, and most of them contain good and helpful advice. But the usual "50 things you can do to save the earth" approach is like trying to cure cancer with a band-aid. Trying to live with powerful technology and having to restrain ourselves from using it fully lest we destroy the world is delusional and does not work. I am talking rather about *choosing simple tools*—landlines instead of cell phones; pen and ink instead of computers; visits and telephone calls instead of emails; voices and musical instruments instead of entertainment centers; feet and bicycles instead of cars; sinks instead of washing machines; shovels instead of bulldozers—and using them to their fullest potential, and thereby using ourselves to our fullest potential and recovering our humanity which we have so stifled. It is the normal use of our technology, not just its abuse, that is destroying nature and ourselves.

I am also emphasizing the personal nature of these choices rather than the political changes others advocate. Political organizing not only postpones the solutions forever into the future, but it relies on faith in what is mostly an illusion: the illusion that one possesses the power to force anyone but oneself to do what one wants. I believe rather that the single most important, effective, and indeed revolutionary thing a person can do on behalf of the planet is to personally stop using violent technology and to support others in doing likewise. Japan once abandoned guns as a society (chapter 28) only because so much of the populace disliked them and were unwilling to use them anymore.

5. It is more important for us to make decisions about the kinds of technologies we choose to have in our lives than to worry about the sources of energy that power them. The speed and violence of some of our technologies, their impact on our lives, their devastation of nature, and their ability to destroy our communities, do not depend on whether their source of power is the sun, the wind, petroleum, or atomic energy. Nor does improving the energy efficiency of our machines decrease our use of energy: it simply allows us to make more machines. The market will not have it any other way.

6. If we are to solve any of our environmental problems at all, business and economics will have to take a back seat to them (chapter 27). We will have in some way to live more cooperatively with our neighbors, and to find other ways of distributing the wealth. As long as everyone's main worry in life is having a job and earning money, this planet hasn't got a prayer. At one time not so long ago in the United States, a single income was enough to support an entire family. Today both husband and wife and also the children must go to work to support themselves. We are doing four times as much work to support the same number of people, and the extra work is helping to destroy the earth.

We must also get out of the habit of thinking we can solve environmental problems by throwing money at them. Truly effective environmental action does not cost anything. Spending money, on the other hand, almost always makes the situation worse.

CHAPTER 35

The Earth and I

Environmentalists are engaged in therapy of the grandest sort. Our purpose is to heal the entire earth, and perhaps it cannot be done. It seems incredibly presumptuous to even try, but the alternative is to give up on life itself, so we accept the long odds and continue the work. Therapy, we know, does not always succeed, and sometimes the patient dies, but we must do our job, because as long as we continue to explore the problem and create space for the patient to discover the cure, recovery is possible. She is still breathing and her heart is still beating strong. If we can see the road to recovery clearly enough, the patient may yet decide to follow it. Science, history, philosophy, psychology, and common sense are at our service in illuminating the path.

I have hinted throughout this book that it is mainly an addiction to technology that is destroying us. One may argue that only certain particular technologies are bad; that although, like drugs, they have come to pervade everything we do, only a relatively few of our most addictive technologies must be abandoned, and that we need not all be thrown back into the Stone Age in order to survive. The atomic bomb is one of the few items we will probably all agree must go. But if modern technology is indeed killing us, we must be willing to look at any or all of it, or we will not solve the problem. And we ought finally to realize that even stone tools are part of a venerable and successful way of life, and although that way of life could not possibly support the number of people presently living on the earth, it could nevertheless support a surprisingly large fraction of them. Unless we can acknowledge the full range of possibilities, we limit our ability to make choices and to solve our problems. And if we do not solve our problems we will end up back in the Stone Age before very long without having chosen to go there.

An addiction, by itself, is a relatively simple thing. But to give one up usually requires recognizing that the dysfunction and disease it is causing are threatening one's life. That has been part of the purpose of this book: to document the disease and connect it home to its causes. The book has also searched below the surface, both inwardly and outwardly, among our sciences and our cultural institutions, for clues to a better understanding of our place in the ecology of the world.

> Who am I upon this earth so green, so blue, so round?
> Man, beast, lover, fire-keeper, maker of horrendous sound,
> contaminator of the ocean, polluter of the summer sky—
> No, no, no! my race does that. I love you, earth, it's they, not I.
> Your breezes crisp, your rivers pure, delight my lungs, refresh my veins,
> my skin, my bones, they are your soil, your ancient woods, your fertile plains.
> Each morning I look forward to one more sunrise, one more day.
> Each day the wilderness grows less, I don't know why, I cannot say.
>
> Who are we, earth, my race and I, what does our future hold?
> A living, vibrant, complex world, or one grown bare and cold?
> I cannot get enough of you, your butterflies, your ocean breezes,
> your starry skies, your gentle rains, your hurricanes, your winter freezes.
> You're all I had when I was born, you're all I have today.
> Each day the wilderness grows less, I don't know why, I cannot say.
>
> I relish every step I walk, each fragrant breath I take,
> each drink, each morsel that I swallow. Every morning that I wake,
> I thank the sun for rising, and the clouds for making rain,
> I greet my furry neighbors, I salute my feathered friends.
> I'm neither tree nor squirrel, I have no gills, I cannot fly,

but in this little human form, as goes the earth, so go I.
Each day the wilderness gets less, but in my prayers—I don't know when,
I don't know how, I cannot say—the wilderness will grow again.

Notes

1. Letter to Charles Francis Adams, Jr., April 11, 1862, in A *Cycle of Adams Letters 1861–1865*, Worthington Chauncey Ford, ed. (Boston: Houghton Mifflin, 1920), Vol. 1, p. 135.
2. Rachel Carson, *Silent Spring* (Boston: Houghton Mifflin 1962), pp. 7–8, 12.
3. Irven De Vore and Richard Borshay Lee, eds., *Man the Hunter* (Chicago: Aldine, 1968), pp. 30–43.
4. Desmond Morris, *The Naked Ape* (New York: McGraw-Hill, 1967), p. 85.
5. Marston Bates, *The Forest and the Sea* (New York: Vintage, 1960), p. 127.
6. Vitus B. Droscher, *The Friendly Beast* (New York: Harper, 1972), pp. 95–125; V. C. Wynne-Edwards, "Population control in animals," *Scientific American*, Aug. 1964, pp. 68–74; V.C. Wynne-Edwards, *Evolution Through Group Selection* (Oxford: Blackwell Scientific, 1986), pp. 84–94.
7. Anthony Huxley, *Plant and Planet* (New York: Pelican, 1975), p. 310; L. J. Audus, *Plant Growth Substances* (London: Leonard Hill, 1959), pp. 391–392.
8. Konrad Lorenz, *On Aggression* (New York: Harcourt, Brace & World, 1966), p. 242.
9. Morris, pp. 175–176.
10. Loren Eiseley, *The Immense Journey* (New York: Random, 1957), p. 14.
11. Huxley, p. 264.
12. Huxley, p. 266.
13. Alfred Russel Wallace, *Tropical Nature and Other Essays* (London: Macmillan, 1878), p. 65.
14. John Daniel, "The long dance of the trees," *Wilderness*, Spring 1988, p. 21.
15. James Hutton, *Theory of the Earth* (Edinburgh: W. Creech 1795), reprinted by Wheldon & Wesley, Codicote, Herts., 1959), Vol. 2, pp. 560, 562.

16. James Lovelock, *Gaia: A New Look at Life on Earth* (Oxford University, 1987), p. 9.
17. Charles Darwin, *Journal of Researches*, 2nd edition (London: J. Murray 1845), pp. 398–399.
18. Charles Darwin, *The Descent of Man* (London: J. Murray, 1871), p. 74.
19. Piotr Kropotkin, *Mutual Aid: A Factor of Evolution* (London: W. Heinemann, 1902), pp. vii-ix.
20. Howard Stansbury, *An Expedition to the Valley of the Great Salt Lake of Utah* (Philadelphia: Lippincott 1852), p. 193.
21. Darwin, *The Descent of Man*, p. 77.
22. Kropotkin, p. 39.
23. Kropotkin, pp. 74–75.
24. Published in *Seattle Star*, Oct. 29, 1887.
25. Most of the oxygen ever created by photosynthesis has already been used up in the oxidation of minerals in the earth's crust.
26. Daniel Martin, Helen McKenna, and Valerie Livina, "The human physiological impact of global deoxygenation," *Journal of Physiological Sciences* 67: 97–106 (2017).
27. Andrew Falle, Ewan Wright, et al., "One million (paper) satellites," *Science* 382(6667): 150–152 (2023).
28. R. G. Ryan, Marais, E. A., Balhatchet, C. J., & Eastham, S. D., "Impact of rocket launch and space debris air pollutant emissions on stratospheric ozone and global climate," *Earth's Future* 10: e2021EF002612 (2022).
29. Qing-Bin Lu, "Observation of large and all-season ozone losses over the tropics," *AIP Advances* 12: 075006 (2022).
30. Ian T. Cousins, Jana H. Johansson, Matthew E. Salter, Bo Sha, and Martin Scheringer, "Outside the safe operating space of a new planetary boundary for per- and polyfluoroalkyl substances (PFAS)," *Environmental Science and Technology* 56: 11172–11179 (2022).
31. B. Goldberg, W.H. Klein, "Comparison of normal incident solar energy measurements at Washington, D.C.," *Solar Energy* 13(3): 311–321 (1971).
32. S. Suraqui, H. Tabor, W.H. Klein, and B. Goldberg, "Solar radiation changes at Mt. St. Katherine after forty years," *Solar Energy* 16(3–4): 155–158 (1974).
33. Gerald Stanhill and Shabtai Cohen, "Global dimming: a review of the evidence for a widespread and significant reduction in global radiation with discussion of its probable causes and possible agricultural consequences," *Agricultural and Forest Meteorology* 107: 255–278 (2001).
34. Ziyan Wang, Ming Zhang, et al., "A comprehensive research on the global all-sky surface solar radiation and its driving factors during 1980–2019," *Atmospheric Research* 265: 105870 (2022); Hejin Fang, Wenmin Qin,

et al., "Solar brightening/dimming over China's nainland: effects of atmospheric aerosols, anthropogenic emissions, and meteorological conditions," *Remote Sensing* 13: 88 (2021).
35. Shuyue Yang, Xiaotong Zhang, et al., "A review and comparison of surface incident shortwave radiation from multiple data sources: satellite retrievals, reanalysis data and GCM simulations," *International Journal of Digital Earth* 16(1): 1332–1357 (2023).
36. *Ibid.*, Fig. 7(c).
37. Anthony Smith, *Mato Grosso* (New York: Dutton, 1971), p. 270.
38. William L. Thomas, Jr., ed., *Man's Role in Changing the Face of the Earth* (University of Chicago, 1956), p. 927.
39. Plato, Critias, 111, translation by H. D. P. Lee (*Plato: Timaeus and Critias*, Baltimore: Penguin, 1971).
40. *Boreal Logging Scars.* Wildlands League 2019. https://wildlandsleague.org/media/LOGGING-SCARS-FINAL-Dec2019-Exec-Summary.pdf
41. *Clearcut Carbon*, Sierra Club B.C., December 2019 https://sierraclub.bc.ca/wp-content/uploads/2019-Clearcut-Carbon-Executive-summary.pdf
42. Lucas Bessire, *Running Out: In Search of Water on the High Plains* (Princeton University Press, 2021).
43. William R. Pearson, *High Plains-Ogallala Aquifer Study: Water Transfer Element*, U.S. Army Corps of Engineers, 1987; Kansas Water Office and the U.S. Army Corps of Engineers, Kansas City District, *Update of 1982 Six State High Plains Aquifer Study, Alternate Route B*, 2015).
44. Bessire, pp. 177–179.
45. Chinchu Mohan, Andrew W. Western, et al., "Global assessment of groundwater stress vis-à-vis sustainability of irrigated food production," *Sustainability* 14: 16896 (2022).
46. *Sierra*, Jan./Feb. 1986, p. 38.
47. U.S. Dept. of Energy, *Reservoir Engineering 1976–2006: A History of Geothermal Energy Research and Development in the United States*, September 2010.
48. Abraham Lustgarten, "Injection wells: The poison beneath us," ProPublica, June 21, 2012, https://www.propublica.org/article/injection-wells-the-poison-beneath-us.
49. "Drilled too far: The perils of injection wells," Food and Water Watch, June 2020, https://www.foodandwaterwatch.org/wp-content/uploads/2021/03/fs_2006_injectionwells-web.pdf; William L. Ellsworth, "Injection-induced earthquakes," *Science* 341: 142–149 (2013).
50. Ground Water Protection Council, "Injection wells: An introduction to their use, operation, and regulation" (2005), https://www.academia.edu/download/35994773/JNGSE-CO2_paper_Xie_Economides.pdf.

51. *Ibid.*
52. Testimony before the House Subcommittee on Investigation and Oversight, March 16, 1983, on "Hazardous Waste Contamination of Water Resources."
53. Environmental Protection Agency, "EPA at 50: Managing waste across the nation." July 20, 2020, https://www.epa.gov/newsreleases/epa-50-managing-waste-across-nation.
54. Environmental Protection Agency. Guide for Industrial Waste Management (2016), https://www.epa.gov/sites/default/files/2016-03/documents/industrial-waste-guide.pdf.
55. David Lazarus, Toxic Technology, *San Francisco Chronicle*, Dec. 3, 2000, https://www.sfgate.com/bayarea/article/TOXIC-TECHNOLOGY-Critics-say-chemicals-used-in-3302899.php.
56. Sibélia Zanon, "Dam-building spree pushes Amazon Basin's aquatic life closer to extinction," *Mongabay*, 22 June 2023.
57. What's Up With Water, *Circle of Blue*, May 24, 2022, https://www.circleofblue.org/2022/world/whats-up-with-water-may-24-2022; One Million Dams Threaten Fish in Europe, *Teller Report*, Jan. 21, 2021, https://www.tellerreport.com/life/2021-01-21-%0A---one-million-dams-threaten-fish-in-europe%0A--.rJw_M73U1_.html.
58. Pinar Büyükakpinar, Simone Cesca, et al., "Reservoir-triggered earthquakes around the Atatürk Dam (Southeastern Turkey)," *Frontiers in Earth Science* 9: 663385 (2021).
59. Jonathan S. Stark, "Effects of lubricant oil and diesel on macrofaunal communities in marine sediments: A five year field experiment in Antarctica," *Environmental Pollution*, 311: 119885 (2022).
60. Jenna R. Jambeck, Roland Geyer, Chris Wilcox, et al., "Plastic waste inputs from land into the ocean," *Science* 347(6223): 768–771 (2015).
61. Speech given at the Long Island Press Distinguished Service Award Dinner, 1970, quoted by the Honorable Otis Pike in the Congressional Record, Oct. 6, 1970.
62. Thor Heyerdahl, *Fagu-Hiva* (Garden City: Doubleday, 1974), p. 267.
63. John and Mildred Teal, *The Sargasso Sea* (Boston: Little, Brown, 1975), p. 180.
64. Captain Charles Moore, *Plastic Ocean* (NY: Avery, 2011), p. 214.
65. *Ibid.*, p. 324.
66. "The New Plastics Economy," January 2016, newplasticseconomy.org.
67. Moore, p. 309.
68. *Ibid.* p. 222.
69. Chris Wilcox, Erik Van Sebille, and Britta Denise Hardesty, "Threat of plastic pollution to seabirds is global, pervasive, and increasing,"

Proceedings of the National Academy of Sciences 112(38): 11899–11904 (2015).
70. Peter G. Ryan, "The effects of ingested plastic on seabirds: Correlations between plastic load and body condition," *Environmental Pollution* 46(2): 119–125 (1987).
71. Moore, pp. 275–276.
72. Chelsea M. Rochman, Eunha Hoh, et al., "Ingested plastic transfers hazardous chemicals to fish and induces hepatic stress," *Scientific Reports* 3: 3263 (2013).
73. Moore, pp. 228–279.
74. *Ibid.*, p. 165.
75. https://outrider.org/nuclear-weapons/articles/chernobyl-slow-motion-under-arctic-seas.
76. Boris Worm, Edward B. Barbier, et al., "Impacts of biodiversity loss on ocean ecosystem services," *Science* 314: 787–790 (2006).
77. Christopher Costello, Daniel Ovando, and Tyler Clavelle, "Global fishery prospects under contrasting management regimes," *Proceedings of the National Academy of Sciences* 113(18): 5125–5129 (2016).
78. Ransom Myers and Boris Worm, "Rapid worldwide depletion of predatory fish communities," *Nature* 423: 280–3 (2003)).
79. Charles Clover, *The End of the Line: How Overfishing Is Changing the World and What We Eat* (Berkeley: University of California Press, 2006), p. 82.
80. *Ibid.*, p. 83.
81. Robert G. Bednarik, "The initial peopling of Wallacea and Sahul," *Anthropos* 92: 355–367 (1997); R. Fullagar, D. Price, and L. Head, "Early human occupation of Northern Australia," *Antiquity* 70: 751–752 (1996); G. Singh and E.A. Geissler, "Late Cainozoic history of vegetation, fire, lake levels, and climate at Lake George, New South Wales, Australia," *Philosophical Transactions of the Royal Society of London (Series B)* 311: 379–447 (1985).
82. Adam Brumm, Gerrit D. van den Bergh, et al., "Age and context of the oldest known hominin fossils from Flores, *Nature* 534: 249–253 (2016); Adam Brumm, Fachroel Aziz, and Gert D. van den Bergh, "Early stone technology on Flores and its implications for *Homo floresiensis*," *Nature* 441: 624–628 (2006); Robert G. Bednarik, "The earliest evidence of ocean navigation," *The International Journal of Nautical Archaeology* 26(3): 183–191 (1997); Robert G. Bednarik, "Seafaring in the Pleistocene," *Cambridge Archaeological Journal* 13(1): 41–46 (2003); Robin W. Dennell, Julien Louys, et al., "The origins and persistence of *Homo floresiensis* on Flores: biogeographical and ecological perspectives," *Quaternary Science Reviews* 96: 98–107 (2014); M. J. Morwood, P. B.

O'Sullivan, et al., "Fission-track ages of stone tools and fossils on the east Indonesian island of Flores," *Nature* 392: 173–176 (1998); Adam Brumm, Gitte M. Jensen, et al., "Hominins on Flores, Indonesia, by one million years ago," *Nature* 464: 748–752 (2010).

83. Paul S. Martin, "Prehistoric overkill," in *Pleistocene Extinctions: The Search for a Cause*, Paul S. Martin and H. E. Wright, eds. (New Haven: Yale University, 1967), p. 111.

84. Paulette F. C. Steeves, *The Indigenous Paleolithic of the Western Hemisphere*, University of Nebraska Press 2021; Steven R. Holen, Thomas A. Demère, et al., "A 130,000-year-old archaeological site in southern California, USA," *Nature* 544: 479–483 (2017); Juan Armenta Camacho, "Vestigios de labor humana en huesos de animales extintos de Valsequillo, Puebla, Mexico," presented at the 35th International Congress of the Americanists, Puebla, 1978; H. de Lumley, H. A. de Lumley, et al., Présence d'outils taillés associés à une faune quaternaire datée du Pleistocene moyen dans la Toca de Esperança, région de Central, État de Bahia, Brésil," *L'Anthropologie (Paris)* 91(4): 917–942 (1987); T. D. Dillehay, *The Settlement of the Americas: A New Prehistory*. NY: Basic Books 2000; Barney J. Szabo and Harold E. Malde, "Dilemma posed by uranium-series dates on archaeologically significant bones from Valsequillo, Puebla, Mexico," *Earth and Planetary Science Letters* 6: 237–244 (1969); Virginia Steen-McIntyre, Roald Fryxell and Harold E. Malde, "Geologic evidence for age of deposits at Hueyatlaco archeological site, Valsequillo, Mexico," *Quaternary Research* 16: 1–17 (1981).

85. Charlevoix, Pierre-François-Xavier de, *Journal d'un Voyage Fait Par Ordre du Rois Dans L'Amérique Septentrionale* (Paris 1744), Vol. V., Septième Lettre, p. 187, my translation.

86. Harry Allen, Michelle C. Langley, and Paul S. C. Taçon, "Bone projectile points in prehistoric Australia: evidence from archaeologically recovered implements, ethnography, and rock art," in Michelle C. Langley (ed.), *Osseous Projectile Weaponry*, Springer, Netherlands 2017, pp. 209–218.

87. Daniel H. Janzen and Paul S. Martin, "Neotropical anachronisms: The fruits the Gomphotheres ate," *Science* 215: 19–27 (1982).

88. Firstenberg, Arthur, *The Invisible Rainbow: A History of Electricity and Life* (White River Junction, VT: Chelsea Green, 2020), Chapters 10, 11, 12, and 13.

89. Herbert L. Ratcliffe, T. G. Yerasimides, and G. A. Elliott, "Changes in the character and location of arterial lesions in mammals and birds in the Philadelphia Zoological Garden," *Circulation* 21: 730–738 (1960); Kathleen J. Rigg, R. Finlayson, et al., "Degenerative arterial disease of animals in captivity with special reference to the comparative pathology of atherosclerosis," *Proceedings of the Zoological Society of London* 135(2):

157–164 (1960); Marcel M. Vastesaeger and R. Delcourt, "The natural history of atherosclerosis," *Circulation* 26: 851–855 (1962); Louise S. Lombard and Ernest J. Witte, "Frequency and types of tumors in mammals and birds of the Philadelphia Zoological Gardens," *Cancer Research* 19(2): 127–141 (1959); "Yann C. Klimentidis, T. Mark Beasley, et al., "Canaries in the coal mine: A cross-species analysis of the plurality of obesity epidemics," *Proceedings of the Royal Society B* 278: 1626–1632 (2011).

90. Augustus D. Imms, "Report on a disease of bees in the Isle of Wight," *Journal of the Board of Agriculture* 14(3): 129–140 (1907).

91. Caspar A. Hallman, Martin Sorg, et al., "More than 75 percent decline over 27 years in total flying insect biomass in protected areas," *PLoS ONE* 22(10): e0185809 (2017).

92. Bradford C. Lister and Andres Garcia, "Climate-driven declines in arthropod abundance restructure a rainforest food web," *Proceedings of the National Academy of Sciences* 115(44): E10397-E10406 (2018).

93. Francisco Sánchez-Bayo and Kris A. G. Wyckhuys, "Worldwide decline of the entomofauna: A review of its drivers," *Biological Conservation* 232 8–27 (2019).

94. Personal communication from New Mexico pigeon racer Larry Lucero, 1999.

95. J. Alan Tanner, "Effect of microwave radiation on birds," *Nature* 210: 636 (1966); J. Alan Tanner, "Bird feathers as sensory detectors of microwave fields," in Stephen F. Cleary, ed., *Biological Effects and Health Implications of Microwave Radiation. Symposium Proceedings* (Rockville, MD, U.S. Dept. of Health, Education and Welfare), Publication BRH/DBE 70-2, pp. 185–187 (1970); Jaime Bigu del Blanco and César Romero-Sierra, "Bird feathers as dielectric receptors of microwave radiation," Laboratory Technical Report LTR-CS-89, Control Systems Laboratory, Division of Mechanical Engineering, National Research Council Canada (1973).

96. Alfonso Balmori and Örjan Hallberg, "The urban decline of the house sparrow (*Passer domesticus*): A possible link with electromagnetic radiation," *Electromagnetic Biology and Medicine* 26:141–151 (2007).

97. Jenny De Laet and James Denis Summers-Smith, "The status of the urban house sparrow *Passer domesticus* in north-western Europe: a review," *Journal of Ornithology* 148 (suppl. 2) S275–278.

98. Colin Galbraith, "The population status of birds in the U.K.: Birds of conservation concern: 2002–2007," *Bird Populations* 7: 173–179 (2002).

99. Sainudeen Pattazhy, "Dwindling number of sparrow," *Karala Calling*, March 2012, pp. 32–33.

100. Benita Sen, "Calling back the sparrow," *Deccan Herald*, November 16, 2012.

101. Svenja Engels, Nils-Lasse Schneider, et al., Anthropogenic electromagnetic noise disrupts magnetic compass orientation in a migratory bird, *Nature* 509: 353–356 (2014).
102. Arthur Firstenberg, *Birds on Texel Island*, Cellular Phone Task Force, 2022, https://cellphonetaskforce.org/birds-on-texel-island-2; Arthur Firstenberg, *Sea Birds' Last Refuges*, Cellular Phone Task Force, 2023, https://cellphonetaskforce.org/sea-birds-last-refuges.
103. Arthur Firstenberg, "Acute electrical illness," in Arthur Firstenberg, *The Invisible Rainbow: A History of Electricity and Life*, White River Junction, VT: Chelsea Green, 2020, pp. 75–93; Arthur Firstenberg, "Mystery on the Isle of Wight," *Ibid.*, pp. 95–112.
104. Arthur Firstenberg, "You mean you can hear electricity?," *Ibid.*, pp. 275–321.
105. Demetria Lee, "Cave in southeast Minnesota sees another dramatic decline in bat population," *Minnpost* March 20, 2019, https://www.minnpost.com/environment/2019/03/cave-in-southeast-minnesota-sees-another-dramatic-decline-in-bat-population.
106. https://pgc.pa.gov/Education/WildlifeNotesIndex/Pages/Bats.aspx.
107. https://inews.co.uk/news/environment/bats-declining-uk-insect-numbers-halloween-advice-tips-356141.
108. https://www.stuff.co.nz/marlborough-express/news/100537330/lack-of-insects-bad-news-for-bats.
109. Jens Rydell, Marcus Elfström, et al., "Dramatic decline of northern bat *Eptesicus nilssonii* in Sweden over 30 years," *Royal Society Open Science* 7: 191754 (2020).
110. Marcia Barinaga, "Where have all the froggies gone?" *Science* 247(4946): 1033–1034 (1990).
111. *Ibid.*; also: David B. Wake and Vance T. Vredenburg, "Are we in the midst of the sixth mass extinction? A view from the world of amphibians," *Proceedings of the National Academy of Sciences* 105(Suppl. 1): 11466–11473 (2008).
112. Barinaga, *op. cit.*
113. *Ibid.*
114. *Ibid.*
115. J. Alan Pound and Martha I. Crump, "Amphibian declines and climate disturbance: The case of the golden toad and the harlequin frog," *Conservation Biology* 8(1): 72–85 (1994).
116. Charles A. Drost and Gary M. Fellers, "Collapse of a regional frog fauna in the Yosemite area of the California Sierra Nevada, USA," *Conservation Biology* 10(2): 414–425 (1996).
117. Pound and Crump, *op. cit.*

118. Sonia L. Ghose, Tiffany A. Yap, et al., *Frontiers in Conservation Science* 4: 1069490 (2023).
119. Robert O. Becker and Gary Selden, *The Body Electric* (NY: William Morrow, 1985).
120. William Souder, "An amphibian horror story," *New York Newsday*, Oct. 15, 1996, p. B19+; Christopher Hallowell, "Trouble in the Lily Pads," *Time*, October 28, 1996, p. 87.
121. Hoperskaya, O.A., L.A. Belkova, M.E. Bogdanov, and S.G. Denisov. "The action of the "Gamma-7N" device on biological objects exposed to radiation from personal computers." In *Electromagnetic Fields and Human Health: Proceedings of the Second International Conference*, Moscow, Sept. 20–24, 1999, pp. 354–355, Abstract.
122. Balmori, Alfonso. "Mobile phone mast effects on common frog (*Rana temporaria*) tadpole: The city turned into a laboratory." *Electromagnetic Biology and Medicine* 29: 31–35 (2010).
123. *The Chautauquan* 22: 203–206 (1895).
124. *Audubon*, September 1990, pp. 50–104.
125. Frans Lanting, "Botswana: A gathering of waters and wildlife," *National Geographic*, Dec. 1990, pp. 5–37.
126. Michael J. Chase, Scott Schlossberg, et al., "Continent-wide survey reveals massive decline in African savannah elephants," *PeerJ* 4:e2354 (2016).
127. Harry Selby, professional hunter and safari leader (in Lanting, *op. cit.*, 1990).
128. Peter O. Thomas, Randall R. Reeves, and Robert L. Brownell, Jr., "Status of the world's baleen whales," *Marine Mammal Science* 32(2): 682–734 (2016).
129. Kent E. Carpenter, Mohammad Abrar, et al., "One-third of reef-building corals face elevated extinction risk from climate change and local impacts," *Science* 321(5888): 560–563 (2008).
130. Daniel G. Boyce, Marlon R. Lewis, and Boris Worm, "Global phytoplankton decline over the past century," *Nature* 466: 591–596 (2010).
131. A. K. Mohammad Mohsin and Mohammad Azmi Ambak, *Freshwater Fishes of Peninsular Malaysia* (Universiti Pertanian Malaysia, 1983).
132. Terry Erwin, "The tropical forest canopy: the heart of biotic diversity," in *Biodiversity*, E. O. Wilson, ed. (Washington: National Academy, 1988), p. 129.
133. Reprinted in *Indiana Historical Society Publications*, Vol. 10, No. 5, 1933.
134. Andrew Falle, Ewan Wright, et al., "One million (paper) satellites," *Science* 382(6667): 150–152 (2023).

135. Canadian Astronomical Society, *Report on Mega-Constellations to the Government of Canada and the Canadian Space Agency*," March 31, 2021, https://arxiv.org/pdf/2104.05733.
136. Jesse Granger, Lucianne Walkowicz, et al., "Gray whales strand more often on days with increased levels of atmospheric radio-frequency noise," *Current Biology* 30 R135–R158 (2020).
137. Wang, Zhanyun, Glen W. Walker, Derek C. G. Muir, and Kakuko Nagatani-Yoshida, "Toward a global understanding of chemical pollution: A first comprehensive analysis of national and regional chemical inventories," *Environmental Science and Technology* 54: 2575–2584 (2020).
138. W. Ross Adey, "Neurophysiologic effects of radiofrequency and microwave radiation," *Bulletin of the New York Academy of Medicine* 55: 1079–1093 (1979); W. Ross Adey, "Some fundamental aspects of biological effects of extremely low frequencies (ELF)," in *Biological Effects and Dosimetry of Non-Ionizing Radiation*, M. Grandolfo et al., eds. (New York: Plenum, 1983), pp, 561–580.
139. S. M. Bawin, R. J. Gavalas-Medici, and W. R. Adey, "Reinforcement of transient brain rhythms by amplitude-modulated VHF fields," in *Biologic and Clinical Effects of Low-Frequency Magnetic and Electric Fields*, J. G. Llaurado et al., eds. (Springfield: C. Thomas, 1974), pp, 172–173.
140. W. R. Klemm, *Animal Electroencephalography* (New York: Academic, 1969), pp. 165–167.
141. L. David Mech and Shannon M. Barber, *A Critique of Wildlife Radio-Tracking and Its Use in National Parks*, U.S. Geological Survey, Northern Prairie Wildlife Research Center (2002); Jon E. Swenson, Kjell Wallin, et al., "Effects of ear-tagging with radio transmitters on survival of moose calves." *Journal of Wildlife Management* 63(1) 354–358 (1999); Tom P. Moorhouse and David W. Macdonald, "Indirect negative impacts of radio-collaring: Sex ratio variation in water voles," *Journal of Applied Ecology* 42: 91–98 (2005); "The snow tiger's last stand," *Reader's Digest*, November 1998; Jason D. Godfrey and David M. Bryant, "Effects of radio transmitters: Review of recent radio-tracking studies," in: M. Williams, ed., *Conservation Applications of Measuring Energy Expenditures of New Zealand Birds Assessing Habitat Quality and Costs of Carrying Radio Transmitters* (Wellington, New Zealand Dept. of Conservation, pp. 83–85 (2003)).
142. Behnoush Maherani, Farah Hossain, et al., "World market development and consumer acceptance of irradiation technology," *Foods* 5: 79 (2016); Tamikazu Kume and Setsuko Todoriki, "Food irradiation in Asia, the European Union and the United States," *Radioisotopes* 62(5): 291–299 (2013).

143. Melanie Bergmann, Sophia Mützel, Sebastian Primpke, et al., "White and wonderful? Microplastics prevail in snow from the Alps to the Arctic," *Science Advances* 2019:5.
144. Steve Allen, Deonie Allen, et al., "Atmospheric transport and deposition of microplastics in a remote mountain catchment," *Nature Geoscience* 12: 339–344 (2019).
145. Janice Brahney, Margaret Hallerud, et al., "Plastic rain in protected areas of the United States," *Science* 368: 1257–1260 (2020).
146. Danyang Tao, Kai Zhang, et al., "Microfibers released into the air from a household tumble dryer," *Environmental Science and Technology Letters* 9(2): 120–126 (2022).
147. Claudia Dessì, Elvis D. Okoffo, et al., "Plastics contamination of store-bought rice," *Journal of Hazardous Materials* 416: 125778 (2021); Mary Kosuth, Sherri A. Mason, and Elizabeth V. Wattenberg, "Anthropogenic contamination of tap water, beer, and sea salt," *PLoS ONE* 13(4): e0194970 (2018).
148. Antonio Ragusa, Alessandro Svelato, et al., "Plasticenta: First evidence of microplastics in human placenta," *Environment International* 146: 106274 (2021).
149. Philipp Schwabl, Sebastian Köppel, et al., "Detection of Various Microplastics in Human Stool," *Annals of Internal Medicine* 171(7): 453–457 (2019).
150. Neda Sharifi Soltani, Mark Patrick Taylor, et al., "International quantification of microplastics in indoor dust: prevalence, exposure and risk assessment," *Environmental Pollution* 312: 119957 (2022); Graham Readfearn, "It's on our plates and in our poo, but are microplastics a health risk?" *The Guardian*, May 15, 2021.
151. H. C. Borghetty and C. A. Bergman, "Synthetic detergents in the soap industry," *Journal of the American Oil Chemists' Society*, March 1950, pp. 88–90.
152. Foster Dee Snell, "Soap vs. synthetic detergents," *Journal of the American Oil Chemists' Society*, July 1949, pp. 338–341.
153. *Real Goods News*, Summer/Fall 1980, p. 20.
154. Alan Devoe, *Our Animal Neighbors* (New York: McGraw-Hill, 1953), pp. 184–185.
155. *The Notebooks of Leonardo Da Vinci*, arranged and translated by Edward MacCurdy (New York: Reynal and Hitchcock, 1938), p. 284.
156. Emrecan Demirors, Jiacheng Shi, Anh Duong, et al., "The SEANet Project: Toward a programmable Internet of Underwater Things," In *Proceedings of the 2018 Fourth Underwater Communications and Networking Conference (UComms)*, Lerici, Italy, 28–30 August 2018; Jose Ilton de Oliveira Filho, Abderrahmen Trichili, Boon S. Ooi et al,

"Towards self-powered Internet of Underwater Things devices," *IEEE Communications Magazine*, July 23, 2019, arXiv: 1907.11652; Syed Agha Hassnain Mohsan, Alireza Mazinani, et al., "Towards the internet of underwater things: a comprehensive survey," *Earth Science Informatics* 15: 735–764 (2022).

157. The calculations were made assuming the following assumptions: (a) a person takes a breath of about half a liter of air about every four seconds; (b) for ever liter of air breathed, 40 cubic centimeters of oxygen are absorbed by the lungs; (c) an average car gets 20 miles to the gallon; (d) a Boeing 747–200B jet airliner with 4 Rolls-Royce turbofans RF 211–254 burns 17,849 pounds of fuel per hour per engine on takeoff, and 5,450 piounds per hour while cruising (FAA data provided in 1993 by Marc Roddin, Manager of Seaport and Airport Planning, Metropolitan Transportation Commission, Oakland, California); (e) each pound of fuel contains 0.86 pounds of carbon and 0.14 pounds of hydrogen; (f) each tree produces 2,500 gallons of oxygen daily, of which half is used for its own respiration; and (g) a one-acre forest contains about 100 mature trees.

158. A Henderson-Sellers, "North American total cloud amount variations this century," *Global and Planetary Change*, Aug. 1989, p. 175.

159. Lewis Mumford, *The Highway and the City* (New York: Harcourt, Brace, and World, 1963), p. 243,

160. World Population Review, "Gun deaths by country 2023," https://www.worldpopulationreview/country-rankings/gun-deaths-by-country.

161. Munshi Naser Ibne Afzal and Jeff Gow, "Electricity consumption and information and communication technology in the next eleven emerging economies," *International Journal of Energy Economics and Policy* 6(3): 381–388 (2016); Mark Mills, *The Internet Begins with Coal.* Greening Earth Society, 1999; Mark Mills, *The Cloud Begins with Coal.* Digital Power Group, August 2013; Mark P. Mills and Peter W. Huber, "Dig more coal – the PCs are coming," *Forbes*, May 31, 1999, pp 70–72.

162. Alisa Zainu'ddin, A *Short History of Indonesia* (New York: Praeger, 1970), p. 116.

163. J. H. Boeke, *The Interests of the Voiceless Far East* (Leiden: Universitaire Pers Leiden, 1948), p. 3.

164. W. Arens, *The Man-Eating Myth* (New York: Oxford University, 1979).

165. Quoted on pages 3–4 of Paul Jacobs and Saul Landau, *To Serve the Devil* (New York: Random, 1971), Vol. 1.

166. Father Chrétien Le Clercq, *New Relations of Gaspesia* (1691, reprinted by The Champlain Society, Toronto, 1910), pp. 103–106.

167. George Bird Grinnell, *Pawnee Hero Stories and Folk Tales* (1889, reprinted by University of Nebraska Press, Lincoln, 1961), pp. 258–259.

168. Quoted in Jacobs and Landau, Vol. 1, p. 50.
169. Quoted in T. L. McLuhan, *Touch the Earth: A Portrait of Indian Existence* (New York: Outerbridge and Dienstfrey, 1971), p. 67.
170. Quoted in Fridtjof Nansen, *Eskimo Life* (London: Longmans, Green, 1893), pp. 180–184.
171. Lucien Bodard, *Green Hell* (New York: Outerbridge and Dienstfrey, 1971), pp. 161–162.
172. Quoted in Emilienne Ireland, "Neither warriors nor victims, the Wauja peacefully organize to defend their land," *Cultural Survival Quarterly*, pp. 58–59.
173. Pertti J. Pelto, *The Snowmobile Revolution: Technology and Social Change in the Arctic* (Prospect Heights, Ill, Waveland, 1987). p. 205.
174. Michael S. Teitelbaum and Jay M. Winter, *The Fear of Population Decline* (Orlando: Academic, 1985), p. 122.
175. *Population and Development Review* 10: 571 (1984).
176. Resolution No. C127/78m 14,5,84 of the European Economic Community, reproduced in Teitelbaum and Winter, p. 162.
177. Thomas Robert Malthus, *Essay on the Principle of Population as It Affects the Future Improvement of Society* (London: J. Johnson, 1798).
178. David Hume, "Of the populousness of ancient nations," in his *Political Discourses* (Edinburgh: R. Fleming, 1752), pp. 155–261.
179. Montesquieu, *The Persian Letters* (1721, reprinted by Bobbs-Merrill, Indianapolis, 1964), letters CXII to CXXII; Robert Wallace, *A Dissertation on the Numbers of Mankind in Ancient and Modern Times* (1753, reprinted by Augustus M. Kelley, New York, 1969).
180. T. H. Hollingsworth, *Historical Demography*. Ithaca: Cornell University Press, 1969.
181. Tertullian, *De Anima*, chapter 30, my translation.
182. Han Fei Tzu, chapter 49. I have used the translation appearing in Garrett Hardin, *Population, Evolution and Birth Control* (San Francisco: Freeman, 1969), p. 18. See also *The Complete Works of Han Fei Tzu*, 2 volumes, translation by W. K. Liao (London: Probsthain, 1939–1959).
183. 2 parents, 4 grandparents, 8 great-grandparents, etc, numbering 32,768 ancestors of the 15th generation. The marriage of cousins will reduce the number of ancestors. For an interesting viewpoint on race, see Jack D. Forbes, "The manipulation of race, caste and identity: Classifying Afroamericans, Native Americans and Red-Black People," *Journal of Ethnic Studies*, Winter 1990, pp. 1–51.
184. Plato, *Symposium*, 175e.
185. Edward Gibbon, *Decline and Fall of the Roman Empire* (1776, reprinted by The Modern Library, New York, 1932), Vol. 1, pp. 43–44.

186. Duarte Lopez, *A Report on the Kingdom of Congo* (1591, reprinted by Negro Universities Press, New York, 1969), pp. 65–70.
187. Basil Davidson, *The Lost Cities of Africa* (Boston: Little, Brown, 1970), pp. 80–81.
188. E. D. Morel, *Great Britain and the Congo: the Pillage of the Congo Basin* (London: Smith, Elder, 1909), p. xiv.
189. Siddharth Kara, *Cobalt Red* (NY: Saint Martin's Press, 2023); Oluwole Ojewale, *Mining and illicit trading of coltan in the Democratic Republic of Congo*, ENACT, Research paper 29, March 2022; Jenna Marie Goldblatt, *Conflict and Coltan: Resource Extraction and Collision in The Democratic Republic of the Congo and Venezuela*, Fordham University student thesis, May 19, 2023; Anna Rosa Juurlink, *Congo, Coltan and Conservation*, Forest and Nature Conservation Policy Group, Master of Science thesis, Wageninen University, Dec. 22, 2021; *Congo's Mining Slaves*, Free the Slaves Investigative Report, June 2013, https://freetheslaves.net/wp-content/uploads/2015/03/Congos-Mining-Slaves-web-130622.pdf; *DRC: Efe Pygmies deprived of their homeland and their livelihood*, World Rainforest Movement, Bulletin 118, May 19, 2007; Blaine Harden, "The dirt in the new machine," *New York Times Magazine*, August 12, 2001; "The forest people: life and death under the Green Revolution," by Stone Age Herbalist, https://www.resilience.org/stories/2021-05-05/the-forest-people-life-and-death-under-the-green-revolution.
190. John Boyd Thacher, *Christopher Columbus, His Life, His Work, His Remains* (New York: G. P. Putnam's Sons, 1903), Vol. 2, Narrative of Third Voyage," pp. 379–380.
191. *De Orbe Novo: The Eight Decades of Peter Martyr d'Anghera*, first published 1516, translation by Francis Augustus MacNutt (New York: G. P. Putnam's Sons, 1912), The Third Decade, Book 1, p. 286.
192. Jean Frédéric Bernard, *Recueil de Voyage au Nord* (Amsterdam, 1731), pp. 128–129.
193. Fray Gaspar de Carvajal, *Relación del Nuevo Descubrimiento del Famoso Río Grande de las Amazonas* (Mexico: Fondo de Cultura Económica, 1955), p. 63, my translation.
194. Al-'Umari, "Masālik Al-Absār fi mamalik al-amsār," 1342–1349, translated into French by Joseph M. Cuoq, *Recueil des Sources Arabes Concernant L'Africue Occidentale du VIIIe au XVIe Siècle* (Paris: Editions du Centre National de la Recherche Scientifique, 1975), pp. 274–275, my translation from the French.
195. Antonio Pigafetta, *Magellan's Voyage: A Narrative Account of the First Circumnavigation*, original publication 1563, translation by R. A. Skelton (New Haven: Yale University, 1969), Vol. I, Chapter X, p. 51. The existence of such a map as early as the 1420s is recorded in Antonio

Galvano, *The Discoveries of the World*, 1563, published in English by Richard Hakluyt, 1601, reprinted by Burt Franklin, New York, Hakluyt Society #30, 1971, p. 66.

196. Montezuma recounted his people's history to Cortés and his men on their meeting on Nov. 8, 1519. See Cortés' second letter to Charles V, in *Cartas de Relación* (Madrid: Hermanos García Noblejas, 1985), pp. 116–117. For English translation see *Letters from Mexico*, translation by Anthony Pagden (New Haven: Yale University, 1986), pp. 85–86. See also Zelia Nuttall, "Some Unsolved Problems in Mexican Archeology," *American Anthropologist*, New Series, 8: 133–149 (1906).
197. Diodorus, Book V, 19, 20; Aristotle, *On Marvelous Things Heard*, 84.
198. Pliny, II, 170.
199. Thor Heyerdahl, *Early Man and the Ocean* (Garden City: Doubleday, 1979), pp. 191–192.
200. Ma Twan-Lin, *Antiquarian Researches*, about 1321, translation by Samuel Wells Williams (New Haven: Tuttle, Morehouse and Taylor, 1881); *Shih Chi*, chapter 6, quoted on p. 115 of Derk Bodde, *China's First Unifier* (Leiden: E. J. Brill, 1938); *Shan Hai Ching*, translation by Hsiao-Chieh Cheng, Hui-Chen Pai Cheng and Kenneth Lawrence Thern (Republic of China, 1985).
201. Lord Kingsborough, *Antiquities of Mexico* (London: R. Havell, 1831), Vol. 6, pp. 156–159, text accompanying Plate I of the Codex Vaticanus.
202. Columbus's journal of Dec. 16 and 26, 1492, in Fray Bartolomé de las Casas, *The Diario of Christopher Columbus's First Voyage to America, 1492–1493*, translation by Oliver Dunn and James E. Kelley, Jr. (Norman: University of Oklahoma, 1989), pp. 233 and 289.
203. Carvajal, pp. 78, 80, 100.
204. R. W. Whiteway, *The Rise of Portuguese Power in India 1497–1550* (Patna: Janaki Prakashan, 1979), pp. 33–35.
205. *Anthropos* 52: 239–262 (1957).
206. Charles K. Meek, *Law and Authority in a Nigerian Tribe* (London: Oxford University, 1937), p. 242.
207. William Wilson, A *Missionary Voyage to the Southern Pacific Ocean Performed in the Years 1796, 1797, 1798 in the Ship Duff, Commanded by Captain James Wilson* (London: T. Chapman, 1799), p. 363.
208. *Oceania* 1: 15–16 (1930–31).
209. *Man* 20(5): 10–12 (1920).
210. *Oceania* 1: 18 (1930–31).
211. Dorothy Shineberg, *The Trading Voyages of Andrew Cheyne 1841–1844* (Honolulu: University of Hawaii, 1971), p. 105.

212. Mervyn Meggitt, *Blood Is Their Argument: Warfare Among the Mae Enga Tribesmen of the New Guinea Highlands* (Palo Alto: Mayfield, 1977), pp. 17–21.
213. Stanton Coblenz, *From Atom to Atom Bomb* (New York: Beechhurst, 1953), p. 20.
214. Frances Densmore, *Papago Music* (Washington: U.S. Government Printing Office, 1929), p. 187.
215. Edward William Nelson, *The Eskimo About Bering Strait* (Washington: U.S. Government Printing Office, 1900), p. 329.
216. Coblenz, p. 18.
217. J. B. Tyrrell, ed., *David Thompson's Narrative of His Explorations in Western America 1784–1812* (Champlain Society Publication 12, reprinted by Greenwood Press, New York, 1968), pp. 328–335.

Bibliography

I am indebted to these authors for the clarity of their thinking or the thoroughness of their vision:

Bates, Marston, *The Forest and the Sea*. New York: Vintage, 1960.
———, *Where Winter Never Comes*. New York: C. Scribner's Sons, 1952.
Bessire, Lucas, *Running Out: In Search of Water on the High Plains*. Princeton University Press, 2021.
Carson, Rachel, *The Edge of the Sea*. Boston: Houghton Mifflin, 1955.
———, *The Sea Around Us*. New York: Oxford University Press, 1951.
———, *Silent Spring*. Boston: Houghton Mifflin, 1962.
———, *Under the Sea-Wind*. New York: Simon & Schuster, 1941.
Clover, Charles, *The End of the Line: How Overfishing Is Changing the World and What We Eat*. University of California Press, 2006.
Commoner, Barry, *The Closing Circle: Nature, Man, and Technology*. New York: Alfred A. Knopf, 1971.
De Grazia, Sebastian, *Of Time, Work, and Leisure*. Garden City: Doubleday, 1962.
Devoe, Alan and Mary Devoe, *Our Animal Neighbors*. New York: McGraw-Hill, 1953.
Dubos, René, *Beast or Angel?* New York: C. Scribner's Sons, 1974.
———, *Celebrations of Life*. New York: McGraw-Hill, 1981.
———, "Conservation, stewardship, and the human spirit," *Audubon*, Sept. 1972, pp. 21–28.
———, *Mirage of Health*. New York: Harper & Row, 1959.
———, *The Wooing of Earth*. New York: C. Scribner's Sons, 1980.
Eiseley, Loren, *The Firmament of Time*. New York: Atheneum, 1960.
———, *The Immense Journey*. New York: Random, 1957.
———, *The Invisible Pyramid*. New York: C. Scribner's Sons, 1970.
———, *The Man Who Saw Through Time*. New York: C. Scribner's Sons, 1973.
———, *The Star Thrower*. New York: Random 1978.
Goldschmidt, Walter, *Man's Way*. New York: Holt, 1959.
Huxley, Anthony, *Plant and Planet*. New York: Pelican, 1975.

Kara, Siddharth, *Cobalt Red: How the Blood of the Congo Powers Our Lives*. New York: St. Martin's Press, 2023.
Marsh, George Perkins, *Man and Nature*. New York: C. Scribner, 1864.
Marx, Wesley, *The Frail Ocean*. New York: Ballantine, 1967.
Moore, Captain Charles, *Plastic Ocean*. New York: Avery, 2011.
Ponting, Clive, A *New Green History of the World*. London: Vintage, 2007.
Roueché, Berton, *What's Left: Reports on a diminishing America*. New York: Little, Brown, 1968.
Steeves, Paulette F. C., *The Indigenous Paleolithic of the Western Hemisphere*. University of Nebraska Press, 2021.
Thomas, William L., Jr., editor, *Man's Role in Changing the Face of the Earth*. University of Chicago Press, 1956.
Tolstoy, Ivan, *The Pulse of a Planet*. New York: New Amsterdam Library, 1971.
Vogt, William, *Road to Survival*. New York: William Sloane Associates, 1948
Wallace, Alfred Russel, *Tropical Nature and Other Essays*. London: Macmillan, 1878.
Wilcove, David S., *No Way Home: The Decline of the World's Great Animal Migrations*. Washington: Island Press, 2008.

Other sources and related materials are listed below:

General

American Chemical Society, *Cleaning Our Environment: A Chemical Perspective*. Washington, D.C., 1978.
Basalla, George, *The Evolution of Technology*. Cambridge University Press, 1988.
Benn, F. R. and C. A. McAuliffe, eds., *Chemistry and Pollution*. London: Macmillan, 1975.
Cailliet, Greg M., Paulette Y. Setzer, and Milton S. Love, *Everyman's Guide to Ecological Living*. New York: Macmillan, 1971.
Center for Science in the Public Interest, *99 Ways to a Simple Lifestyle*. Garden City: Anchor, 1977.
Daumas, Maurice, ed., A *History of Technology and Invention*. New York: Crown, 1962.
De Bell, Garrett, ed., *The Environmental Handbook*. New York: Ballantine, 1970.
———, *The New Environmental Handbook*. San Francisco: Friends of the Earth, 1980.
Diamant, R. M. E., *The Prevention of Pollution*. London: Pitman, 1974.
Ellul, Jacques, *The Technological Bluff*. Grand Rapids: W. Eerdmans, 1990.
Forbes, R. J., *The Conquest of Nature: Technology and Its Consequences*. New York: Praeger, 1968.

Hughes, J. Donald, *Ecology in Ancient Civilizations*. Albuquerque: University of New Mexico Press, 1975.

Johnson, Huey D., ed., *No Deposit—No Return*. Reading, Mass.: Addison-Wesley, 1970.

Kroeber, Alfred, *Anthropology*. New York: Harcourt, Brace, and World, 1948.

Laurence, William F., "The Anthropocene," *Current Biology* 29: R942–R995 (2019).

Love, Glen A. and Rhoda M. Love, eds., *Ecological Crisis: Readings for Survival*. New York: Harcourt, Brace, Jovanovich, 1970.

McNeil, Ian, ed., *An Encyclopedia of the History of Technology*. London: Routledge, 1990.

McRobie, George, *Small Is Possible*. New York: Harper & Row, 1981.

Meadows, Donella H., Dennis L. Meadows, et al., *The Limits to Growth*. New York: New American Library, 1972.

Mumford, Lewis, *The Myth of the Machine*, 2 volumes. New York: Harcourt, Brace, Jovanovich, 1967 and 1970.

Murchie, Guy, *The Seven Mysteries of Life*. Boston: Houghton Mifflin, 1978.

Nobile, Philip and John Deedy, eds., *The Complete Ecology Fact Book*. Garden City: Doubleday, 1972.

Novick, Sheldon and Dorothy Cottrell, *Our World in Peril: An Environmental Review*. Greenwich, Conn.: Fawcett, 1971.

Osborn, Fairfield, *Our Plundered Planet*. Boston: Little, Brown, 1948.

Pyke, Magnus, *The Science Myth*. London: J. Murray, 1962.

Raven, P. H., *We're Killing Our World: The Global Ecosystem in Crisis*. Chicago: MacArthur Foundation, 1987.

Schumacher, E. F., *Small Is Beautiful: Economics As If People Mattered*. New York: Harper & Row, 1973.

Sears, Paul B., *Lands Beyond the Forest*. Englewood Cliffs, N.J.: Prentice-Hall, 1969.

Shaheen, Esber I., *Environmental Pollution: Awareness and Control*. Mahomet, Ill.: Engineering Technology, 1974.

Singer, Charles, E. J. Holmyard, and A. R. Hall, eds., *A History of Technology*, 5 vols. Oxford: Clarendon, 1954–1958.

Singer, S. Fred, ed., *The Changing Global Environment*. Dordrecht, Netherlands: D. Reidel, 1975.

Ternes, Alan, ed., *Ants, Indians, and Little Dinosaurs*. New York: C. Scribner's Sons, 1975.

White, Lynn, Jr., "The historical roots of our ecological crisis," *Science* 155: 1203–1207 (1967).

"Will earth survive man? A planetary life or death struggle is unfolding," *United Nations Chronicle*, June 1988, pp. 40–51.

Which Way Is Forward? (pp. 1–3)
Carson, Rachel, *Silent Spring*. Boston: Houghton Mifflin, 1962.
Ford, Worthington Chauncey, ed., A *Cycle of Adams Letters 1861–1865*. Boston: Houghton Mifflin, 1920.

1. Road to the Present (pp. 9–20)
Bender, Barbara, *Farming in Prehistory*. New York: St. Martin's 1975.
Day, Michael H., *Guide to Fossil Man*, 4th edition. University of Chicago Press, 1986.
Diamond, Arthur Sigismund, *The Evolution of Law and Order*. London: Watts, 1951.
Heiser, Charles B., Jr., *Seed to Civilization*. San Francisco: Freeman, 1973.
Jacobs, Jane, *The Economy of Cities*. New York: Random, 1969.
Leakey, Richard E., *The Making of Mankind*. New York: Dutton, 1981.
Leakey, Richard E. and Roger Lewin, *People of the Lake*. Garden City: Anchor, 1978.
Lee, Richard B., "What hunters do for a living, or How to make out on scarce resources," in Lee, R. B. and I. De Vore, *Man the Hunter*. Chicago: Aldine, 1968, pp. 30–48, on the !Kung.
Vavilov, N. I., *The Origin, Immunity and Breeding of Cultivated Plants. Chronica Botanica*, Vol. 13, Waltham, Mass., 1949/50.

2. Self-Portrait (pp. 21–28)
Audus, L. J., *Plant Growth Substances*. London: Leonard Hill, 1959. Chapter XV," Natural Plant Growth Inhibitors," pp. 384–404.
Bright, Michael, *Animal Languages*. Ithaca: Cornell University Press, 1984.
Clarke, James, *Man Is the Prey*. New York: Stern and Day, 1969.
Dröscher, Vitus B., *The Friendly Beast: Latest Discoveries in Animal Behavior*. New York: Harper & Row, 1972.
Griffin, Donald R., *Animal Thinking*. Cambridge, Mass.: Harvard University Press, 1984.
———, *The Question of Animal Awareness*. New York: Rockefeller University Press, 1976.
Lorenz, Konrad, *On Aggression*. New York: Harcourt, Brace, and World, 1966.
Masson, Jeffrey Moussaieff, *When Elephants Weep: The Emotional Lives of Animals*. New York: Dell, 1995.
McIntyre, Joan, *Mind in the Water*. Toronto: McClelland and Stewart, 1974.
Morris, Desmond, *The Naked Ape*. New York: McGraw-Hill, 1967.
Rivers, W. H. R., *Social Organization*. New York: Knopf, 1924.
Wynne-Edwards, V. C., "Population control in animals," *Scientific American*, Aug. 1964, pp. 68–74.
———, *Evolution Through Group Selection*. Oxford: Blackwell Scientific, 1986.

4. A Living Universe (pp. 32–36)

Alfvén, Hannes, *Atom, Man, and Universe*. San Francisco: Freeman, 1969.
———, *Cosmic Plasma*. Dordrecht: D. Reidel, 1981.
———, "Cosmology in the plasma universe: An introductory exposition," *IEEE Transactions on Plasma Science* 18(1): 5–10 (1990).
———, "The Plasma Universe," *Physics Today*, Sept. 1986, pp. 22–27.
———, *Worlds – Antiworlds*. San Francisco: Freeman, 1966.
Bering, Edgar A., III, Arthur A. Few, and James R. Benbrook, "The Global Electric Circuit," *Physics Today*, October 1998, pp. 24–30.
Burr, Harold Saxton, "Moon Madness," *Yale Journal of Biology and Medicine*, 16: 249–256 (1943/44). Describes electrical measurements on trees.
Chalmers, J. Alan, *Atmospheric Electricity*. New York: Pergammon, 1967.
———, "Point-discharge through a living tree during a thunderstorm," *Journal of Atmospheric And Terrestrial Physics* 24: 1059–1063 (1962).
Chevalier, Gaetan, *The Earth's Electrical Surface Potential*. Encinitas, CA: California Institute for Human Science (2007).
Dolezalek, Hans, "Atmospheric Electricity," *CRC Handbook of Chemistry and Physics*, 72nd edition. Boca Raton: CRC Press, 1991/92, sec. 14–23 to 14–25.
Houghton, Henry C., *Physical Meteorology*. Cambridge, Mass.: MIT Press, 1985. "Atmospheric Electricity," pp. 361–407.
King, J. W. and W. S. Newman, eds., *Solar-Terrestrial Physics*. New York: Academic, 1967.
Lerner, Eric J., *The Big Bang Never Happened*. New York: Times Books, 1991.
Marmet, Paul and Grote Reber, "Cosmic matter and the nonexpanding universe," *IEEE Transactions on Plasma Science* 17: 264–269 (1989).
Peratt, Anthony L., "Not with a Bang," *The Sciences*, Jan./Feb. 1990, pp. 24–32.
———, "Plasma Cosmology, Part I. Interpretations of the visible universe," *The World and I*, Aug. 1989, pp. 294–301.
———, "Plasma Cosmology. Part II. The universe is a sea of electrically charged particles," *The World and I*, Sept. 1989, pp. 307–317.
———, *Physics of the Plasma Universe*. New York: Springer, 1992.
Roble, Raymond G. and Israel Tzur, "The Global Atmospheric-Electrical Circuit," in National Research Council, *The Earth's Electrical Environment*, Washington, D.C.: National Academy Press 1986, pp. 206–231.
Volland, Hans, ed., *Handbook of Atmospheric Electrodynamics*, 2 vols. Boca Raton, FL: CRC Press (1995).
Williams, Erle R., "The Global Electrical Circuit: A review," *Atmospheric Research* 91(2-4): 140–152 (2009).
Yourgrau, Wolfgang and Allen D. Breck, ed., *Cosmology, History, and Theology*. New York: Plenum, 1977.

5. Of Marvelous Form (pp. 37–45)

Becker, Robert O. and Gary Selden, *The Body Electric: Electromagnetism and the Foundation of Life*. New York: Morrow, 1985.

Becker, Robert O. and Andrew A. Marino, *Electromagnetism and Life*. Albany: SUNY Press, 1982.

Brown, Glenn H. and Jerome J. Wolken, *Liquid Crystals and Biological Structures*. New York: Academic, 1979.

Crile, George, *Phenomena of Life: A Radio-Electric Interpretation*. New York: Norton, 1936.

Fisch, Michael R., *Liquid Crystals, Laptops and Life*. Singapore: World Scientific, 2004.

Ho, Mae-Wan, *The Rainbow and the Worm: The Physics of Organisms*, 3rd ed. Singapore: World Scientific (2008).

Marx, Jean, "A 'mitey' theory for gene jumping," *Science* 253: 1092–1093 (1991).

6. Ecology (pp. 46–56)

Bates, Marston, *The Forest and the Sea*. New York: Vintage, 1960.
Daniel, John, "The Long Dance of the Trees," *Wilderness*, Spring 1988, p. 21.
Huxley, Anthony, *Plant and Planet*. New York: Pelican, 1978.
Maser, Chris, *Forest Primeval*. San Francisco: Sierra, 1989.
Storer, John H., *The Web of Life*. New York: New American Library, 1956.

7. Gaia (pp. 57–60)

Hutton, James, *Theory of the Earth*. Edinburgh: W. Creech, 1795. Codicote, Herts.: Wheldon and Wesley, 1959.

Joseph, Lawrence E., *Gaia: The Growth of an Idea*. New York: St. Martin's Press, 1991.

Lovelock, James, *Gaia A New Look at Life on Earth*. New York: Oxford University Press, 1987.

Margulis, Lynn and Dorion Sagan, *Slanted Truths: Essays on Gaia, Symbiosis, and Evolution*. New York: Springer-Verlag, 1997.

8. Being and Becoming (pp. 61–70)

Cairns-Smith, A. G., *Genetic Takeover and the Mineral Origins of Life*. Cambridge University Press, 1982.

———, *The Life Puzzle*. University of Toronto Press, 1971.

———, *Seven Clues to the Origin of Life*. Cambridge University Press, 1985.

Cairns-Smith, A. G. and H. Hartman, eds., *Clay Minerals and the Origins of Life*. Cambridge University Press, 1986.

Darwin, Charles, *Journal of Researches*, 2nd edition. London: J. Murray, 1845.

———, *The Descent of Man*. London: J. Murray, 1871.

Dauvillier, A., *The Photochemical Origins of Life*. New York: Academic, 1965.
Forel, Auguste, *Social World of the Ants*. New York: Boni, 1930.
Hardin, Garret, ed., *Population, Evolution, and Birth Control*. San Francisco: Freeman, 1969.
Kropotkin, Piotr, *Mutual Aid: A Factor of Evolution*. London: W. Heinemann, 1902.
Miller, Stanley L. and Leslie E. Orgel, *The Origins of Life on Earth*. Englewood Cliffs, N.J.: Prentice-Hall, 1974.
Shapiro, Robert, *Origins: A Skeptic's Guide to the Creation of Life on Earth*. New York: Summit, 1987.
Stansbury, Howard, *An Expedition to the Valley of the Great Salt Lake of Utah*. Philadelphia: Lippincott, 1852.
Wynne-Edwards, V. C., *Evolution through Group Selection*. Oxford: Blackwell Scientific, 1986.

9. We Are All Indians (pp. 73–78)

Chief Seattle, 1854. Speech published in *Seattle Star*, Oct. 29, 1887. A copy was obtained from the Library of the Museum of the American Indian, New York.

10. Air (pp. 79–82)

Breuer, Georg, *Air in Danger*. Cambridge University Press, 1980.
Faure, Hugues, "Changes in the global continental reservoir of carbon," *Global and Planetary Change* 2: 47–52 (1990).
Firor, John, *The Changing Atmosphere*. New Haven: Yale University Press, 1990.
Foley, Gerald, *The Energy Question*. New York: Viking, 1987.
Ho, Mae-Wan, Institute of Science in Society, "More CO_2 could mean less biodiversity and worse," ISIS Special Miniseries - Life of Gaia, Oct. 8, 2003.
———, "O_2 dropping faster than CO_2 rising." ISIS Report, August 19, 2009.
———, O_2 "Diving towards danger point," ISIS Report, May 18, 2015.
Huang, Jinpiang, Jiping Huang, Xiaoyue Liu, et al., "The global oxygen budget and its future projection," *Science Bulletin* 63: 1180–1186 (2018).
Huang, Jinpiang, Xiaoyue Liu, Yongsheng He, et al., "The oxygen cycle and a habitable earth," *Science China Earth Sciences* 64(4): 511–528 (2021).
Körner, Christian, "Ecological impacts of atmospheric CO_2 enrichment on terrestrial ecosystems," in *Proceedings of the Royal Society conference, Abrupt climate change: evidence, mechanisms and implications*, held February 4–5, 2003. Republished in *Philosophical Transactions of the Royal Society A* 361: 2023–2041 (2003).
Liu, Xiaoyue, Jianping Huang, Jiping Huang, et al., "Estimation of gridded atmospheric oxygen consumption from 1975 to 2018," *Journal of Meteorological Research* 34(3): 646–658 (2020).

Livina, V. N., Vaz Martins, and A. Forbes, "Tipping point analysis of atmospheric oxygen concentration," *Chaos* 25: 036403 (2015).

Luce, Larry, ed., *The Breather's Guide to Invisible Air Pollution*. Berkeley: Ecology Information Group, 1970.

Martin, Daniel, Helen McKenna, and Valerie Livina, "The human physiological impact of global deoxygenation," *Journal of Physiological Sciences* 67: 97–106 (2017).

Pettijohn, F. J., *Sedimentary Rocks*. New York: Harper & Row, 1975.

Scientific American, *Energy and Power*. San Francisco: Freeman, 1971.

Stonehouse, B., ed., *Arctic Air Pollution*. Cambridge University Press, 1986.

Sundquist, E. T. and G. A. Miller, "Oil shales and carbon dioxide," *Science* 208: 740 (1980).

Wei, Jun, Jianguo Wu, Jianping Huang, et al., "Declining oxygen level as an emerging concern to global cities," *Environmental Science and Technology* 55: 7808-7817 (2021).

In a Greenhouse (pp. 82–95)

Budyko, M. I., *The Earth's Climate: Past and Future*. New York: Academic, 1982.

Cheng, Lijing, John Abraham, Kevin E. Trenberth, et al., "Another year of record heat for the oceans," *Advances in Atmospheric Sciences* 40: 963–974 (2023).

Haeberli, Wilfried, Frank Paul, and Michael Zemp. "Vanishing glaciers in the European Alps," Zurich Open Repository and Archive, University of Zurich (2011), https://www.zora.uzh.ch/id/eprint/83973/1/2013_HaeberliW
_etal_PASSV_2013.pdf

Massachusetts Institute of Technology, *Inadvertent Climate Modification: Report of the Study of Man's Impact on Climate*. Cambridge, Mass.: MIT Press, 1971.

National Centers for Environmental Information, monthly and annual *Global Climate Reports*, https://www.ncei.noaa.gov/access/monitoring/monthly-report.

National Oceanic and Atmospheric Administration, "Arctic report card: climate change transforming Arctic into 'dramatically different state,'" https://www.noaa.gov/news-release/arctic-report-card-climate-change-transforming-arctic-into-dramatically-different-state

Nunn, Patrick and Paul A. Williams, "Drowning islands: climate change imperatives in the Asia-Pacific region," *TEXT* 22: 1–15 (2018).

Rantanen, Mika, Alexey Yu. Karpechko, Antti Lipponen, et al., "The Arctic has warmed nearly four times faster than the globe since 1979," *Communications Earth and Environment* 3: 168 (2022).

Revkin, Paolo Antonio, "Endless summer: Living with the greenhouse effect," *Discover*, Oct. 1989, pp. 50–61.

Rounce, David R., Regine Hock, Fabien Maussion, et al., "Global glacier change in the 21st century: Every increase in temperature matters," *Science* 379(6627): 78–83 (2023).

Shepherd, Andrew, Erik Ivins, et al., "Mass balance of the Greenland Ice Sheet from 1992 to 2018," *Nature*, 579 (7798): 233–239 (2020).

Under a Blazing Sun (pp. 95–101)

Gribbin, John, *The Hole in the Sky*. New York: Bantam, 1988

Petkov, Boyan H., Vito Vitale, Piero Di Carlo et al., "An unprecedented Arctic ozone depletion event during Spring 2020 and its impacts across Europe" *Journal of Geophysical Research: Atmospheres* 128(3): e2022JD037581 (2023).

Qing-Bin Lu, "Observation of large and all-season ozone losses over the Tropics," *AIP Advances* 12, 075006 (2022).

Ryan, R. G., Marais, E. A., Balhatchet, C. J., & Eastham, S. D. (2022). "Impact of rocket launch and space debris air pollutant emissions on stratospheric ozone and global climate," *Earth's Future*, 10: e2021EF002612).

Smith, R. C. et al., "Ozone depletion: Ultraviolet radiation and phytoplankton biology in Antarctica," *Science* 255: 952 (1992).

Thévenot, Roger, A *History of Refrigeration*. Paris: International Institute of Refrigeration, 1979.

The Gentle Rain (pp. 101–104)

Cousins, Ian T., Jana H. Johansson, Matthew E. Salter, et al., "Outside the safe operating space of a new planetary boundary for per- and polyfluoroalkyl substances (PFAS)," *Environmental Science and Technology* 56: 11172-11179 (2022).

Garnett, Jack, Crispin Halsall, Holly Winton, et al., "Increasing accumulation of perfluorocarboxylate contaminants revealed in an Antarctic firn core (1958-2017)," *Environmental Science and Technology* 56: 11246-11255 (2022).

Li, Can, Chris McLinden, Vitali Fioletov, et al., "India Is overtaking China as the world's largest emitter of anthropogenic sulfur dioxide," *Nature Scientific Reports* 7: 14304 (2017).

Prakash, Jigyasa, Shashi Bhushan Agrawal, and Madhoolika Agrawal, "Global trends of acidity in rainfall and its impact on plants and soil," *Journal of Soil Science and Plant Nutrition* 23:398–419 (2023).

Zhao, Xiaode, Zhifang Xu, Wenjing Liu, et al., "Chemical composition of precipitation in Shenzhen, a coastal mega-city in South China: influence of urbanization and anthropogenic activities on acidity and ionic composition," *Science of the Total Environment* 662: 218–226 (2019).

Chemical Soup (pp. 104–108)

Fang, Hejin, Wenmin Qin, Lunche Wang, et al., "Solar brightening/dimming over China's mainland: effects of atmospheric aerosols, anthropogenic emissions, and meteorological conditions," *Remote Sensing* 13(1): 88 (2021).

Goldberg, B. and W. H. Klein, "Comparison of normal incident solar energy measurements at Washington, D.C.," *Solar Energy* 13(3): 311–321 (1971).

Stanhill, Gerald and Shabtai Cohen, "Global dimming: a review of the evidence for a widespread and significant reduction in global radiation with discussion of its probable causes and possible agricultural consequences, *Agricultural and Forest Meteorology* 107: 255–278 (2001).

Suraqui, S., H. Tabor, W. H. Klein, and B. Goldberg, "Solar radiation changes at Mt. St. Katherine after forty years, *Solar Energy* 16(3–4): 155–158 (1974).

Wang, Ziyan, Ming Zhang, Lunche Wang, and Wenmin Qin, "A comprehensive research on the global all-sky surface solar radiation and its driving factors during 1980-2019," *Atmospheric Research* 265: 105870 (2022).

Yang, Shuyue, Xiaotong Zhang, Shikang Guan, et al., A review and comparison of surface incident shortwave radiation from multiple data sources: satellite retrievals, reanalysis data and GCM simulations, *International Journal of Digital Earth* 16(1): 1332–1357 (2023).

11. Wood (pp. 109–123)

Breuer, Georg, *Air in Danger*. Cambridge University Press, 1980.

Bryson, Reid A. and Thomas J. Murray, *Climates of Hunger*. Madison: University of Wisconsin Press, 1977.

Claiborne, Robert. *Climate, Man, and History*. New York: Norton, 1970.

Day, Gordon M., "The Indian as an ecological factor in the northeastern forest," *Ecology* 34: 329–346 (1953).

Eckholm, Erik and Lester R. Brown, *Spreading Deserts—The Hand of Man*, Worldwatch Paper 13. Washington: Worldwatch Institute, August 1977.

Grainger, Alan, "The state of the world's tropical forests," *The Ecologist*, Jan./Feb. 1980, pp. 6–52.

Hughes, J. Donald, *Ecology in Ancient Civilizations*. Albuquerque: University of New Mexico Press, 1975.

Keast, A., R. L. Crocker, and C. S. Christian, eds., *Biogeography and Ecology in Australia*. The Hague: Junk, 1959. See especially pp. 36–51, Normal B. Tindale, "Ecology of Primitive Man in Australia."

Lamb, Robert, *World Without Trees*. London: Paddington, 1979.

Maxwell, Hu, "The use and abuse of forests by the Virginia Indians," *William and Mary Quarterly Historical Magazine*, Oct. 1910, pp. 86–103.

McHugh, William P., "Cattle pastoralism in Africa—A model for interpreting archaeological evidence from the Easten Sahara," *Arctic Anthropology* XI - Suppl.: 236–244 (1974).

Michelmore, Peter, "Our trees Are dying," *Reader's Digest*, Nov. 1984, pp. 157–161.

Nicholson, Robert, "A far plateau," *Natural History*, Sept. 1991, pp. 22–29, on the Sahara cypress.

Plato, *Critias*, 111.

Putnam John J., "Timber: how much is enough?" *National Geographic* 145: 484–511 (1974).

Pyne, Stephen J., "Fire down under: how the first australians put a continent to the torch," *The Sciences*, March/April 1991, pp. 39–44.

Raikes, Robert L. and Robert H. Dyson, Jr., "The prehistoric climate of baluchistan and the Indus Valley," *American Anthropologist* 63: 265–281 (1961).

Ritchie, J. C., C. H. Eyles, and C. V. Haynes, "Sediment and pollen evidence for an early to mid-Holocene humidperiod in the Eastern Sahara," *Nature* 314: 352–355 (1985).

Sears, Paul B., *Lands Beyond the Forest*. Englewood Cliffs, N.J.: Prentice-Hall, 1969.

Smith, Anthony, *Mato Grosso*. New York: Dutton, 1971.

12. Water (pp. 124–161)

Canby, Thomas Y., "Water, our most precious resource," *National Geographic* 158: 144–179 (1980).

Coulton, Frederick and Emil Mrak, eds., *Water Quality: Proceedings of an International Forum*. New York: Academic, 1977.

Little, Arthur D., *Water Quality Criteria Data Book*. Washington: Environmental Protection Agency, 1970.

McKee, J. E. and A. W. Wolf, eds., *Water Quality Criteria*, 2nd edition. Sacramento: State Water Quality Control Board, 1963.

Groundwater (pp. 125–129)

Bessire, Lucas, *Running Out: In Search of Water on the High Plains* (Princeton University Press, 2021

Borrell, John, "A plan to make the desert gush," *Time*, Sept. 29, 1986, p. 63.

Groundwater Protection Council, *Groundwater Report to the Nation: A Call to Action*, 2007.

Hastings, James Rodney and Raymond M. Turner, *The Changing Mile*. Tucson: University of Arizona Press, 1965.

Issar, Arie, "Fossil water under the Sinai-Negev peninsula," *Scientific American*, July 1985, pp. 104–110.
Jorgensen, Eric P., ed., *The Poisoned Well*. Washington: Island Press, 1989.
Kansas Water Office and the U.S. Army Corps of Engineers, Kansas City District, *Update of 1982 Six State High Plains Aquifer Study, Alternate Route B*, 2015.
Little, Jane Braxton, "Saving the Ogallala aquifer," *Scientific American Special Editions* 19(1s): 32–29 (March 2009).
Mohan, Chinchu, Andrew W. Western, et al., "Global assessment of groundwater stress vis-à-vis sustainability of irrigated food production," *Sustainability* 14: 16896 (2022).
Moss, Senator Frank E., *The Water Crisis*. New York: Praeger, 1967.
Pearson, William R., *High Plains-Ogallala Aquifer Study: Water Transfer Element*, U.S. Army Corps of Engineers, 1987.
Stengel, Richard, "Ebbing of the Ogallala," *Time*, May 10, 1982, p. 98.
Tevis, James H., *Arizona in the '50s*. Albuquerque: University of New Mexico Press, 1954.
United Nations Economic and Social Commission for Asia and the Pacific, *Water Resources Development in Asia and the Pacific—Some Issues and Concerns*, Water Resources Series #62. New York: United Nations, 1987.
Webb, Walter Prescott, *The Great Plains*. New York: Ginn, 1931.
———, "The American West—perpetual pmirage," *Harper's*, May 1957, pp. 25–31.

Endangered Springs (pp. 129–130)

Christopher, H. Armstead, ed., *Geothermal Energy*. Paris: Unesco, 1973.
Kerr, Richard A., "Geothermal tragedy of the commons," *Science* 253: 134–135 (1991).
U.S. Department of Energy, *Reservoir Engineering 1976-2006: A History of Geothermal Energy Research and Development in the United States*, September 2010.

Toxic Wastes (pp. 133–136)

Crawford, Mark, "Hazardous waste: where to put it?" *Science* 235: 156 (1987).
Ellsworth, William L., "Injection-induced earthquakes," *Science* 341: 142–149 (2013).
Epstein, B. J. W., L. O. Brown, and C. Pope, *Hazardous Wastes in America*. San Francisco: Sierra, 1982.
Food and Water Watch, "Drilled too far: The perils of injection wells," Food and Water Watch, June 2020, https://www.foodandwaterwatch.org/wp-content/uploads/2021/03/fs_2006_injectionwells-web.pdf

Groundwater Protection Council, *Injection Wells: An Introduction to Their Use, Operation, and Regulation*, 2013, https://www.academia.edu/download/35994773/JNGSE-CO2_paper_Xie_Economides.pdf.

Lave, Lester and Arthur C. Upton, eds., *Toxic Chemicals, Health, and the Environment*. Baltimore: The Johns Hopkins University Press, 1987.

Lustgarten, Abraham, *Injection Wells: The Poison Beneath Us*. ProPublica, June 21, 2012, https://www.propublica.org/article/injection-wells-the-poison-beneath-us.

Regenstein, Lewis, *America the Poisoned*. Washington: Acropolis, 1982.

Hi-Tech Poisons (pp. 136–137)

Lazarus, David, "Toxic technology," *San Francisco Chronicle*, Dec. 3, 2000.

National Institute for Occupational Safety and Health, *Hazard Assessment of the Electronic Component Manufacturing Industry*, Dept. of Health and Human Services Publication #85-100. Washington, D.C., Feb. 1985.

Sherry, Susan, *High Tech and Toxics*. Sacramento: Golden Empire Health Planning Center, 1985.

Dams (pp. 138–148)

Fradkin, Philip L., A *River No More: The Colorado River and the West*. Tucson: University of Arizona Press, 1984.

Goldsmith, Edward and Nicholas Hildyard, *The Social and Environmental Effects of Large Dams*. San Francisco: Sierra, 1984.

Morris, John W., *The Southwestern United States*. New York: Van Nostrand Reinhold, 1970.

One Million Dams Threaten Fish in Europe, *Teller Report*, Jan. 21, 2021, https://www.tellerreport.com/life/2021-01-21-%0A---one-million-dams-threaten-fish-in-europe%0A--.rJw_M73U1_.html

Palmer, Tim, *Endangered Rivers and the Conservation Movement*. Berkeley: University of California Press, 1986.

Picard, Andre, "James Bay: A power play," *The Globe and Mail*, Toronto, five-part series, April 13, 14, 16, 17, 18, 1990.

Sattaur, Omar, "India's troubled waters," *New Scientist*, May 27, 1989, pp. 46–51.

Udall, Stewart and W. Kent Olson, "Too many dammed rivers," *New York Times*, Nov. 25, 1989, p. A23.

Walton, Susan, "Egypt: After the Aswan Dam," *Environment*, May 1981, pp. 31–36.

What's Up With Water, *Circle of Blue*, May 24, 2022, https://www.circleofblue.org/2022/world/whats-up-with-water-may-24-2022.

Zanon, Sibélia, "Dam-building spree pushes Amazon Basin's aquatic life closer to extinction," *Mongabay*, 22 June 2023.

The Sea (pp. 150–159)

Batisse, Michael, "Probing the future of the Mediterranean Basin," *Environment*, June 1990, pp. 4–9.
Boucher, Julian and Damien Friot, *Primary Microplastics in the Oceans: A Global Evaluation of Sources* (International Union for Conservation of Nature, 2017).
Carpenter, E. J. and K. L. Smith, Jr., "Plastics on the Sargasso Sea surface," *Science* 175: 1240–1241 (1972).
Goldberg, Edward D., *The Health of the Oceans*. Paris: Unesco, 1976.
Heyerdahl, Thor, *Fagu-Hiva*. Garden City: Doubleday, 1974.
Jambeck, Jenna R., Roland Geyer, Chris Wilcox, et al., "Plastic waste inputs from land into the ocean," *Science* 347(6223): 768–771 (2015).
La Vallee, Douglas, "Mediterranean Sea: Just how much can it take?" *Greenpeace Examiner*, June 1986, pp. 13–16.
Marx, Wesley, *The Frail Ocean*. New York: Ballantine, 1967.
McLean, Joyce, "Great Lakes: An abused ecosystem," *Greenpeace Examiner*, June 1986, pp. 17–18.
Moorcraft, Colin, *Must the Seas Die?* Boston: Gambit, 1973.
Moore, Captain Charles, *Plastic Ocean* (NY: Avery, 2011).
Morello, Lauren, "Phytoplankton population drops 40 percent since 1950," *Scientific American*, July 29, 2010.
"The New Plastics Economy," January 2016, newplasticseconomy.org.
Simon, Anne W., *Neptune's Revenge*. New York: Watts, 1984.
Teal, John and Mildred Teal, *The Sargasso Sea*. Boston: Little, Brown, 1975.
Weisskopf, Michael, "Plastic reaps a grim harvest in the oceans of the world," *Smithsonian*, March 1988, pp. 58–67.
Wilcox, Chris, Erik Van Sebille, and Britta Denise Hardesty, "Threat of plastic pollution to seabirds is global, pervasive, and increasing," *Proceedings of the National Academy of Sciences* 112(38): 11899–11904 (2015).
Wilderness Society, *100 Spills, 1000 Excuses*. Washington, 1990.

They Nuked Paradise (p. 159)

Rossi, Melissa, "A 'Chernobyl in slow motion' under Arctic seas," *Outrider*, Nov. 28, 2022, https://outrider.org/nuclear-weapons/articles/chernobyl-slow-motion-under-arctic-seas.

Strip-Mining the Seas (pp. 160–161)

Carver, Edward, "Illegal bottom trawling widespread inside Mediterranean marine protected areas," *Global Fishing Watch*, Nov. 2, 2022, https://globalfishingwatch.org/press-release/new-data-reveals-bottom-trawling-in-protected-areas.

Clover, Charles, *The End of the Line: How Overfishing Is Changing the World and What We Eat* (Berkeley: University of California Press, 2006).

Costello, Christopher, Daniel Ovando, and Tyler Clavelle, "Global fishery prospects under contrasting management regimes," *Proceedings of the National Academy of Sciences* 113(18): 5125–5129 (2016).

Glavin, Terry, "North Pacific being stripped by Asian fishery nets," *Vancouver Sun*, Sept. 15, 1989, p. A1.

Myers, Ransom and Boris Worm, "Rapid worldwide depletion of predatory fish communities," *Nature* 423: 280–283 (2003)).

Neslen, Arthur, "Europe's fishing industry to battle with conservationists over bottom trawling," *The Guardian*, 17 Feb. 2022.

Shabecoff, Philip, "Huge drifting nets raise fears for an ocean's fish," *New York Times*, March 21, 1989, p. C1.

Thomas, Karli, "New Zealand's bottom trawling isolation continues in the South Pacific," *Deep Sea Conservation Coalition*, Sept. 29, 2022, https://savethehighseas.org/2022/09/29/new-zealands-bottom-trawling-isolation-continues-in-the-south-pacific/

von Brandt, Andres, *Fish Catching Methods of the World*. Farnham, Surrey: Fishing News, 1984.

Worm, Boris, Edward B. Barbier, Nicola Beaumont, et al., "Impacts of biodiversity loss on ocean ecosystem services," *Science* 314: 787–790 (2006);

Yeung, Peter. "'Walls of Death': Surge in illegal drift nets threatens endangered species," *The Guardian*, 18 Aug. 2020.

14. Plants and Animals (pp. 165–194)

Allen, Harry, Michelle C. Langley, and Paul S. C. Taçon, "Bone projectile points in prehistoric Australia: Evidence from archaeologically recovered implements, ethnography, and rock art," in Michelle C. Langley (ed.), *Osseous Projectile Weaponry*, pp. 209–218, Springer, Netherlands, 2017.

Bakker, Robert T., *The Dinosaur Heresies*. New York: Morrow, 1986.

Bednarik, Robert G., "The earliest evidence of ocean navigation, *The International Journal of Nautical Archaeology* 26(3): 183–191 (1997)

———, "The initial peopling of Wallacea and Sahul, *Anthropos* 92: 355–367 (1997).

———, "Seafaring in the Pleistocene," *Cambridge Archaeological Journal* 13(1): 41–46 (2003).

Boyce, Daniel G., Marlon R. Lewis and Boris Worm, "Global phytoplankton decline over the past century," *Nature* 466: 591–596 (2010).

Brody, Jane E., "Water-based animals are becoming extinct faster than others," *New York Times*, April 23, 1991, p. C4.

———, "Far from fearsome, bats lose ground to ignorance and greed," *New York Times*, Oct. 29, 1991, p. C1.

———, "Boom in poaching threatens bears worldwide," *New York Times,* May 1, 1990, p. C1.

Brumm, Adam, Fachroel Aziz, and Gert D. van den Bergh, "Early stone technology on Flores and its implications for *Homo floresiensis, Nature* 441: 624–628 (2006).

Brumm, Adam, Gitte M. Jensen, Gert D. van den Bergh, et al., "Hominins on Flores, Indonesia, by one million years ago," *Nature* 464: 748–752 (2010).

Brumm, Adam, Gerrit D. van den Bergh, Michael Storey, et al., "Age and context of the oldest known hominin fossils from Flores," *Nature* 534: 249–253 (2016).

Camacho, Juan Armenta, "Vestigios de labor humana en huesos de animales extintos de Valsequillo, Puebla, Mexico," presented at the 35th International Congress of the Americanists, Puebla, 1978.

Carpenter, Kent E., Mohammad Abrar, Greta Aeby, et al., "One-third of reef-building corals face elevated extinction risk from climate change and local impacts," *Science* 321(5888): 560–563 (2008).

Curry-Lindahl, Kai, *Let Them Live: A Worldwide Survey of Animals Threatened with Extinction.* New York: Morrow, 1972.

Day, David, *The Doomsday Book of Animals.* New York: Viking, 1981.

de Lumley, H., H. A. de Lumley, M. C. M. C. Beltrão, et al., "Présence d'outils taillés associés à une faune quaternaire datée du Pleistocene moyen dans la Toca de Esperança, région de Central, État de Bahia, Brésil," *L'Anthropologie (Paris)* 91(4): 917–942 (1987).

Dennell, Robin W., Julien Louys, Hannah J. O'Regan, et al., "The origins and persistence of *Homo floresiensis* on Flores: Biogeographical and ecological perspectives, *Quaternary Science Reviews* 96: 98–107 (2014).

Diamond, Jared, "The Golden Age that never was," *Discover,* Dec. 1988, pp. 71–79.

Dillehay, T. D., *The Settlement of the Americas: A New Prehistory.* NY: Basic Books 2000.

Ehrlich, Paul, "Extinction: What is happening now and what needs to be done," in *Dynamics of Extinction*, David Elliott, ed. (New York: Wiley, 1986),
pp. 157–164.

Eley, Thomas J. and T. H. Watkins, "In a sea of trouble," *Wilderness,* Fall 1991, pp. 18–26.

Fisher, James, Noel Simon, and Jack Vincent, *Wildlife in Danger.* New York: Viking, 1969.

Fullagar, R., D. Price, and L. Head, "Early human occupation of Northern Australia," *Antiquity* 70: 751–752 (1996).

Holen, Steven R., Thomas A. Demère, Daniel C. Fisher, et al., "A 130,000-year-old archaeological site in southern California, USA," *Nature* 544: 479–483 (2017).

Holmes, Kathryn, "Death among the lowlife," *Sierra*, March/April 1986, pp. 22–23.

Janzen, Daniel H. and Paul S. Marin, "Neotropical anachronisms: The fruits the Gomphotheres ate," *Science* 215: 19–27 (1982).

Koopowitz, Harold and Hilary Kaye, *Plant Extinctions: A Global Crisis*. Washington: Stone Wall, 1983.

Larsen, Al and Jeannette Larsen, "Flying flowers: appreciating butterflies," *Wilderness*, Fall 1991, p. 16.

Livingston, John A., *The Fallacy of Wildlife Conservation*. Toronto: McClelland and Steward, 1981.

Martin, Paul S. and H. E. Wright, eds., *Pleistocene Extinctions*. New Haven: Yale University Press, 1967.

Mohlenbrock, *Where Have All the Wildflowers Gone?* New York: Macmillan, 1983.

Mohsin, A. K. Mohammad and Mohammad Azmi Ambak, *Freshwater Fishes of Peninsular Malaysia*, Universiti Pertanian Malaysia 1983.

Morwood, M. J., P. B. O'Sullivan, F. Aziz, and A. Raza, "Fission-track ages of stone tools and fossils on the East Indonesian island of Flores, *Nature* 392: 173–176 (1998).

Mowat, Farley, *Sea of Slaughter*. Toronto: McClelland and Stewart, 1984.

Myers, Norman, "Threatened biotas," *The Environmentalist* 8: 187–208 (1988).

Pinney, Roy, *Vanishing Wildlife*. New York: Dodd, Mead, 1963.

Pokagon, Simon, *The Red Man's Rebuke*, 1893. Reprinted in *Indiana Historical Society Publications*, Vol. 50, No. 5, 1933.

Raven, P. H., *We're Killing Our World: The Global Ecosystems in Crisis*. Chicago: MacArthur Foundation, 1987.

Singh, G. and E.A. Geissler, "Late Cainozoic history of vegetation, fire, lake levels, and climate at Lake George, New South Wales, Australia," *Philosophical Transactions of the Royal Society of London (Series B)* 311: 379–447 (1985).

Soule, M. E. and B. A. Wilcox, eds., "Conservation biology: An evolutionary-ecological perspective." Sunderland, Mass.: Sinauer, 1980.

Steen-McIntyre, Virginia, Roald Fryxell, and Harold E. Malde, "Geologic evidence for age of eeposits at Hueyatlaco archeological site, Valsequillo, Mexico, *Quaternary Research* 16: 1-17 (1981).

Steeves, Paulette F. C., *The Indigenous Paleolithic of the Western Hemisphere*, University of Nebraska Press, 2021.

Szabo, Barney J. and Harold E. Malde, "Dilemma posed by uranium-series dates on archaeologically significant bones from Valsequillo, Puebla, Mexico," *Earth and Planetary Science Letters* 6: 237–244 (1969).

Usher, Ann Danaiya, "The wild flesh trade," *Free Spirit*, Summer/Fall 1990, pp. 57–59.

Wilson, E. O., ed., *Biodiversity*. Washington: National Academy Press, 1988.

Oxygen (pp. 174–184)

Firstenberg, Arthur, *The Invisible Rainbow: A History of Electricity and Life* (White River Junction, VT: Chelsea Green, 2020).

Klimentidis, Yann C., T. Mark Beasley, Hui-Yi Lin, et al., "Canaries in the coal mine: A cross-species analysis of theplurality of obesity epidemics," *Proceedings of the Royal Society B* 278: 1626–1632 (2011).

Lombard, Louise S. and Ernest J. Witte, "Frequency and types of tumors in mammals and birds of the Philadelphia Zoological Gardens," *Cancer Research* 19(2): 127–141 (1959).

Ratcliffe, Herbert L., T. G. Yerasimides, and G.A. Elliott, Changes in the character and location of arterial sesions in mammals and birds in the Philadelphia Zoological Garden, *Circulation* 21 730–738 (1960)

Rigg, Kathleen J., R. Finlayson, C. Symons, et al., "Degenerative arterial disease of animals in captivity with special reference to the comparative pathology of atherosclerosis," *Proceedings of the Zoological Society of London* 135(2): 157–164 (1960).

Vastesaeger, Marcel M. and R. Delcourt, "The natural history of atherosclerosis," *Circulation* 26: 851–855 (1962).

Insects (pp. 175–176)

Carringon, Damian, "Plummeting insect numbers 'threaten collapse of nature,'" *The Guardian*, 10 Feb. 2019.

Cucurachi, S., W. L. M. Tamis, M. G. Vijver, et al., "A review of the ecological effects of radiofrequency electromagnetic fields (RF-EMF)," *Environment International* 51: 116–140 (2013).

Goulson, Dave, *Insect Declines and Why They Matter*, Somerset Wildlife Trust, 2019.

Hallman, Caspar A., Martin Sorg, Eelke Jongejans, et al., "More than 75 percent decline over 27 years in total flying insect biomass in protected areas, *PLoS ONE* 22(10): e0185809 (2017).

Imms, Augustus D., "Report on a disease of bees in the Isle of Wight," *Journal of the Board of Agriculture* 14(3): 129–140 (1907).

Lázaro, Ámparo, Antonia Chroni, Thomas Tscheulin, et al., "Electromagnetic radiation of mobile telecommunication antennas affects the abundance and composition of wild pollinators," *Journal of Insect Conservation* 20: 315–324 (2016).

Lister, Bradford C. and Andres Garcia, "Climate-driven declines in arthropod abundance restructure a rainforest food web," *Proceedings of the National Academy of Sciences* 115(44): E10397-E10406 (2018).

Pincher, Michael Chapman, The Insect Inspector, "Insects are disappearing all over the world. The usual suspects are pesticides and habitat loss but is EMR the real villain of the piece?" October 7, 2019, https://www.fcc.gov/ecfs/document/10914205092143/3

Sánchez-Bayo, Francisco and Kris A.G. Wyckhuys, "Worldwide decline of the entomofauna: A review of its drivers," *Biological Conservation* 232: 8–27 (2019).

Thill, Alain, "Biologische Wirkungen elektromagnetischer Felder auf Insekten," *Umwelt Medizin Gesellschaft* 33, March 2020.

Birds (pp. 176–178)

Balmori, Alfonso and Örjan Hallberg, "The urban decline of the house sparrow (*Passer domesticus*): A Possible Link with Electromagnetic Radiation, *Electromagnetic Biology and Medicine* 26:141–151 (2007).

Bigu del Blanco, Jaime and César Romero-Sierra, "Bird feathers as dielectric receptors of microwave radiation," Laboratory Technical Report LTR-CS-89, Control Systems Laboratory, Division of Mechanical Engineering, National Research Council Canada (1973).

De Laet, Jenny and James Denis Summers-Smith, "The status of the urban house sparrow *Passer domesticus* in north-western Europe: A Review, *Journal of Ornithology* 148 (suppl. 2) S275-278 (2007).

Engels, Svenja, Nils-Lasse Schneider, Nele Lefeldt, et al., "Anthropogenic electromagnetic noise disrupts magnetic compass orientation in a migratory bird," *Nature* 509: 353–356 (2014).

Firstenberg, Arthur, "Acute electrical illness," in Arthur Firstenberg, *The Invisible Rainbow: A History of Electricity and Life*, White River Junction, VT: Chelsea Green 2020, pp. 75–93.

———, Mystery on the Isle of Wight, in Arthur Firstenberg, *The Invisible Rainbow: A History of Electricity and Life*, White River Junction, VT: Chelsea Green 2020, pp. 95–112).

Galbraith, Colin, "The population status of birds in the U.K.: Birds of conservation concern: 2002-2007," *Bird Populations* 7: 173–179 (2002).

Sainudeen Pattazhy, "Dwindling number of sparrow," *Karala Calling*, March 2012, pp. 32–33.

Sen, Benita, "Calling back the sparrow," *Deccan Herald*, November 16, 2012.

Tanner, J. Alan, "Effect of microwave radiation on birds," *Nature* 210: 636 (1966).

———, "Bird feathers as sensory detectors of microwave fields, in Stephen F. Cleary, ed., *Biological Effects and Health Implications of Microwave Radiation. Symposium Proceedings* (Rockville, MD: U.S. Dept. of Health, Education and Welfare), Publication BRH/DBE 70-2, pp. 185–187 (1970).

Wee, Eric L., "Homing pigeons race off to oblivion," *Washington Post*, Oct. 8, 1998, p. A3.

Bats (p. 179)

Firstenberg, Arthur, "You mean you can hear electricity?" in Arthur Firstenberg, *The Invisible Rainbow: A History of Electricity and Life*, White River Junction, VT: Chelsea Green 2020, pp. 275–321

Winifred F. Frick, Winifred F., Tigga Kingston, and Jon Flanders, "A review of the major threats and challenges to global bat conservation," *Annals of the New York Academy of Sciences* 1469(1): 5–29 (2020).

Lee, Demetria, "Cave in southeast Minnesota sees another dramatic decline in bat population," *Minnpost*, March 20, 2019, https://www.minnpost.com/environment/2019/03/cave-in-southeast-minnesota-sees-another-dramatic-decline-in-bat-population.

Pennsylvania Game Commission, *Bats Wildlife Note*, https://pgc.pa.gov/Education/WildlifeNotesIndex/Pages/Bats.aspx.

Rydell, Jens, Marcus Elfström, Johan Eklöf, and Sonia Sánchez-Navarro, "Dramatic decline of northern bat Eptesicus nilssonii in Sweden over 30 years," *Royal Society Open Science* 7: 191754 (2020).

Amphibians (pp. 179–183)

Balmori, Alfonso, "Mobile phone mast effects on common frog (*Rana temporaria*) tadpole: The city turned into a laboratory," *Electromagnetic Biology and Medicine* 29: 31–35 (2010).

Baringa, Marcia, "Where have all the froggies gone? *Science* 247(4946): 1033–1034.

Collins, James P., "Amphibian decline and extinction: What we know and what we need to learn," *Diseases of Aquatic Organisms* 92: 93–99 (2010).

Drost, Charles A. and Gary M. Fellers. "Collapse of a regional frog fauna in the Yosemite area of the California Sierra Nevada, USA," *Conservation Biology* 10(2): 414–425 (1996).

Ghose, Sonia L., Tiffany A. Yap, Allison Q. Byrne, et al., "Continent-wide recent emergence of a global pathogen in African amphibians," *Frontiers in Conservation Science* 4: 1069490 (2023).

Hallowell, Christopher, "Trouble in the lily pads," *Time*, October 28, 1996, p. 7.

Houlahan, Jeff. E., C. Scott Findlay, Benedikt R. Schmidt, et al., "Quantitative evidence for global amphibian population declines," *Nature* 404: 752–755 (2000).

Lips, Karen R., Patricia A. Burrowes, Joseph R. Mendelson III, and Gabriela Parra-Olea, "Amphibian declines in Latin America: Widepread population declines, extinctions, and impacts," *Biotropica* 37(2): 163–165 (2005).

McCallum, Malcolm L., "Amphibian decline or extinction? current declines dwarf background extinction rate," *Journal of Herpetology* 4(3): 483–491 (2007).

Norris, Scott, "Ghosts in our midst: Coming to terms with amphibian extinctions," *BioScience* 57(4): 311–316 (2007).

Pound, J. Alan and Martha I. Crump, "Amphibian declines and climate disturbance: The case of the golden toad and the harlequin frog," *Conservation Biology* 8(1): 72–85 (1994).

Stuart, Simon N., Janice S. Chanson, Neil A. Cox, et al., "Status and trends of amphibian declines and extinctions worldwide," *Science* 306(5702): 1783–1784 (2004).

Souder, William, "An amphibian horror story," *New York Newsday*, Oct. 15, 1996, p. B19+.

Toledo, Luís Felipe, Sergio Potsch de Carvalho-e-Silva, Ana Maria Paulino Telles de Carvalho-e-Silva, et al., "A retrospective overview of amphibian declines in Brazil's Atlantic Forest," *Biological Conservation* 277: 109845 (2023).

Wake, David B. and Vance T. Vredenburg, "Are we in the midst of the sixth mass extinction? A view from the world of amphibians," *Proceedings of the National Academy of Sciences* 105(Suppl. 1): 11466–11473 (2008).

Forgotten Fliers (pp. 184–187)

Diamond, Jared, "World of the living dead," *Natural History*, Sept. 1991, pp. 30–37.

Greenway, James C., Jr., *Extinct and Vanishing Birds of the World*, American Committee for International Wild Life Protection, Special Publication #13. New York, 1958.

Pokagon, Chief Simon, "The wild pigeon of North America," *The Chautauquan* 22: 202–206 (1895).

Shorger, A. W., *The Passenger Pigeon: Its Natural History and Extinction.* Madison: University of Wisconsin Press, 1955.

Wallace, Joseph, "Where have all the songbirds gone?' *Sierra*, March/April 1986, pp. 44–47.

Wilcove, David, "Empty skies," *The Nature Conservancy Magazine*, Jan./Feb. 1990, pp. 4–13.

The Last Refuge (pp. 187–189)

Chase, Michael J., Scott Schlossberg, Curtice R. Griffin, et al., "Continent-wide survey reveals massive decline in African savannah elephants," *PeerJ* 4: e2354 (2016).
Jones, Robert F., "Farewell to Africa," *Audubon*, Sept. 1990, pp. 50–104.
Lanting, Frans, "Botswana: A gathering of waters and wildlife," *National Geographic*, Dec. 1990, pp. 5–37.
Lee, Douglas and Frans Lanting, "Okavango Delta: Old Africa's last refuge," *National Geographic*, Dec. 1990, pp. 38–69.
Stevens, William K., "Huge conservation effort aims to save vanishing architect of the savanna," *New York Times*, Feb. 28, 1989, p. C1.

A World Park (pp. 189–194)

Antarctica Special, *Greenpeace Examiner*, Dec. 1985, pp. 11–20.
Bonner, W. N. and D. W. H. Walton, eds., *Key Environments: Antarctica*. New York: Pergammon, 1985.
Douglis, Carole, "The last global commons," *Wilderness*, Summer 1990, pp. 12–17+.
Ellis, Richard, *The Book of Whales*. New York: Knopf, 1980.
King, Judith E., *Seals of the World*. Ithaca, N.Y.: Cornell University Press, 1983.
Lemonick, Michael D., "Antarctica," *Time*, Jan. 15, 1990, pp. 56–62.
Thomas, Peter O., Randall R. Reeves, and Robert L. Brownell, Jr., "Status of the world's baleen whales," *Marine Mammal Science* 32(2): 682–734 (2016).

15. Space (pp. 195–198)

Boyle, Rebecca, "The threat of satellite constellations," *Scientific American*, Feb. 2023, pp. 46–51.
Canadian Astronomical Society, *Report on Mega-Constellations to the Government of Canada and the Canadian Space Agency*," March 31, 2021, https://arxiv.org/pdf/2104.05733.
Falle, Andrew, Ewan Wright, Aaron Boley, and Michael Byers, "One million (paper) satellites," *Science* 382(6667): 150–152 (2023).
Granger, Jesse, Lucianne Walkowicz, Robert Fitak, and Sönke Johnsen, Gray whales strand more often on days with increased levels of atmospheric radio-frequency noise, *Current Biology* 30 R135–R158 (2020).
Lawrence, Andy, Meredith L. Rawls, Moriba Jah, et al., "The case for space environmentalism," *Nature Astronomy* 6: 428–435 (2022).

Wofsey, Michael, "Space, the cluttered frontier," *Spheric*, Aug. 24, 1990, pp. 1+.

Wood-Kaczmar, Barbara, "The junkyard in the sky," *New Scientist*, Oct. 13, 1990, pp. 37–40.

17. Chemistry (pp. 206–207)

Center for Science in the Public Interest, *The Household Pollutants Guide*. Garden City: Anchor, 1978.

"Chemicals: A double edged sword," *United Nations Chronicle*, June 1988, p. 47.

Lave, Lester and Arthur C. Upton, eds., *Toxic Chemicals, Health, and the Envirionment*. Baltimore: The Johns Hopkins University Press, 1987.

Maugh, Thomas H. II, "Chemicals: How many are there?" *Science* 199: 162 (1978).

Regenstein, Lewis, *America the Poisoned*. Washington: Acropolis, 1982.

Wang, Zhanyun, Glen W. Walker, Derek C. G. Muir, and Kakuko Nagatani-Yoshida, "Toward a global understanding of chemical pollution: A first comprehensive analysis of national and regional chemical inventories," *Environmental Science and Technology* 54: 2575-2584 (2020)

Young, Bambi Batts and Kirk A. Johnson, "Toxic Home Syndrome," *New Age Journal*, April 1986, pp. 46–49+.

18. Radiation (pp. 208–215)

Adey, W. Ross, "Neurophysiologic effects of radiofrequency and microwave radiation," *Bulletin of the New York Academy of Medicine* 55: 1079–1093 (1979).

Becker, Robert O., *Cross Currents: The Promise of Electromedicine, the Perils of Electropollution*. New York: Tarcher, 1990.

Becker, Robert O. and Andrew Marino, *Electromagnetism and Life*. Albany: SUNY Press, 1982.

Becker, Robert O. and Gary Selden, *The Body Electric: Electromagnetism and the Foundation of Life*. New York: Morrow, 1985.

Brodeur, Paul, *Currents of Death: Power Lines, Computer Terminals, and the Attempt to Cover Up Their Threat to Your Health*. New York: Simon & Schuster, 1989.

———, *The Zapping of America*. New York: Norton, 1977.

Dubrow, A. P., *The Geomagnetic Field and Life*. New York: Plenum, 1978.

Firstenberg, Arthur, *The Invisible Rainbow: A History of Electricity and Life*. White River Junction, VT: Chelsea Green, 2020.

Godfrey, Jason D. and David M. Bryant, "Effects of radio transmitters: Review of recent radio-tracking studies," in M. Williams, ed., *Conservation Applications of Measuring Energy Expenditures of New Zealand Birds Assessing*

Habitat Quality and Costs of Carrying Radio Transmitters (Wellington, New Zealand Dept. of Conservation, 2003), pp. 83–85.

Gofman, John W., *Radiation and Human Health*. San Francisco: Sierra, 1981.

Grandolfo, M., S. M. Michaelson, and A. Rindi, eds., *Biological Effects and Dosimetry of Non-Ionizing radiation*. New York: Plenum, 1983.

Klemm, W. R., *Animal Electroencephalography*. New York: Academic, 1969.

Kume, Tamikazu and Setsuko Todoriki, "Food irradiation in Asia, the European Union and the United States," *Radioisotopes* 62(5): 291–299 (2013).

Lee, Jack M., Jr., *Electrical and Biological Effects of Transmission Lines: A Review*. Portland, OR: U.S. Dept. of Energy, Bonneville Power Administration, Oct. 1985.

Levitt, B. Blake, Henry C. Lai, and Albert M. Manville III, "Effects of non-ionizing electromagnetic fields on flora and fauna, Part 2 impacts: how species interact with natural and man-made EMF," *Reviews on Environmental Health* 37(3): 327–406 (2022).

Lin, James C., ed., *Electromagnetic Interaction with Biological Systems*. New York: Plenum, 1989.

Llaurado, J. G., A. Sances, Jr., and J. H. Battocletti, eds., *Biologic and Clinical Effects of Low-Frequency Magnetic and Electric Fields*. Springfield: C. Thomas, 1974.

Luce, Gay Gaer, *Biological Rhythms in Human and Animal Physiology*. New York: Dover, 1971.

Maherani, Behnoush, Farah Hossain, et al., "World market development and consumer acceptance of irradiation technology," *Foods* 5: 79 (2016).

Mech, L. David and Shannon M. Barber, *A Critique of Wildlife Radio-Tracking and Its Use in National Parks*. U.S. Geological Survey, Northern Prairie Wildlife Research Center (2002).

Moorhouse, Tom P. and David W. Macdonald, "Indirect negative impacts of radio-collaring: Sex ratio variation in water voles," *Journal of Applied Ecology* 42: 91–98 (2005).

Ott, John N. *Light, Radiation, and You*. Greenwich, CT: Devin-Adair, 1985.

Polk, Charles and Elliot Postow, *CRC Handbook of Biological Effects of Electromagnetic Fields*. Boca Raton: CRC Press, 1986.

Sheppard, Asher R. and Merril Eisenbud, *Biological Effects of Electric and Magnetic Fields of Extremely Low Frequency*. New York: NY Press, 1977.

"The Snow Tiger's Last Stand," *Reader's Digest*, November 1998.

Swenson, Jon E., Kjell Wallin, et al., "Effects of ear-tagging with radio transmitters on survival of moose calves," *Journal of Wildlife Management* 63(1) 354–358 (1999).

Symposium on Health Aspects of Nonionizing Radiation, *Bulletin of the New York Academy of Medicine* 55: 973–1296 (1979).

United Nations Scientific Committee on the Effects of Atomic Radiation (UNSCEAR), *Sources and Effects of Ionizing Radiation*, Report to the General Assembly. New York: United Nations, 1977. Table of radioactive consumer products is on pages 95–99.

Wertheimer, Nancy and E. D. Leeper, "Electrical wiring configurations and childhood cancer," *American Journal of Epidemiology* 109: 273–284 (1979).

Whiteford, Gary T., "Deadly combination: H-bombs and earthquakes," *Globe and Mail*, Toronto, Oct. 4, 1989, p. A7.

Young, Louise, *Power Over People*. New York: Oxford University Press, 1973.

19. Plastic (pp. 216–224)

"Additives in the environment: A challenge for the nineties," *Plastics Technology*, July 1990, pp. 48–59.

Allen, Steve, Deonie Allen, Vernon R. Phoenix, et al., Atmospheric transport and deposition of microplastics in a remote mountain catchment, *Nature Geoscience* 12:339-344 (2019).

Bergmann, Melanie, Sophia Mützel, Sebastian Primpke, et al., "White and wonderful? Microplastics prevail in snow from the Alps to the Arctic," *Science Advances* 2019:5.

Bischoff, Fritz, "Organic polymer biocompatibility and toxicology," *Clinical Chemistry* 18: 869 (1972).

Brahney, Janice, Margaret Hallerud, Eric Heim, et al., "Plastic rain in protected areas of the United States," *Science* 368: 1257–1260 (2020)

Briston, John Herbert and C. C. Gosselin, *Introduction to Plastics*. Feltham, Middlesex: Newnes, 1968.

Briston, John Herbert and Leonard L. Katan, *Plastics in Contact with Food*. London: Food Trade, 1974.

"Chemicals and additives," *Modern Plastics*, Vol. 66, Mid-October Encyclopedia Issue, 1990, pp. 159–212.

Crosby, N. T., *Food Packaging Materials: Aspects of Analysis and Migration of Contaminants*. London: Applied Science, 1981.

Dessì, Claudia, Elvis D. Okoffo, Jake W. O'Brien, et al., "Plastics contamination of store-bought rice," *Journal of Hazardous Materials* 416: 125778 (2021).

Driver, Walter E., *Plastics Chemistry and Technology*. New York: Litton, 1979.

Giam, C. S. et al., "Phthalate esters, PCB, and DDT residues in the Gulf of Mexico atmosphere," *Atmospheric Environment* 14: 65–69 (1980).

Gosselin, R. E., R. P. Smith, and H. Hodge, *Clinical Toxicology of Commercial Products*, 5th Edition. Baltiore: Williams and Wilkins, 1984.

Kosuth, Mary, Sherri A. Mason, and Elizabeth V. Wattenberg, "Anthropogenic contamination of tap water, beer, and sea salt," *PLoS ONE* 13(4): e0194970 (2018).
Mascia, L, *The Role of Additives in Plastics*. New York: Wiley, 1974.
Moore, John W., "The vinyl chloride story," *Chemistry*, June 1975, pp. 12–16.
Oswin, C. R., *Plastic Films and Packaging*. New York: Wiley, 1975.
Ragusa, Antonio, Alessandro Svelato, Criselda Santacroce, et al., "Plasticenta: First evidence of microplastics in human placenta," *Environment International* 146: 106274 (2021).
Readfearn, Graham, "It's on our plates and in our poo, but are microplastics a health risk?" *The Guardian*, May 15, 2021.
Sacharow, Stanley and Roger C. Griffin, Jr., *Basic Guide to Plastics in Packaging*. Boston: Cahners, 1973.
Schwabl, Philipp, Sebastian Köppel, Philipp Königshofer, et al., "Detection of various microplastics in human stool," *Annals of Internal Medicine* 171(7): 453–457 (2019).
Soltani, Neda Sharifi, Mark Patrick Taylor, Scott Paton Wilson, et al., "International quantification of microplastics in indoor dust: prevalence, exposure and risk assessment," *Environmental Pollution* 312: 119957 (2022).
Tao, Danyang, Kai Zhang, Shaopeng Xu, et al., "Microfibers released into the air from a household tumble dryer," *Environmental Science and Technology Letters* 9(2): 120–126 (2022).
United States Code of Federal Regulations, Title 21, "Food and Drugs," Parts 175, 176, 177, 178, 180, 181, and 184, listing additives from plastic packing allowed to enter foods.
Vainio, H., P. Pfäffli and A. Zitting, "Chemical hazards in the plastics industry," *Journal of Toxicology and Environmental Health* 6: 1179–1186 (1980).

20. Detergents (pp. 225–229)

Balsam, H. S. and Edward Sagarin, eds., *Cosmetics: Science and Technology*. New York: Wiley, 1972.
Borghetty, H. C. and C. A. Bergman, "Synthetic detergents in the soap industry," *Journal of the American Oil Chemists' Society*, March 1950, pp. 88–90.
Clute, W. N., *The Useful Plants of the World*. Indianapolis: Clute, 1943.
Davidsohn, A. and B. M. Milwidsky, *Synthetic Detergents*, 7th edition. London: Leonard Hill, 1987.
Schwartz, A., J. Perry, and J. Berch, *Surface Active Agents and Detergents*. Huntington, N.Y.: Krieger, 1977.
Sisley, J. P. and P. J. Wood, *Encyclopedia of Surface Active Agents*. New York: Chemical, 1952.

Snell, Foster Dee, "Soap vs. synthetic detergents," *Journal of the American Oil Chemists' Society*, July 1949, pp. 338–341.
Swisher, R. D., *Surfactant Biodegradation*. New York: Dekker, 1970.

21. Biocides (pp. 230–232)

Carson, Rachel, *Silent Spring*. Boston: Houghton Mifflin, 1962.
van den Bosh, Robert, *The Pesticide Conspiracy*. Garden City: Doubleday, 1978.
Ware, George W., *The Pesticide Book*. Fresno: Thomson, 1989.
White-Stevens, Robert, ed., *Pesticides in the Environment*. New York: Dekker, 1976.

22. Farewell to Silence (pp. 233–237)

Baggeroer, Arthur and Walter Munk, "The Heard Island feasibility test," *Physics Today*, Sept. 1992, pp. 22–30.
Burdic, William S., *Underwater Acoustic System Analysis*. Englewood Cliffs, N.J.: Prentice-Hall, 1984, pp. 298–301.
Coates, Rodney F. W., *Underwater Acoustic Systems*. New York: Wiley, 1989, pp. 153–170.
Committee on Underwater Telecommunication National Research Council, *Present and Future Civil Uses of Underwater Sound*. Washington: National Academy of Sciences, 1970.
da Vinci, Leonardo, *Notebooks*, arranged and translated by Edward McCurdy. New York: Reynal and Hithcock, 1938, p. 284.
Demirors, Emrecan, Jiacheng Shi, Anh Duong, et al., "The SEANet Project: Toward a programmable Internet of Underwater Things," In *Proceedings of the 2018 Fourth Underwater Communications and Networking Conference (UComms)*, Lerici, Italy, 28–30 August 2018.
Ewing, Maurice and J. Lamar Worzel, "Long-range sound transmission," in *Propagation of Sound in the Ocean*, Geological Society of America Memoir 27, 1948.
Filho, Jose Ilton de Oliveira, Abderrahmen Trichili, Boon S. Ooi et al, "Towards self-powered Internet of Underwater Things devices," IEEE Communications Magazine, July 23, 2019, arXiv: 1907.11652.
Fletcher, John and R. G. Busnel, eds., *Effects of Noise on Wildlife*. New York: Academic, 1978.
Griffin, Donald, *Listening in the Dark: The Acoustic Orientation of Bats and Men*. New York: Dover, 1974.
King, Judith E., *Seals of the World*. Ithaca, N.Y; Cornell University Press, 1983.
Mohsan, Syed Agha Hassnain, Alireza Mazinani, Nawaf Qasem Hamood Othman, and Hussain Amjad, "Towards the internet of underwater things: a comprehensive survey," *Earth Science Informatics* 15: 735–764 (2022).

Tavolga, William N., ed., *Marine Bio-Acoustics*. New York: Pergammon, 1967. See especially pp. 159–186, William E. Evans, "Vocalization among marine mammals."

White, Frederick A., *Our Acoustic Environment*. New York: Wiley, 1975.

23. Transportation (pp. 238–239)

Burwell, David, "The end of the road," in *The New Environmental Handbook*, Garrett de Bell, ed. (San Francisco: Friends of the Earth, 1980), pp. 67–78.

Henderson-Sellers, A., "North American total cloud amount variations this century," *Global and Planetary Change* 1: 175–194 (1989).

Mumford, Lewis, *The Highway and the City*. New York: Harcourt, Brace, and World, 1963.

Rattner, Robert, "Mermaids among us," *Animals*, July/Aug. 1990, pp. 26–31, on manatees and boats.

Schneider, Kenneth R., *Augokind vs. Mankind*. New York: Norton, 1971.

24. Guns (pp. 240–241)

World Population Review, *Gun Deaths by Country 2023*, https://www.worldpopulationreview/country-rankings/gun-deaths-by-country.

25. The Internet (pp. 242–244)

Afzal, Munshi Naser Ibne ans Jeff Gow, "Electricity consumption and information and communication technology in the next eleven emerging economies," *International Journal of Energy Economics and Policy* 6(3): 381–388 (2016).

Mills, Mark, *The Internet Begins with Coal*. Greening Earth Society, 1999.

———, *The Cloud Begins with Coal*. Digital Power Group, August 2013.

Mills, Mark P. and Peter W. Huber, "Dig more coal – the PCs are coming," *Forbes*, May 31, 1999, pp 70–72.

27. Economics, and Human Diversity (pp. 253–298)

Arndt, H. W., *The Rise and Fall of Economic Growth*. Melbourne: Longman Cheshire, 1978.

Barnes, Harry, *Economic History of the Western World*. New York: Harcourt, Brace, 1937. See especially pp. 377–415 on the industrialization of non-Western countries.

Beckhart, Benjamin Haggott, *Federal Reserve System*. New York: Columbia University Press, 1972.

Cole, G. D. H., *Introduction to Economic History 1750-1950*. New York: Macmillan, 1965.

Del Mar, Alexander, *A History of Money in Ancient Countries*, 1885, reprinted by Franklin, New York,, 1968.

Diamond, Arthur Sigismund, *The Evolution of Law and Order*. London: Watts, 1951.
Galbraith, John Kenneth, *Money: Whence It Came, Where It Went*. Boston: Houghton, Mifflin, 1975.
Herkovits, Melville J., *Economic Antrhopology: The Economic Life of Primitive Peoples*. New York: Norton, 1952.
Hilton, Rodney, ed., *The Transition from Feudalism to Capitalism*. London: Verson, 1976.
Jarrett, Henry, ed., *Environmental Quality in a Growing Economy*. Baltimore: The Johns Hopkins Press, 1966.
Kenwood, A. G. and A. L. Lougheed, *Technological Diffusion and Industrialization Before 1914*. New York: St. Martin's, 1982.
Keyfitz, Nathan, "World resources and the world middle class," *Scientific American*, July 1976, pp. 28–35.
Keynes, John Maynard, *The Pure Theory of Money*, 1930, reprinted by Macmillan, London, 1971.
Lappé, Frances Moore and Joseph Collins, *Food First: Beyond the Myth of Scarcity*. New York: Ballantine, 1978.
Miller, J. Innes, *The Spice Trade of the Roman Empire, 29 B.C. to A.D. 641*. Oxford: Clarendon, 1969.
Mishan, Ezra J., *The Costs of Economic Growth*. New York: Praeger, 1967.
Saint Phalle, Thibaut de, *The Federal Reserve, An Intentional Mystery*. New York: Praeger, 1985.
Schumacher, E. F., *Small Is Beautiful: Economics as if People Mattered*. New York: Harper & Row, 1973.
Warsh, David, "The great hamburger paradox: Things John Maynard Keynes never told us," *Forbes*, Sept. 15, 1977, pp. 166–239.

Agricultural Societies (pp. 255–265)

Boeke, J. H., *The Interests of the Voiceless Far East*. Leiden: Universitaire Pers Leiden, 1948.
Contemporary Nomadic and Pastoral People: Asia and the North, Studies in Third World Societies, Publication #18. Williamsburg, Va., 1982.
Croxall, Harold E. and Lionel P. Smith, *The Fight for Food: Factors Limiting Agricultural Producation*. London: Allen, 1984.
Zainu'ddin, Ailsa, *A Short History of Indonesia*. New York: Praeger, 1970.

Foragers (pp. 265–278)

Arens, W., *The Man-Eating Myth*. New York: Oxford University Press, 1979.
Aspinall, Ron, "Sarawak: The human consequences of logging," *Cultural Survival Quarterly* 14(4): 71–78 (1990).
Bodley, John H., *Victims of Progress*. Palo Alto: Mayfield, 1990.

Carlisle, Richard, ed., *Illustrated Encyclopedia of Mankind.* New York: Cavendish, 1984.
Egan, Timothy, "Alaska natives sadly ponder whether to spend their land," *New York Times,* June 1, 1990, p. A1.
Gaisford, John, ed., *Atlas of Man.* New York: St. Martin's, 1978.
Grinnell, George Bird, *Pawnee Hero Stories and Folk Tales,* 1889, reprinted by University of Nebraska Press, Lincoln, 1961.
Hahn, C. H. L., H. Vedder, and L. Fourie, *The Native Tribes of Southwest Africa.* New York: Barnes and Noble, 1966.
Hall, Sam, *The Fourth World: The Heritage of the Arctic and Its Destruction.* London: The Bodley Head, 1987.
Hoffman, Carl, *The Bunan: Hunters and Gatherers of Borneo.* Ann Arbor: UMI Research Press, 1986. Relevant critiques of this book are found in *Borneo Research Bulletin,* Sept. 1988, pp. 31–129.
Jacobs, Paul and Saul Landau, *To Serve the Devil.* New York: Random, 1971.
Kroeber, Alfred L., *Handbook of the Indians of California,* Bulletin 78 of the Bureau of American Ethnology, 1925, reprinted by California Book Company, 1953.
Le Clercq, Father Chrétien, *New Relations of Gaspesia,* 1691, reprinted by Champlain Society, Toronto, 1910.
Lee, Richard Borshay and Irven De Vore, eds., *Man the Hunter.* Chicago: Aldine, 1968. See especially pp. 13–20, George Peter Murdock, "The current status of the world's hunting and gathering peoples."
McLuhan, T. L., *Touch the Earth: A Portrait of Indian Existence.* New York: Outerbridge, 1971.
Mowat, Farley, *Canada North Now: The Great Betrayal.* Toronto: McClelland and Stewart, 1976.
Murdock, George Peter, "Ethnographic Atlas," *Ethnology* 6: 109–236 (1967).
Nansen, Fridtjof, *Eskimo Life.* London: Longmans, Green, 1893.
Pinney, Roy, *Vanishing Tribes.* New York: Crowell, 1968.
Schapera, I., *The Khoisan Peoples of South Africa.* London: Routledge, 1930.
Seligman, C.G., *The Veddas.* Cambridge University Press, 1911.

Central Africa (pp. 278–279)

Bailey, Rober C., "The Efe, Archers of the African Rain forest," *National Geographic,* 76 (1989): 664–686.
Cavalli-Sforza, Luigi Luca, ed., *African Pygmies.* Orlando: Academic, 1986.
Hallet, Jean-Pierre, *Pygmy Kitabu.* Greenwich, CT: Fawcett, 1973.
Peterson, Richard B., "Searching for life on Zaire's Ituri Forest frontier, *Cultural Survival Quarterly* 14(4): 56–62 (1990).

Turnbull, Colin, *The Forest People: A Study of the Pygmies of the Congo*. New York: Simon & Schuster, 1962.

———, "The Mbuti Pygmies: An ethnographic survey," *Anthropological Papers of the American Museum of Natural History*, Vol. 50, Part 3, 1965, pp. 137–282.

Wilkie, David S. and Gilda A. Morelli, "Pitfalls of the Pygmy hunt," *Natural History*, Dec. 1988, pp. 32–41.

South America (pp. 279–286)

"The Aché of Paraguay," *Cultural Survival Quarterly* 8(3): 11–12 (1984).

Arcand, Bernard, *The Urgent Situation of the Cuiva Indians of Colombia*. Copenhagen: International Work Group for Indigenous Affairs (AWGIA), 1973.

Bodard, Lucien, *Green Hell*. New York: Outerbridge, 1971.

Carrera, Nicolás Iñigo, *Violence as an Economic Force: The Process of Proletarianisation among the Indigenous People of the Argentine Chaco, 1884-1930*. Copenhagen: IWGIA, 1982.

Escobar, Ticio, *Ethnocide: Mission Accomplished?* Copenhagen: IWGIA, Aug. 1989, on the Ayoreode.

Ireland, Emilienne, "Neither warriors nor victims, the Wauja peacefully organize to defend their land," *Cultural Survival Quarterly* 15(1): 54 (1991).

Münzel, Mark, *The Aché Indians: Genocide in Paraguay*. Copenhagen: IWGIA, 1972.

Reed, Richard, "The Mataco of Argentina," *Cultural Survival Quarterly* 8(3): 9 (1984).

Rodrigues, Nemesio J., *Oppression in Argentina*. Copenhagen: IWGIA, 1975, on the Mataco.

Steward, Julian H., ed., *Handbook of South American Indians*, 7 volumes. Washington: U.S. Government Printing Office, 1947.

Villas Boas, Ortlando and Claudio Villas Boas, *Xingu: The Indians, Their Myths*. New York: Farrar, 1973.

Europe (pp. 286–289)

Pelto, Pertti J., *The Snowmobile Revolution: Technology and Social Change in the Arctic*, 2nd edition. Prospect Heights, Ill.: Waveland, 1987.

Industrial Societies (pp. 289–294)

Barnet, Richard J. and Ronald E. Müller, *Global Reach: The Power of the Multinational Corporations*. New York: Simon & Schuster, 1974.

De Grazia, Sebastian, *Of Time, Work, and Leisure*. Garden City: Doubleday, 1962.

The Great Depression of the 1930s (pp. 295–296)
Kindleberger, Charles P., *The World in Depression*. Berkeley: University of California Press, 1986.

28. Population, and a Tour of the World (pp. 299–357)
Bibby, Geoffrey, *Four Thousand Years Ago*. New York: Knopf, 1963.
Carr-Saunders, A. M., *World Population*. Oxford: Clarendon, 1936.
Darwin, Sir Charles, *The Problems of World Population*. Cambridge University Press, 1958.
Ehrlich, Paul R., *The Population Bomb*. New York: Ballantine, 1968.
Hardin, Garrett, ed., *Population, Evolution and Birth Control*. San Francisco: Freeman, 1969.
Hollingsworth, T. H., *Historical Demography*. Ithaca: Cornell University Press, 1969.
Hume, David, "Of the populousness of ancient nations," in his *Political Discourses*, R. Fleming, Edinburgh, 1752, pp. 155–261.
Malthus, Thomas Robert, *Essay on the Principle of Population as It Affects the Future Improvement of Society*. London: J. Johnson, 1798.
McEvedy, Colin and Richard Jones, *Atlas of World Population History*. New York: Facts on File, 1979.
Montesquieu, Charles Louis de Secondat, baron de, *The Persian Letters*, 1721. Indianapolis: Bobbs-Merrille, 1964, letters CXII to CXXII.
Organski, Katherine and A. F. K. Organski, *Population and World Power*. New York: Knopf, 1961.
Rienow, Robert and Leona Rienow, *Moment in the Sun*. New York: Dial, 1967.
Rivers, W. H. R., "The contact of peoples," in *Essays and studies Presented to William Ridgeway*. London: Cambridge University Press, 1913, pp. 474–492.
Stangeland, Charles Emil, *Pre-Malthusian Doctrines of Population*. New York: Kelley, 1966.
Teitelbaum, Michael S. and Jay M. Winter, *The Fear of Population Decline*. Orlando: Academic, 1985.
Tertullian, *De Anima*, Chapter 30.
Trewartha, Glenn T., *A Geography of Population*. New York: Wiley, 1969.
Waldron, Ingrid and Robert E. Ricklefs, *Environment and Population*. New York: Holt, 1973.
Wallace, Robert, A *Dissertation on the Numbers of Mankind in Ancient and Modern Times*, 1753. Reprinted by Kelley, New Yorik, 1969.
Wrigley, E. A., *Population and History*. New York: McGraw-Hill, 1969.

Java (p. 306)

Van der Kroef, Justus M., "Populatin pressure and economic development in Indonesia," *American Journal of Economics and Sociology* 12: 355–371 (1953).

China (pp. 306–309)

Durand, John D., "The population statistics of China, A.D. 2 - 1953," *Population Studies* 13: 209–256 (1960).

Legge, James, *The Chinese Classics*. Hong Kong University Press, 1960. Vol. 3, pp. 76–77 gives the population of China in the 23rd century B.C.

Peffer, Nathaniel, *China: The Collapse of a Civilization*. New York: Day, 1930.

India (pp. 309–311)

See Hollingsworth, pp. 228–229 and Trewartha, pp. 15, 32.

Sri Lanka (pp. 311–313)

Taeuber, Irene B., "Ceylon as a demographic laboratory," *Population Index* 15: 293–304 (1949).

The Fertile Crescent (pp. 313–315)

See Hollingsworth, pp. 307–311, on the history of Egypt's population.

Japan (pp. 315–319)

Perrin, Noel, *Giving up the Gun*. Boston: Hall, 1979.
Taeuberg, Irene B., *The Population of Japan*. Princeton University Press, 1958.
Reischauer, Edwin O., *Japan: The Story of a Nation*. New York: Knopf, 1970.

Mongolia and Turkestan (pp. 319–324)

Area Handbook for Mongolia. Washington: U.S. Government Printing Office, March 1970.

Fox, Ralph, *Genghis Khan*. London: The Bodley Head, 1936.

Grousset, René, *A History of Asia*. New York: Walker, 1963.

Heissig, Walther, *A Lost Civilization: The Mongols Rediscovered*. New York: Basic, 1966.

McGovern, William Montgomery, *The Early Empires of Central Asia*. Chapel Hill: University of North Carolina Press, 1939.

Sokolewicz, Zofia, "Traditional worldview in contemporary Mongolia," in *Contemporary Nomadic and Pastoral People: Asia and the North*. Williamsburg, VA: Studies Third world Societies, 1982, pp. 125–139.

Siberia (pp. 324–326)

Jochelson, Waldemar, *Peoples of Asiatic Russia*. New York: American Museum of Natural History, 1928.

Kolarz, Walter, *The Peoples of the Soviet Far East*. New York: Praeger, 1954.

Norton, Henry Kittredge, *The Far Eastern Republic of Siberia*. New York: Holt, 1923.

Europe (pp. 326–328)

Gibbon, Edward, *The Decline and Fall of the Roman Empire*, 1776. Reprinted by Modern Library, New York, 1932. See especially p. 37 and pp. 43–44 on ancient Europe's population.

Gottfried, Robert S., *The Black Death*. New York: Free Press, 1983.

Ireland (pp. 328–329)

Kennedy, K., T. Gibbin, and D. Mchugh, *The Economic Development of Ireland in the Twentieth Century*. New York: Routledge, 1988.

Africa (pp. 330–339)

Congo's Mining Slaves, Free the Slaves Investigative Report, June 2013, https://freetheslaves.net/wp-content/uploads/2015/03/Congos-Mining-Slaves-web-130622.pdf.

Davidson, Basil, *The Lost Cities of Africa*. Boston: Little, Brown, 1970.

DRC: Efe Pygmies deprived of their homeland and their livelihood, World Rainforest Movement, Bulletin 118, May 19, 2007.

Goldblatt, Jenna Marie, *Conflict and Coltan: Resource Extraction and Collision in The Democratic Republic of the Congo and Venezuela*, Fordham University student thesis, May 19, 2023.

Hall, R. N. and W. G. Neal, *The Ancient Ruins of Rhodesia*, 1904. Reprinted by Negro Universities Press, New York, 1969.

Harden, Blaine, "The dirt in the new machine," *New York Times Magazine*, August 12, 2001.

Juurlink, Anna Rosa, *Congo, Coltan and Conservation*, Forest and Nature Conservation Policy Group, Master of Science thesis, Wageninen University, Dec. 22, 2021.

Kara, Siddharth, *Cobalt Red*. NY: Saint Martin's Press, 2023.

Legum, Colin, *Congo Disaster*. Gloucester, Mass.: P. Smith, 1972.

Lopez, Duarte, *A Report on the Kingdom of Congo*, Rome, 1551. Reprinted by Negro Universities Press, 1969.

Mallows, Wilfrid, *The Mystery of the Great Zimbabwe*. London: Hale, 1985.

Middleton, Lamar, *The Rape of Africa*. New York: Negro Universities Press, 1969.
Morel, E. D., *Red Rubber*. New York: Nassau, 1906.
———, *Great Britain and the Congo*. London: Smith, Elder, 1909.
Ojewale, Oluwole, *Mining and Illicit Trading of Coltan in the Democratic Republic of Congo*, ENACT, Research paper 29, March 2022.
Peters, Carl, *King Solomon's Golden Ophir*, 1899. Reprinted by Negro Universities Press, 1969.
Schuyler, Philippa, *Who Killed the Congo?* New York: Devin-Adair, 1962.
Sheriff, Abdul, *Slaves, Spices and Ivory in Zanzibar*. London: Curry, 1987.
Stone Age Herbalist, "The Forest People: life and feath under the Green Revolution," https://www.resilience.org/stories/2021-05-05/the-forest-people-life-and-death-under-the-green-revolution.
Vansina, Jan, *Kingdoms of the Savannas*. Madison: University of Wisconsin Press, 1968.
Walker, Eric A., A *History of South Africa*. London: Longmans, Green, 1935.

America (pp. 341–352)

Allman, William, "Lost Empires of the Americas," *U.S. News & World Report*, April 2, 1990, pp. 46–54.
Bancroft, Hubert Howe, *Native Races of the Pacific States*, 5 vols. San Francisco: The History Co., 1886.
Barbar, John, "Oriental enigma," *Equinox*, Jan./Feb. 1990, pp. 83–95.
Barrau, Jacques, ed., *Plants and the Migratins of Pacific Peoples*. Honolulu: Bishop Museum, 1963.
Bernard, Jean Frédic, *Recueil de Voyage au Nord*. Amsterdam, 1731.
Cain, Stanley A., *Foundations of Plant Geography*. New York: Harper, 1944.
Carter, George F., "Plant evidence for early contact with America," *Southwestern Journal of Anthropology* 6: 161–181 (1940.
———, "Plants Across the Pacific," *Memoirs of the Society for American Archaeology* 9: 62–71 (1953).
Carvajal, Fray Gaspar de, *Relación del nuevo Descubrimiento del famoso Río Grande de las Amazonas*. Mexico: Fondo de Cultura Económica, 1955.
Colinvaux, Paul A., "The past and future Amazon," *Scientific American*, May 1989, pp. 102–108.
Cook, Sherburne and Woodrow Borah, *Essays in Population History*. Berkeley: University of California Press, 1971.
Cortés, Hernán, *Cartas de Relación*. Madrid: García Noblejas, 1985.
Cuoq, Joseph M., translator, *Recueil des Sources Arabes Concernant L'Afrique Occidentale du VIIIe Siècle*. Paris: Centre National de la Recherche Scientifique, 1975.

de Candolle, Alphonse, *Origin of Cultivated Plants*, 1885. Reprinted by Hafner, New York, 1959.

Denevan, William M., *The Natie Population of the Americas in 1492*. Madison: University of Wisconsin Press, 1976.

Dobyns, Henry F., *Their Number Become Thinned: Native American Population Dynamics in Eastern North America*. Knoxville: University of Tennessee Press, 1983.

———, "Estimating Aboriginal American population 1: An appraisal of techniques with a new hemispheric estimate," *Current Anthropology* 7: 395–416 (1966).

Dostal, W., ed., *The situation of the Indian in South America*. Geneva: World Council of Churches, 1972.

Durán, Fray Diego, *Book of the gods and rites and the Ancient Calendar*, about 1576–1579. Translation by F. Horcasitas and D. Heyden, University of Oklahoma Press, 1971.

Eckholm, Gordon F., "A possible focus of Asiatic influence in the Late Classic cultures of Mesoamerica," *Memoirs of the Society for American Archaeology* 9: 72–89 (1953).

Galvano, Antonio, *The Discoveries of the World*, 1563. Published in English by Richard Hakluyt, 1601. Reprinted by Franklin, New York, 1971.

Gibbons, Ann, "New view of early Amazonia," *Science* 248: 1488–1490 (1990).

Hagar, Stansbury, "The problems of the unity or plurality and the probable place of origin of the American Aborigines: The bearing of astronomy on the subject," *American Anthropologist* 14: 43–48 (1912).

Harrisse, Henry, *the Discovery of North America*. Amsterdam: N. Israel, 1961.

Herkovits, Melville J., *Economic Anthropology*. New York: Norton, 1952, on trade routes of the Hopi, etc.

Heyerdahl, Thor, *Early Man and the Ocean*. Garden City: Doubleday, 1979.

———, *Kon-Tiki: Across the Pacific by Raft*. Chicago: Rand McNally, 1950.

Hornell, James, *Water Transport, Origins and Early Evolution*. Cambridge University Press, 1946.

Hultkranz, Ake, *The North American Indian Orpheus Tradition*. Stockholm: Caslon, 1957.

Imbelloni, J., "On the diffusion in America of Patu Onewa, Okewa, Patu Paraoa, Miti, and other relatives of the Mere family," *the Journal of the Polynesian Society* 39: 322–345 (1930).

Jeffreys, D. W., "Pre-Columbian Negroes in America," *Scientia* 88: 202–218 (1953).

Kingsborough, Lord, *Antiquities of Mexico*. London: Havell, 1831.

Kroeber, Alfred, *Anthropology*. New York: Harcourt, Brace, and World, 1948, p. 535 on the diffusion of culture to and from America.

las Casas, Fray Bartolomé de, *The Diario of Christopher Columbus's First Voyage to America, 1492-1493*, translation by Oliver Dunn and James E. Kelley, Jr. Norman: University of Oklahoma, 1989.

Leland, Charles G., *Fusang: Or the discovery of America by Chinese Buddhist priests in the fifth century.* London: Curzon, 1973.

Ma Twan-Lin, *Antiquarian Researches*, about 1321. Samuel Wells Williams, translator (New Haven: Tuttle, 1881).

Marcus, G. J., *The Conquest of the North Atlantic*. New York: Oxford University Press, 1981.

Martyr, Peter, *De Orbe Novo*, 1516. Francis MacNutt, translator (New York: Putnam, 1912).

Maudsley, Alfred Percival, *Archaeology*. vols. 58–63 of *Biologia Centrali-Americana*, Frederick DuCane Godman and Osbert Salvin, eds. London: Porter, 1889–1902.

Mertz, Henriette, *Pale Ink: Two Ancient Records of Chinese Exploration in America*. Chicago: Swallow, 1972.

Moore, Thomas, "Thor Heyerdahl: Sailing against the curren," *U.S. News & World Report*, April 2, 1990, pp. 55–60.

Morison, Samuel Eliot, *The European Discovery of America*. New York: Oxford University Press, 1971.

———, *Portuguese Voyages to America in the Fifteenth Century.* Cambridge, Mass.: Harvard University Press, 1950.

Nuttall, Zelia, "The fundamental principles of Old and New World civilizations," *Archaeological and Ethnological Papers of the Peabody Museum*, Vol. 2, 1900, reprinted by Kraus Reprint Co., N.Y., 1970.

———, "Some unsolved problems in Mexican archaeology," *American Anthropologist*, New Series, 8: 133–149 (1906).

———, *A curious Survival in Mexico of the Use of the Purpura Shell-Fish for Dyeing*, Putnam Anniversary Volume. Cedar Rapids, Iowa: The Torch Press, 1909.

Owen, J. Deetz and A. Fisher, eds., *North American Indians: A Sourcebook*. New York: Macmillan, 1967.

Pigafetta, Antonio, *Magellan's Voyage*, 1563. R. A. Skelton, translator (New Haven: Yale University Press, 1969).

Pohl, Fred, *Atlantic Crossings Before Columbus*. New York: Norton, 1961.

Quinn, David B., "The argument for the English discovery of America between 1480 and 1494," *Geographical Journal* 127: 277–285 (1961).

———, ed., *New American World: A Documentary History of North America to 1612*. New York: Arno, 1979.

Riley, Carroll L. et al., eds., *Man Acrodd the Sea: Problems of Pre-Columbian Contacts*. Austin: University of Texas Press, 1971.

Roosevelt, Anna, "Lost civilizatins of the Lower Amazon," *Natural History*, Feb. 1898, pp. 74–83.

Sahagún, Fray Bernardino de, *General History of the things of New Spain*, also known as the *Florentine Codex*. Published in English by the School of American Research, Santa Fe, New Mexico, 1982, in 12 volumes.

Sauer, Carl O., *Agricultural Origins and Dispersals*. New York: American Geographical Society, 1952.

———, *Northern Mists*. Berkeley: University of California Press, 1968.

Shan Hai Ching, translation by Hsiao-Chieh Cheng et al. Republic of China: National Institute fo Compilation, 1985.

Thacher, John Boyd, *Christopher Columbis: His Life, His Work, His Remains*. New York: Putnam, 1903.

Thronton, Russell, *American Indian Holocaust and Survival: A Population History Since 1492*. Norman: University of Oklahoma Press, 1987.

Tout, Charles Hill, "Report on the ethnology of the Okanák•en of British Columbia, an interior division of the Salish stock," *The Journal of the Royal Anthropological Society of Great Britain and Ireland* 41: 133–135 (1911).

Tylor, Edward B., *Researches into the Early History of Mankind and the Development of Civilization*. London: J. Murray, 1878.

———, "On the game of patolli in ancient Mexico and its probably Asiatic origin," *The Journal of the Anthropological Institute of Great Britain and Ireland* 8: 116–131 (1879).

———, "On the diffusion of mythical beliefs as evidence in the history of culture," *British Association for the Advancement of Science. Report of the Meetings* 64: 774 (1894).

The Pacific Islands (pp. 352–353)

Beaglehole, J. C., ed., *The Journals of Captain James Cook*. Cambridge University Press, 1969. See Vol. II, p. 409 on the population of Tahiti in 1774.

Bugotu, F., "The culture clash: A Melanesian" view," *New Guinea and Australia, The Pacific and South-East Asia*, June/July 1968, pp. 65–70.

Howard, Alan, "Polynesia origins and migrations," in *Polynesian Culture History*, Bernice P. Bishop Museum Special Publication 56. Honolulu: bishop Museum Press, 1967, pp. 45–101.

McCarthy, F. D., "'Trade' in Aboriginal Australia, and 'trade' relationships with Torres Strait, New Guina, and Malaya," *Oceania* 9: 405–438 (1939) and 10: 81–104, 171–190 (1939).

Shineberg, Dorothy, *They Came for Sandalwood*. Carlton, Victoria: Melbourne University Press, 1967.

Stannard, David E., *Before the Horror: The Population of Hawaii on the Eve of Western Contact*. Honolulu: University of Hawaii Press, 1989.

Turnbull, Clive, *Black War: The Extermination of the Tasmanian Aborigines*. South Melbourne: Sun Books, 1974.

Contraception (pp. 355–357)

Henshaw, Paul S., "Physiologic control of fertility," *Science* 117: 572–582 (1953).

Hines, Norman E., *The Medical History of Contraception*. Baltimore: Williams and Wilkens, 1936.

Laszlo, Henry de and Paul S. Henshaw, "Plant materials used by primitive peoples to affect fertility," *Science* 119: 626–630 (1954).

Noonan, John T., Jr., *Contraception*. Cambridge, Mass.: Harvard University Press, 1965.

Riddle, John M., J. Worth Eates, and Josiah C. Russell, "Ever since Eve: Birth control in the ancient world," *Archaeology*, March/April 1994, pp. 29–35.

Zahl, Paul A., "Some characteristics of the anti-estrous factor in *Lithospermum*," *Proceedings of the Society for Experimental Biology and Medicine*, 67: 405–410 (1948).

29. War (pp. 358–367)

Brodie, Bernard and Fawn Brodie, *From Crossbow to H-bomb*. New York: Dell, 1962.

Chidsey, Donald Barr, *Goodbye to Gunpowder*. New York: Crown, 1963.

Coblenz, Stanton Arthur, *From Arrow to Atom Bomb: The Psychological History of War*. New York: Beechhurst, 1953.

Densmore, Frances, *Papago Music*. Washington: U.S. Government Printing Office, 1929.

Evans-Pritchard, E. E., "Zande warfare," *Anthropos* 52: 239–262 (1957).

Hall, A. R., "A note on military pyrotechnics," in Charles Singer et al., *A History of Technology*. Oxford: Clarendon, 1956, Vol. 3, pp. 374–382.

———, "military technology," *ibid.*, pp. 695–730.

Malinowski, Bronislaw, "War and weapons among the natives of the Trobriand Islands," *Man* 20(5): 10–12 (1920).

Meek, Charles K., *Law and Authority in a Nigerian Tribe*. London: Oxford University Press, 1937.

Meggitt, Mervyn, *Blood Is Their Argument: Warfare Among the Mae Enga Tribesmen of the New Guinea Highlands*. Palo Alto: Mayfield, 1977.

Nansen, Fridtjof, *Eskomo Life*. London: Longmans, Green, 1893.

Nelson, Edward William, *The Eskimo About Bering Strait*. Washington: U.S. Government Printing Office, 1900.

O'Neil, Brian Hugh St. John, *Castles and Cannon*. Oxford: Clarendon, 1960.

Shineberg, Dorothy, ed., *The Trading voyages of Andrew Cheyne 1841-1844*. Honolulu: University of Hawaii Press, 1971.

Tyrrell, J. B., ed., *David Thompson's Narrative of His Explorations in Western America, 1784-1812*. Toronto: Champlain Society, 1916. Reprinted by Greenwood Press, New York, 1968.

Wedgewood, Camilla H., "Some aspects of warfare in Melanesia," *Oceania* 1: 5–33 (1930).

Whiteway, R. S., *The Rise of Purtuguese Power in India 1497-1550*. Patna: Janaki Prakashan, 1979.

Wilson, William, A *Missionary Voyage to the Southern Pacific Ocean Performed in the Years 1796, 1797, 1798 in the Ship Duff, Commanded by Captain James Wilson*. London: T. Chapman, 1799.

30. Slavery (pp. 368–371)

Macleod, William Christie, "Debtor and chattel slavery in Aboriginal North America," *American Anthropologist* 27: 370–380 (1925).

Nieboer, H. J., *Slavery as an Industrial System*, 1910. Reprinted by Franklin, New York, 1971.

Pinney, Roy, *Slavery, Past and Present*. New York: Nelson, 1972.

Rodney, Walter, "African slavery and other forms of social oppression on the Upper Guinea Coast in the context of the Atlantic slave trade," *Journal of African History* 3: 431–443 (1966).

Sheriff, Abdul, *Slaves, Spices and Ivory in Zanzibar*. London: J. Curry, 1987.

Watson, James L., ed., *Asian and African Systems of Slavery*. Berkeley: University of California Press, 1980.

Winks, Robin W., ed., *Slavery: A Comparative Perspective*. New York: NYU Press, 1972.

31. Religion (pp. 372–381)

Ferguson, John, *War and Peace in the World's Religions*. New York: Oxford University Press, 1978.

Parrinder, Geoffrey, *Sex in the World's Religions*. New York: Oxford University Press, 1980.

Thompson, Henry O., *World Religions in War and Peace*. Jefferson, N.C.: McFarland, 1988.

White, Lynn, Jr., "The historical roots of our ecological crisis," *Science* 155:1203-7 (1967).

32. Sex (pp. 382–387)

Bettelheim, Bruno, *Symbolic Wounds: Puberty Rites and the Envious Male*. New York: Collier, 1962.

Bryk, Felix, *Circumcision in Man and Woman*, 1934. Reprinted by American Ethnological Press, New York, 1974.

El Dareer, Asma, *Women, Why Do You Weep? Circumcision and Its Consequences.* London: Zed, 1982.

Encyclopedia Judaica. New York: Macmillan, 1971, Vol. 5, pp. 567–575.

Romberg, Rosemary, *Circumcision: The Painful Dilemma.* South Hadley, Mass.: Bergin & Garvey, 1985.

Singer, Isidore, ed., *the Jewish Enclopedia.* New York: Funk & Wagnalls, 1903, Vol. 4, pp. 92–102.

Wallerstein, Edward, *Circumcision: An American Health Fallacy.* New York: Springer, 1980.

34. Slowing Down (pp. 393–398)

Good, Merle, *Who Are the Amish?* Intercourse, Pa.: Good Books, 1985. Passage quoted is from page 88.

Index

aardvark, 171
Aché, 268, 280–81
acid rain, 101–104, 364
Ackerman, John, 179
acupuncture, 37–38, 175. *See also* meridians
Adams, Henry Brooks, 2, 3
Adirondack State Park, 102
aerosol cans, 96
Africa, 187–89; agriculture, 14, 278–79, 271; dams, 142–43; deforestation, 103–4, 111–15, 118–19; depletion of groundwater, 130; early trans-Atlantic voyages from, 259, 342, 343; foragers, 15, 266, 271, 278–79; human origins in, 11–15; native warfare, 358–59; nomadic herders, 256; pesticide use, 105, 143, 231; population history, 330–39; slavery in, 334–39, 368, 371; wildlife, 25, 69, 267–72
aggression, 26–27
Agnolo, 326–27
agricultural economies, 255–265
Agricultural Revolution, 14–16
agriculture, 255–65, 370, 375–76, 380; mechanized, 256, 295; on Nile, 142; origins, 14–16; plastics in, 216; soil erosion and, 109–110, 114, 115–17, 125–26; swidden, 255–57; water use by, 145–46; *see also* irrigation

ahimsa, 377–78
Ainu, 267, 315
air, *see* atmosphere
air conditioning, *see* refrigeration
air pollution, 1, 2, 101–107, 145, 163, 220, 246, 247
Air Quality Index, 106
airplanes, 1, 18, 76, 83, 123, 166–67, 238–39, 246, 247; air pollution from, 105; birds killed by, 238, 294; cause of cloudiness, 77, 238; effect on warfare, 364; and greenhouse effect, 83; noise pollution from, 192, 201, 237; speed and, 239; supersonic, and ozone depletion, 97–98
airports, 47, 144
Akosombo Dam, 138, 142, 148
Al-'Umarī, 343
Alaska: native warfare in, 362; Russians in, 325, 351
Alaska Native Claims Settlement Act, 283
Aleuts, 346, 351
Alfvén, Hannes, 30
Algae, 40, 58, 59, 61, 84, 109, 209, 227; blue-green, 41–42
Allen, John, 179
aloes wood, 269–71
alpaca, 15
Alps, 85
aluminum smelting, 145, 325

462   THE EARTH AND I

Amazon Basin, 118–19: burning of, 103, 118–19; cities of, 283, 349–51; dams in, 138, 141; deforestation, 103, 111, 118–19; native peoples, 255, 283–85; rubber hunters in, 273, 283, 338; Trans-Amazon Highway, 107
Amazon River, 107, 193
America; extinct animals, 166–72; pre-Columbian contacts, 342–48; population history, 341–52; settlement of, 12
amino acids, 40, 44, 45, 62
Amish, 246, 374, 375, 381, 393
amphibians, 179–83; electromagnetic radiation and, 182–83; evolution of, 70; radio tracking devices on, 213
Anabaptists, 381
Ancestral Pueblo, 118, 128
Andaman Islanders, 267, 363
Andean condor, 132
Andes Mountains, 282, 285, 350, 351
animals: domestication of, 14–15, 16–17; endangered, 19, 174–93; extinct, 1, 14, 45, 121, 166–72; number of species, 192–93, 302; size, 172–74. *See also* biodiversity; migration
Antarctica: ice pack, 57, 84, 85, 124; International Antarctic Treaty, 191; noise pollution in, 192; oil spills in, 152–53, 191; ozone hole over, 99–100, 191; pollution of, 191–92; wildlife, 190–92; as world park, 189–91
antelope, 25, 67, 113, 125, 151, 170, 187, 188; pronghorn, 166
antibiotics, 18, 202, 206
ants, 49, 51, 66, 68
Aquinas, Thomas, 292

Arabia: forests of, 112; Incense Road, 313; traders from, 306, 307, 311, 330, 333
Aral Sea, 151
Arapesh, 28
Arcand, Bernard, 280
Arctic: air pollution, 105; native peoples, 28, 144, 190, 273, 278, 283, 286–89, 346, 351, 363; ozone hole, 100–101; snowmobiles in, 190, 247, 287–89, 405–408; warming of, 85
arctic jellyfish, 173
Argentina, 186, 191, 196, 273, 281, 285, 383; groundwater depletion in, 130
Aristotle, 292, 344
Arizona, groundwater depletion in, 126–29
armadillo, giant, 166
artisans, 271, 272. *See also* crafts
asbestos mining, 247
Asia: early agriculture, 14; settlement of, 12
Asia Minor: forests of ancient, 112
asphalt, 81, 156, 193, 239
astronomy, 30–34
Aswan High Dam, 138, 142
Atamai, Chief, 286
Atlantic Ocean, 91, 104, 141, 155, 163, 166, 185, 221, 314, 331, 341, 343, 344, 349
atmosphere, 79–108; composition, 58; origin of, 58; source of life's elements, 39. *See also* oxygen; ozone layer
atmospheric electricity, 208–211
atomic bomb, 159, 162, 196, 213, 214, 319, 364
aura, 38
aurochs, 170
aurora, 34

Australia, 352, 353; deforestation, 113; extinct animals, 168, 170; foragers, 272; gold rush, 290; groundwater depletion in, 129, 130; population history of, 353; settlement of, 13, 167; warfare in, 363
automation. *See* factories
automobiles, 1, 10, 19, 62, 218, 295, 394, 395, 396; and acid rain, 102, 103, 105, 106; and greenhouse effect, 83, 92, 95; noise pollution from, 201, 237; and ozone depletion, 97; and ocean pollution, 153; and plastics, 221; radiation from, 137; and sexuality, 386, 388; and slavery (the rubber trade) 338; smog from, 105, 106; and wildlife, 163, 238–39, 294
Azores, 14
Aztecs, 129, 138, 269, 347, 348, 365, 368, 380

Babylonia, 138, 313; population history of, 313–14; *see also* Iraq
bacteria, 41–43, 44–45, 49, 52, 62, 231; nitrogen-fixing, 40, 55
Badag, 271
Baghdad, 313, 314
Baha'i, 279–80
Baikal, Lake, 321, 322; pollution of, 151
Balboa, Vasco Nuñez de, 343
Balkh, 306
Balmori, Alfonso, 177
Bangladesh, 310, 311; groundwater depletion in, 130
banks and bankers, 254, 290–92, 294
Bantu, 173
basking shark, 173
Bates, Marston, 25
bats, 9, 10, 51, 55, 176, 179, 231–32, 236; and radio frequency radiation, 179; radio tracking devices on, 213
Battuta, Ibn, 334
Bay of Fundy. 103
bear, 14, 17, 21, 22, 67, 139; black, 165, 166; brown, 166; cave, 170; giant, 14; grizzly, 120, 125, 172; white, 19, 166, 172, 247
beaver, 22, 68, 125, 127, 139, 140, 150, 165, 276; giant, 14, 166
Becker, Robert O., 37–38
Bedouin, 129, 256
beech, 109
Belgians in the Congo, 336–38
Bergman, C.A., 225
Bering, Vitus, 325, 342
Bering land bridge, 167, 342
Berry, Wendell. 246
Bhutan, 257, 291, 307
Biancho, Andréa, 344
Big Bang theory, 30
Biobío River, 140
biocides, 3, 19, 189, 191, 396, 397; about, 230–32; in agriculture, 143, 230–31, 264, 295; air pollution by, 104–105; along rights-of-way of power lines, roads and railways, 211, 231; DDT, 18, 231; green revolution and, 264; history of, 18, 295; in forestry, 230, 231; in the home, 231; in hospitals, 202; in plastics, 220; in reservoirs, 142, 143, 230
biodegradability: of detergents, 227; of plastics, 216, 219, 221, 223–24
biodiversity, 45, 53–54, 63, 165–93; and human diversity, 274–75, 393–94. *See also* endangered species; extinctions
bioelectricity, 35, 36, 37–39, 62, 115
biorhythms, 209, 210

birds, 76, 139, 165; of Antarctica, 190; biocides and, 231; cooperation among, 65–68; of Douglas fir forest, 54–56; electrocuted by power lines, 211–12; electromagnetic fiels and, 176–78, 212, 213; endangered, 183–84, 192; evolution of, 69, 173; extinct, 170–71 ; of Java, 260; killed by airplanes, 238, 294; killed by automobiles, 163; of Okavango delta, 188; population control among, 25; radio tracking devices on, 213; sonar in, 236; of Sri Lanka, 311. *See also* seabirds. *See also* names of specific birds: hawk, pelican, etc.
birds' nests, edible, 236, 269–70
birth control: in animals, 24–26; contraceptive herbs, 355–57 *table*; traditional methods, 354–357
birth defects, electromagnetic radiation and, 215
bison, *see* buffalo
black bear, 165, 166
blue-green algae, 41–42
bluefin tuna, 173
bobcat, 165, 279
Boccaccio, 326
Bodard, Lucien, 284, 338
Boeke, J.H., 263–64
Boletus, 50
Borghetty, H.C., 225
Borneo, 28, 243, *map* 258, 265, 267, 268–271, 291, 383; dams, 140–41; oilfields, 262
Bornu, 334
Botswana, 188–89
Boucher, John, 159
Brazil, native peoples of, 283–86. *See also Amazon Basin*
Brazil Current, 340 *map*, 341

Brethren, 374, 381
British: in Australia, 352, 353; in China, 308; in Egypt, 315; in Fiji, 352; in India, 310–11; in Ireland, 328–29; in Japan, 316; in Java, 260, 306; in Sri Lanka, 312–13
Brown, Jerry, 147
brown bear, 166
Buddhism, 379
buffalo, 68, 165, 166, 170, 260; Cape, 188; giant, 14 ; North American, 67, 165, 166, 170, 173, 174, 194, 273
Bugotū, F., 352
Bukhara, 324
burial, 13
buses, 239
butterflies, 76, 164, 260; radio tracking devices on, 213

cacti, 26, 129; organ-pipe, 51; sahuaro, 29, 127
cahow, 155
California condor, 131–33
Calvaria major, 171
camel, 14, 113, 166, 173, 256, 306, 333; radio collars on, 213
camphor, 269, 271
Canada: acid rain in, 102–103; dams in, 144–45; deforestation, 47, 119–20, 137; extinct elephants, 170; forest fires, 90; motorboats in, 150. *See also* James Bay; Meares Island; Ontario; Queen Charlotte Islands; Yukon
cancer, 19, 180, 182; causes, 214–15; electromagnetic radiation and, 175, 205, 214–15, 395; nuclear testing and, 159; plastics and, 216, 220–21
cannibalism, myth of, 268–69
capitalism, 263–65, 290–96

carbohydrates, 39, 40, 41, 50, 62, 83, 85, 109
carbon: cycle, 40, 58, 79–80, 82–84, 109; and life, 39
carbon dioxide, 40, 58, 82–85, 109, 121
caribou, 67, 144, 165, 166, 292
carnivores, 52, 67
Carolina parakeet, 185–86
carrying capacity, 26, 63, 299; defined, 25
Carson, Rachel, 2, 3, 230, 232
Carthaginians, 326, 333, 344, 345, 364
Carvajal, Friar Gaspar de, 343, 349–50
Caspian Sea, 151, 319
cats, 67, 279; bobcat, 165, 279
cattle, 14, 16, 113, 114, 127–28, 256; as money, 289; replacing wildlife, 54. *See also* herders
cave bear, 170
cedar: of ancient Sahara, 113; Lebanese, 112; western red, 46–47, 55
Celebes, 258 *map*, 272
cell phones, 18, 176, 178, 180, 181, 183, 190, 203, 204, 212, 216, 242, 274, 279, 298, 389, 397, 395–96
cell towers, 177–78, 181, 182, 183, 184, 191, 212, 395
cells, 41–43, 44–45, 63; cancerous, 214–15
Central Africa, 103, 265, 266, 271, 278–79, 331, 332
Central Arizona Project, 128
Central Asia, 273, 306, 307, 308, 309, 325, 327, 364, 368; population history of, 319–24. *See also* Silk Road
CFCs, *see* chlorofluorocarbons

Chaco Canyon, 118
chain saw, 47, 76, 241, 246, 395
chakras, 37
Charlevoix, 170
chemicals, synthetic: air pollution by, 104–107; groundwater pollution by, 133–36; in rain, 104; number of, 206
Chernobyl, 213
Cheyne, Andrew, 361
chi, *see* qi
chicken, 14, 27, 176, 210, 346
child mortality, 12, 18, 19, 353
Chile, 90, 118, 140, 173, 191, 196, 285, 341, 346; dams in, 140; groundwater depletion in, 130
China, and Central Asia, 319–20, 321, 322; dams, 140–41; deforestation, 115; groundwater depletion in, 129; early agriculture, 15; population history of, 306–309; war with Japan, 318, 319
chlorinated compounds, 2, 97
chlorination, 18, 74, 150, 231
chlorofluorocarbons (CFCs), 83, 96, 97, 99
chloroplasts, 41
Christianity, 372–76, 383
chromosomes, 41, 42, 44, 45
cinnamon, 262, 312, 313
circadian rhythms, 209, 210
circumcision, 382–85
cities: migration to, 264, 318; origins of, 16
clam, giant, 173
clay: bentonite, as soap substitute, 229; and global cycles, 59; and origin of life, 62
Clayoquat Sound, 48
climatic change: caused by airplanes, 238; caused by deforestation,

35–36, 110, 114–15, 278; greenhouse effect and, 82–86; Ice Ages, 12, 58, 167, 169, 259
clothing, 17, 22, 389; origins of, 12, 24
cloudiness: effect on climate, 58; increased by airplanes, 77, 238
coal, 80, 117, 262, 282; acid rain and, 102. See also fossil fuels
Coblenz, Stanton A., 362, 363
coconut, 313, 346
coffee, 262, 310, 312, 313
Colombia, dams in, 142
colonialism, 293–94; in Africa, 334–39; in China, 308–309; European, 327, 354; in India, 310; in Ireland, 260–65; Japanese, 318–19; in Java, 260–65; in the Pacific, 352–53; in Sri Lanka, 312–13
colony collapse disorder, 176
Colorado River, 125, 128, 139, 351
Colombia, 119, 142, 196, 268, 280, 282, 383
Columbus, Christopher, 167, 193, 259, 260, 269, 341, 342, 349, 351; myth of, 343–48
commerce, see trade
communes, 122, 246
Communism, 10, 291, 293, 300, 309
community: biological, 14, 48, 56; human, 162–63, 245–46, 369–70; impact of technology on, 245–46
competition, 53, 375; economic, 253; as a force of evolution, 53, 63–64, 68, 208; vs. predation, 63
computers, 1, 10, 29, 96, 136–37, 161, 175, 183, 203, 216, 242–44, 246, 263, 274, 298, 300, 389, 394, 395, 396, 397; as slaves, 386. See also electronics; information superhighway

condor: Andean, 132; California, 131–33
Confucianism, 292, 380
Congo, Democratic Republic of: minerals for cell phones, 279; enslavement of for minerals for cell phones, 338–39; enslavement of for rubber, 336–38; population history, 337
Congo, Kingdom of, 331–32
Congo River, 142, 332, 333, 334
consciousness, 37
contraception, see birth control
contraceptive plants, 355–57 table
Cook, Captain James, 353
cooperation: economic, 253, 364; as a force of evolution, 63–64, 68, 208; human, 27, 245–46, 375; in nature, 63–69
coral, 50, 51, 58, 159, 160, 192
Cornell University, 77–78, 245, 253
corporations, 263, 291, 295, 319
Cortés, Hernán, 269
cosmic rays, 34, 57, 209
cotton, 113, 230, 259, 262, 270, 281, 282, 295, 306, 310, 324, 346, 350, 352, 396
cougar, see mountain lion
Cousteau, Jacques, 150, 155
crab, giant, 173
crafts: in Africa, 333; development of, 16; displaced by manufactured goods, 263, 293, 295, 310, 318; in Japan, 316, 318
Crazy Horse, 277
credit, 290
Cree, 144, 365–67
Crete, 112, 117, 345
crocodiles, 45, 173, 174, 260, 311
Crow (tribe), 273
Crump, Martha, 181
crystals, 38–39, 62; liquid, 39, 62–63

Cuiva, 268, 280
Curly Chief, 276–77
cypress, Sahara, 113

da Gama, Vasco, 259, 310
da Vinci, Leonardo, 234
dams, 47, 118, 127, 138–48; and disease, 139, 142; earthquakes caused by, 139; largest, 148 *table*; opposition to, 146–48; in Saamiland, 287; in Siberia, 325
dandelion, 55
Daniel, John, 56
Danish, in India, 310
Darfur, 334
Darwin, Charles, 61, 64, 65, 67, 68
DDT, 18, 231
de Gamboa, Sarmiento, 345
de Laet, Jenny, 177
decomposers, 52
deep injection wells, 134–35
deep sound channel, 234
deer, 54, 65, 67, 125, 165, 170, 194, 367; fallow-, 65; giant, 14
deer mouse, 52
deforestation, 1, 2, 14, 47, 109–21, 171, 281, 283; in ancient Greece, 2, 116; effects on climate, 35–36; rainforest destruction, 10, 54, 84, 103–104, 110–11, 118–20, 284. *See also* forests; logging
Democratic Republic of Congo, *see* Congo, Democratic Republic of
Densmore, Frances, 362
Derosière, Michel, 161
desertification, 110–16, 118, 121, 323
D'Estaing, Valerie Giscard, 300
detergents, 225–29; additives in, 226; history of, 225, 227; toxicity, 227–28; vs. soap, 225–26; water pollution by, 77, 150, 225, 226–27

Devoe, Alan, 232
diamonds, 283
digging stick, 11, 15, 255
Dilawar, Mohammed, 177
dinosaurs, 173
Diodorus, 344
dire wolf, 14, 166
disappearing disease, 176
Djenne, 333–34
DNA, *see* nucleic acids
dodo, 171
dog, 14, 171, 268, 279, 346. *See also* wolf
dolphin, 23, 67, 141, 166, 213, 234, 236; radio tracking devices on, 213
domestication of animals, 14–15, 17. *See also* herders. *See also* names of domestic animals, such as sheep, cattle, etc.
donkey, 14, 15, 306, 333
Douglas fir, 48, 52–53, 54–56
Doyle, Arthur Conan, 337
driftnets, 160
dry cleaning, 96, 105
dualistic economics, 263–64
Dubois, Mark, 147
ducks, 15, 144, 150, 188, 194, 232
Dust Bowl, 125, 126
Dutch: in China, 308; in India, 310; in Japan, 316; in Java, 260–65, 306; in Sri Lanka, 312

eagles, 188, 211
earth: about, 33–36; as a living being, 57–60; magnetic field, 34, 208, 209; productivity of, 302–303; surface temperature, 57, 84–85
earthquakes, 134, 135, 139, 304; caused by dams, 139; caused by injection wells, 134–35
Easter Island, 118, 346

ecology, 46–56; defined, 45, 46
economics, 2, 95, 121, 253–98;
   defined, 253; dualistic, 263–64;
   population growth and, 263–64,
   290, 292, 299, 303, 354; war and,
   296, 386
economy: agricultural, 255–65;
   foraging, 265–89; industrial,
   253–55, 289–96
ecosystem, defined, 48, 51
Edmonton, 247–48
eels: electric, 209; migration of, 155, 209
Efe, 266, 271
Egede, Paul, 278
Egler, Frank, 115
Egypt, 112, 196, 255, 260, 333,
   345, 347, 368, 371, 382, 383,
   384; dams in, 138, 142–43, 148;
   population history of, 313–15
Eiseley, Loren, 36
electric eel, 209
electric power, 10, 107; and
   electromagnetic pollution,
   211–12; and greenhouse effect,
   83, 92; lines, 144, 211–12, 287.
   See also dams
electricity and magnetism, 29–38,
   208–215; bioelectricity, 35, 36,
   37–39, 62, 115, 209–11; global
   electrical circuit, 35, 35 *illus.*, 115,
   182, 208–209
electrification, 95
electromagnetic fields, 33, 201,
   205, 208, 210, 211, 215; *see also*
   electromagnetic radiation
electromagnetic pollution, 136–37,
   202–205, 211–15; in the home,
   209; in the hospital, 202–203
electromagnetic radiation, 137,
   208–15; ionizing vs. non-ionizing,
   202, 213, 214, 215. See also radio
   waves

electron transport chain, 174,
   175, 215
electronics: and electromagnetic
   radiation, 136–37; and
   groundwater pollution, 137; and
   ozone depletion, 96; radioactive
   components. *See also* computers
elephants, 22, 23, 69, 174; ancient
   trade in, 307; of Botswana, 188,
   189; communication among, 23;
   extinct, 14, 166, 170, 172; in
   the Congo, 338, of Java, 260; of
   Kenya, 187; population control
   among, 26; radio collars on, 213;
   rock art depictions of, 113; size,
   173; of Sri Lanka, 311, 312; of
   Uganda, 26, 172
elephant bird, 26, 171
extra low frequency (ELF) radiation,
   202, 204
elk, 54, 125, 165, 170, 194, 247, 271
Elwha Hot Springs, 147
Elwha Riber, 147
emu, giant, 170
endangered species, 19, 173–93. *See
   also* biodiversity; extinctions
energy: available from photosynthesis,
   302; consumption, 146;
   geothermal, 132–33; nuclear, 17;
   renewable, 92. *See also* fossil fuels
England, forests of, 117
Eno, Amos, 132
environmental illness, 201–205
Erikson, Leif, 341
Erwin, Terry, 192–93
Eskimo curlew, 186
Ethiopia, 196, 266, 294, 331, 343,
   382, 384
eukaryotes, 41–42
Euphrates River, 143, 313
Europe: extinct animals, 170;
   dams in, 138, forests, 116–17;

indigenous peoples, 286–89; population history of, 303, 326–28; settlement of, 12; *See also* colonialism
European Economic Community, Resolution on population, 300–301
European Space Agency, 196
Evans-Pritchard, E. E., 359
evolution: biological, 45, 53, 60, 61–70; of stars, 33
extinctions, 1, 45, 166–72; Pleistocene, 14, 121, 168–70. *See also* biodiversity; endangered species. *See also* names of extinct animals, such as dodo, mammoth, etc.
Exxon Valdez, 152, 153, 280

factories: displacement of crafts by, 263–64, 293, 295, 310, 318; mechanization of, 264.
Falkland Islands, 64, 346
fallout, 102, 105
Federal Reserve System, 254
feminism, 386
Fertile Crescent, 143, 313–15
fertilizer, 97, 106, 111, 143, 206, 264, 295
feudalism, 257, 259–63, 318
fibers, synthetic, 218
fig, 50–51
Fiji, 352; groundwater depletion in, 129
fire, human use of, 12, 20, 22. *See also* forest fires
fire drive, 14
fire extinguishers, *see* CFCs
fish: acid rain and, 102; in the Congo, 336; cooperation in, 67; dams as barriers to, 146–47; detergents and, 227; evolution of, 61; driftnets and, 160; extinct, 192; hearing of, 236; population control among, 26; radio tracking devices on, 213; sonar and, 161, 190, 235; in Sargasso Sea, 155; spawning of, 165, 265. *See also* names of specific species: bluefin tuna, etc.
fisher (mammal), 165
fishers, of Pacific Northwest, 265. *See also* foragers
flush toilet, 145, 149, 388
food chain, 52
foragers, 11, 15, 265–89, 266–68 *table*, 370, 375, 381. *See also* specific groups: Penan, San, etc.
foraging economy, 265–66
forest fires, 14, 90, 103–104, 111, 113, 117–18, 118–19, 121
forests, 46–48, 109–21; acid rain and, 102–103; biocides used in, 230; burning of, 14, 103–104, 111, 113, 117–18, 118–19, 121; and climate, 110; Douglas fir, 48, 52–53, 54–56; forest people, 121, 268–86, 332; impact of automobiles on, 239; of ancient Greece, 2, 116; of Java, 260; oxygen production by, 109; tropical, 53–54, 63, 110–11, 118–19, 260, 268–71, 278–79, 311. *See also* deforestation; logging
formaldehyde, 105, 201
Formosa, 308, 316, 318
fossil fuels, 17; and acid rain, 102; and greenhouse effect, 82–85, 91–92; and oxygen depletion, 58, 79–82. *See also* coal; natural gas; petroleum
Fourier, Joseph, 83
fox, 67, 165, 247
freedom, 1, 16, 369–70, 395

French: in China, 309; in the Congo, 337; in Egypt, 338; in India, 310; in Vietnam, 309
French Guiana, dams in, 140
French Polynesia, nuclear testing in, 159
Freud, Sigmund, 389
Friends (Quakers), 374, 381
Friot, Damien, 159
frogs, 25–26, 179–83, 188, 231, 232
Fu-sang, 345–46
Fundy, Bay of, 103
Fukushima, 213

Gaia, 57–60
Galápagos Islands, 64, 168, 173
gamma radiation, 209, 223
Gandhi, Mahatma, 310
Ganges River, 140
garbage, 1, 154, 156, 157, 192, 195, 217, 224; quantities produced, 136
Gaspesians, 276
geese, 14, 150, 188, 194
genes, 41, 43–45; global pool of, 45
genetic code, 44
genetic engineering, 29, 44
genetic recombination, 64
genocide, 269, 351; and ecocide 279–80
Genghis Khan, 321–22, 323
George, Lake, 77
geothermal energy, 132–33
Germans, in China, 309
Ghana, 196, 255, 333, 479–80; Akosombo Dam, 138, 142, 148 *table*
giardia, 150
Gibbon, Edward, 326
giraffe, 113, 173, 174, 307
Glacier National Park, 85
glaciers, 12, 85

global dimming, 106–107
global electrical circuit, 35, 35 *illus.*, 115, 182, 208–209
global warming, *see* greenhouse effect
global weather report 2021–2023, 87–91
goat, 14, 15, 115, 170, 256, 323
Gobi Desert, 115–16, 233
gold, 254, 269, 278, 289–90; mining, 118, 247, 282, 283, 331, 333, 334, 342–43
gophers, 55
gorillas, 22, 68, 338
government, origins of, 16
Gran Chaco, 273, 281
Grand Canal, 307
grand fir, 55
grasses: desert, 114, 127; of Douglas fir ecosystem, 55–56; prairie, 166, 170
grassland, *see* prairie; savannah
gravity, 30–31, 208
great apes, 22. See also orangutans, gorillas
Great Depression, 295–96
Great Lakes, African, 330
Great Lakes, American, pollution of, 151
Great Peace March for Global Nuclear Disarmament, 162–64, 223, 245
Great Smoky Mountains, 102–103, 237
Great Zimbabwe, 331
Greek gods, 380
Greece, 2, 116, 169, 196, 356, 357, 364, 368, 371
green revolution, 264
greenhouse effect, 2, 82–95; and Heard Island experiment, 233–34
Greenland, 166, 190, 278, 340, 344; acid rain in, 102; black people in, 343; ice cap, 57, 85

Greenlander, Paul, 278
Grinnell, George Bird, 276–77
grizzly bear, 120, 125, 172
ground sloth, giant, 14, 166, 172
groundwater, 125–29, 137
growth, economic, 290, 292, 300, 354
Guatemala, 107–108, 196, 348, 349
guinea pig, 15
Gulf Stream, 57, 92, 155, 341, 345
guns, 73, 74, 240, 306, 334, 353, 354, 389–90; abandonment of, by Japan, 317; and the Columbus myth, 343; effects on warfare, 322, 351, 360, 363–64, 365–67; invention of, 293–94; and sex, 382, 386; trade in, 261, 294, 332; and wildlife, 163, 170
gutta percha, 269, 270–71

Han Fei Tzu, 2, 304
hand, 22
hand axe, 11
Hanford Nuclear Reservation, 213
Harappan, 113
Hawai'i, 102, 158, 346, 368; population history of, 353
hawk, 52, 64, 67
Hayes, Marc, 181
hazardous wastes, *see* toxic wastes
Heard Island experiment, 233–34
Heitzman, Charles, 176
hemlock, 41, 55, 56
Herat, 306, 307, 324
herbicides, *see* biocides
herbivores, 14, 52, 69, 168
herders, 15, 256, 257, 273, 286–89, 324–25
heredity, *see* genes; nucleic acids
Hernandez, John, 135
Heyerdahl, Thor, 156
Hidatsa, 273

Himalayas, 95. 256, 257, 291, 307
Hinduism, 377
hippies, 122, 246, 297–98, 386
hippopotamus, 113, 170, 172–73, 174; pygmy, 171
Hispaniola, population history of, 349
history, human, 11–20
Hohokam, 126–29
Hollingsworth, T.H., 303
honey bees, 32, 38, 49, 175–76, 209, 231
Hoover Dam, 139
Hopi trade routes, 351
horse, 170, 246, 256, 273; domestication of, 14; extinct, 14, 166, 172; intelligence, 23; in warfare, 320, 351, 364, 366
hot springs, 130–33, 147
houses, early, 12, 15
huckleberries, 50, 172
human beings: evolution, 61–62, 64; history, 11–20; portrait of, 21–28
Humboldt Current, 340 *map*, 341
Hume, David, 303, 326, 327
hummingbirds, 51
Huns, 309, 320, 321
hunters aned gatherers, *see* foragers
Hutterites, 374, 381
Hutton, James, 59
Huxley Anthony, 51
Hwui-Shin, 345
hydroelectric projects, *see* dams
hyena, 113, 169; giant, 14

Ibn Battuta, 334
Ice Ages, 12, 58, 168, 259
Iceland, 161, 344; geothermal energy, 133; whaling, 190
Illipe nuts, 269, 270
Imms, Augustus, 175
Inca empire, 138, 345, 349, 365, 368
incense, 270

Incense Road, 313
Inco, Ltd., 102
India: deforestation, 113–14; dams in, 140, 146; foragers of, 271; groundwater depletion in, 129; population history of, 309–11; warfare in, 358–59
Indian Ocean trade, 305 *map*, 306–307, 311–12, 313–14, 330
Indians, 73–76, 117–18, 166, 273, 279–86, 342–52, 390
indigenous peoples, 73–74; defined, 73
Indonesia, 171, 173, 190, 262–65, 313, 355; groundwater depletion in, 129. *See also* Borneo; Celebes; Java; Madagascar; Maluku; Sumatra; Sunda Islands; Timor
Indus River valley, 14, 113–14
industrial economy, 254–55, 289–96
Industrial Revolution, 17
industrial solvents: groundwater pollution by, 133; and ozone depletion, 96
infant mortality, 12, 18, 19, 353
information superhighway, 212–13
infrared, 32, 213
insecticides, *see* biocides
insects, 51, 52, 53, 54, 55, 61, 66, 175–76. *See also* ants; biocides; honey bees
intelligence, 23
interest, 291–92
Internet, 242–44, 395
Internet of Things, 212
Internet of Underwater Things, 235
International Antarctic Treaty, 191
International Union for Conservation of Nature, 159
International Whaling Commission, 190
Internet, 242–44, 395, 396

Internet of Things, 212
Internet of Underwater Things, 235–36
Interoceanic Highway, 285
Inuit, 28, 144, 186, 268, 287, 325, 346, 363; whaling by, 190
investment, 253, 263, 264, 292, 300
ionizing radiation, 202, 213, 214
ionosphere, 34–36, 208, 209
Iraq, 112, 143, 169, 313–14; Persian Gulf War, 152, 364
Ireland, 107, 137, 196; population history of, 328–29
iron mining and smelting, 118
irrigation, 16, 146, 255; in Africa, 335; Anasazi, 118; in Baylonia, 313; in Central Asia, 320, 323; and dams, 138–43; effect on Caspian and Aral Seas, 151; Harappan, 113–14; impact on groundwater, 126–30; in the Roman Empire, 327; in Sri Lanka, 312
Irula, 267, 271
Islam, 376, 380
Isle of Wight: forests of ancient, 112
Isle of Wight disease, 175
Israel, 107, 137, 169, 196, 313, 347, 368; 130; forests of ancient, 112; groundwater depletion in, 129, 130; population history of, 314, 330
Ithaca, New York, 77–78, 138, 247
Ituri forest, 271, 278–79, 363
ivory, 113, 243, 261, 271, 306, 330, 332, 336; as money, 289

Jainism, 377–78
Jakarta, 259, 261
James Bay, 47, 144–45, 146
Janzen, Daniel, 171–72
Japan, 263, 308, 309; abandonment of gun-making, 317; dams, 140;

deforestation, 116; groundwater depletion in, 129; population history of, 315–19; whaling, 190
Japan Current, 340 *map*, 341
Java, 258 *map*, 257–65, 357; population history of, 260, 306
Jefferson, Thomas, 292
jellyfish, arctic, 173
Jericho, 16, 373
Jones, Robert, 187
Judaism, 372–75, 383, 384

kabragoya, 311
Kalahari, 15, 189
Kalimantan, *see* Borneo
Kamuk, 271
Kanem, 334
kangaroo, giant, 170
Karakorum, 323
Kashgar, 306
Khara-khoto, 115–16, 323
Khayyam, Omar, 324
Khoekhoe, 271
Kiuh Yuen, 345
kiwi, 171
Klamath River, 147
koala, giant, 170
Komodo dragon, 173
Korea, 137, 160, 196, 231, 294, 308, 315, 316, 318, 321, 368, 379; groundwater depletion in, 129
Kota, 271
Kropotkin, Piotr, 65–66, 67, 68, 325
Kublai Khan, 307, 322
Kubu, 28, 267, 363
!Kung San, 15
Kurumba, 267, 271

Lake Baikal, 151, 321, 322; pollution of, 151
Lake Brokopondo, 141–42
Lake George, 77

land ownership, 257; in Ireland, 328–29
landlord/tenant systems, 257, 328–29
language, 23
Lanting, Frans, 188
Las Casas, Bartolomé de, 349
law, origins of, 16
Le Clercq, Father Chrétien, 276
Leakey, Richard, 188
Lebanon, deforestation of, 112
leisure, 19, 265
lemming, 25, 26
lemur, 171
leopard, 11, 19, 170
Lese, 271
Libya, groundwater depletion in, 129
lichen, 50, 65, 68, 287, 288
life: evolution of, 61–70; interconnectedness of, 46–56; nature of, 29–45, 62
life force, 29–30, 34, 36, 37–39
light: and health, 95–96, 210–11, 213; natural vs. artificial, 96 ; polarized, 38; pollution, 211
lightning, 11, 19, 29–31, 34–36, 40, 46–47, 61, 111, 182, 208, 209
lion, 11, 25, 113, 166, 168–69, 170, 172, 307; marsupial, 170
lipids, 39, 41, 62
liquid crystals, 39, 62, 63
Livingstone, David, 332, 335
lizards, 29, 150, 232, 307, 311
llama, 15, 172
logging, 47–48, 122–23. *See also* deforestation; forests
Lopez, Duarte, 331–32
Lorenz, Konrad, 26–27
Lovelock, James, 59
Luddites, 1
lupine, 55

Madagascar, 256, 260, 266, 330, 383: deforestation, 171; extinct animals, 168, 171; Indonesian colony, 306, 342
Magellan, Ferdinand, 260, 341, 344
magnetic field of the earth, 33–34, 208, 209
magnetism, *see* electricity and magnetism
magnetosphere, 33–34
Maidu, 262
malaria, 127, 261, 278, 294, 312
Mali, 333–34, 343, 384
Malinowski, Bronislaw, 360
Malthus, Thomas, 303
Maluku, 258 *map*, 259–61, 306
mammals, 61, 70, 336; defined, 22. *See also* marine mammals. *See also* specific mammals: antelope, buffalo, etc.
mammoth, 14, 166, 170, 172
manatee, 173, 238
manufactured goods, *see* factories
Mao Zedong, 309
Marconi, Guglielmo, 175
Marion Island, 157
Maori, 353
Mapuche-Pehuenche, 140
marine mammals: driftnets and, 169; noise pollution and, 233–34, 236; ocean pollution and, 153. *See also* specific mammals: dolphin, whale, etc.
market economy, 263–64
Marquesas, 346
Marshall Islands, 213
marsupial lion, 170
marten, 55, 165
Martin of Bohemia, 344
Martin, Paul, 168, 171–72
Marx, Karl, 293
mastodon, 14

Mataco, *see* Wichi
Mato Grosso, 284, 285–86
Mauritius, extinct animals of, 168, 171
Mayans, 103, 108, 348
Mbuti, 28, 266, 278–79, 338
Meares Island, 46–47, 48
Meau, 271
medical school, 43, 130, 201–203
medicine: electromagnetic radiation in, 202–203, 213; nuclear, 213; plastics in, 216, 221–22
meditation, 210
Mediterranean Sea: pollution, 151; trade, 313–14
Meek, Charles K., 359
Meggit, Mervyn, 361–62
Mekong River, dams on, 140
Melanesia, 352; warfare in, 260–61
Mendocino, 50, 122–23, 172, 203–204, 247
Mennonites, 374, 375, 381
merchants, 261, 291–94
meridians, 175, 182
Merv, 306–307
Mesopotamia, forests of ancient, 112
metabolism, 175, 179, 214–15, 302
metal, early use of, 16
Mexico, 15, 269, 344; groundwater depletion in, 129; population history of, 348. *See also* Aztecs
Mexico City, groundwater depletion in, 129, 130
Meyer, Major Otto, 176
mice, 9–10, 20, 25, 54
microplastics, 102, 158–59, 160, 216–17, 396
microwaves, 176–77, 184, 198, 202, 203, 204, 209. *See also* radio waves
migration: animal, 59, 65, 68; bat, 236; bird, 67, 184–85, 186–87;

buffalo, 67, 166; caribou, 67, 144, 166; eel, 155, 209; fish, 165, 265; human, 12, 14; reindeer, 287; rivers as corridors for, 56
military-industrial complex, 263, 293–94, 296, 365
Milky Way, 33
minerals, global cycles of, 58–59
mining: asbestos, 247; cause of deforestation, 118, 119; coal, 282; gold, 247, 282, 290, 331, 333; groundwater pollution from, 133; iron, 118; uranium, 282; water use in, 145
mink, 165
miscarriage: electromagnetic radiation and, 215
missionaries, 281, 312, 316, 325, 334, 353, 371, 374, 379
mitochondria, 41–42, 174, 175, 215
mobile phones, *see* cell phones
Mogollon, 128
money, 162, 164, 253–54, 256, 257, 292, 369; history of, 289–91, 331; origin of, 254
Mongol Empire, 308, 315, 322
Mongolia, 115; population history of, 319–24
Mongols, 320, 321–22, 364
monkey, 22, 65, 108, 171, 210, 260, 307, 311
monsoon, 113–14, 126
Montesquieu, Charles Louis de Secondat, baron de, 303, 326, 327
Monteverde Cloud Forest Biological Preserve, 181–82, 182 *illus.*
Montezuma II, 344
Montreal Protocol, 96
Moore, Captain Charles, 156–57
moose, 165, 172, 194, 247, 276, 394
Morel, E. D., 336

Morris, Desmond, 24, 27
Moruroa, 159, 213
motorboats, 46, 76, 150, 246; impact on marine life, 238
motorcycle, 107
mountain lion, 19, 52, 54, 125, 165
mouse, 9–10, 20, 25, 52, 54, 357; deer, 52
Muhammad, Sultan of Mali, 343–44
multinational corporations, 263
Munk, Walter, 233–34
Münzel, Mark, 280
Murphy, Robert Cushman, 155–56
Murngin, 363
musk ox, 166, 169, 394; radio collars on, 213
mutation, 64, 221

Nagara River, 140, 146
Narmada River, 140, 146
natural gas, 17, 79, 80, 81, 83, 85, 119, 263, 396
Navajo, 282–83, 356, 368
navigation, *see* oceans
Nelson, Edward William, 362–63
New Guinea, 28, 196, 267, 269, 272, 347, 368, 383, 384; warfare in, 361–62
New Melones Reservoir, 147
New York City, 66–67, 76, 138, 163
New Zealand: extinct animals, 168, 171; groundwater depletion in, 129; population history, 353
niche, 46, 52, 53, 168, 190, 242
Nile River, 142–43, 331
Nishapur, 324
nitric acid: in rain, 103–104
nitrogen cycle, 40
nitrogen-fixing bacteria, 40, 55
nitrogen oxides, 97–99, 106, 197
noise pollution, 150, 192, 201, 233–37

nomadism, 11, 15, 162, 291, 293, 323; among farmers, 255, 256, 257; among foragers, 189, 265–89
non-ionizing radiation, 174, 214–15
non-violence, 377, 378, 379
Norse gods, 380
North America: extinct animals, 166; native warfare, 362; trade routes, 351. *See also* Arctic; Indians
North Equatorial Current, 155, 340 *map*, 341
Northern Ireland, 374; population history of, 329
Norway, whaling by, 190
Nubians, 143
nuclear medicine, 213
nuclear power, 17, 102, 133, 134–35, 195, 213, 214; by-products, 214. *See also* radioactive wastes; radioactive fallout
nuclear submarines, 159, 214
nuclear testing, 159, 196, 213–14
nuclear wastes, *see* radioactive wastes
nuclear weapons, *see* atomic bomb
nucleic acids, 38–45, 62

oceans: deep sound channel, 234; and driftnets, 160; navigation of, 13, 167, 259, 272, 306, 311–12, 315, 330, 341–8; noise pollution of, 233–36; nuclear wastes, 159; origin of life, 39, 62–63; origins of, 58; pollution, 10, 151–59, 191, 221; salt content, 57, 58; sonar and, 161, 190, 235. *See also* Atlantic Ocean; fish; marine mammals; plankton; seabirds; Indian Ocean; Pacific Ocean; Sargasso Sea
ocean currents, 57, 63, 92, 155, 340 *map*, 341
ocean sunfish, 173

octopus, 23
Ogalalla aquifer, 125–26
oil, *see* petroleum
oil spills, 151–53, 154, 156, 191
Okavango delta, 188, 189
Old Tassel, Chief, 277
olive, 112
Ontario: author's stay in, 9–10; Sudbury, 102, 103
opium, 261, 310
opposable thumb, 22
orangutans, 22, 68, 260
organ-pipe cactus, 51
ostrich, 113
Outer Mongolia, 322, 323
overpopulation, *see* population
owls, 29, 231; extinct, 171; spotted, 52, 56, 185
oxygen: cycle, 40, 58, 79–80, 82–84; depletion, and fossil fuels, 79–82; deprivation, and electromagnetic fields, 174–84; production by forests, 109; supply of, 79–82. *See also* atmosphere
ozone layer, 95; depletion, 2, 3, 10, 79, 95–101, 104, 191, 197; in Antarctic, 99–100; in Arctic, 100–101; in tropics, 101

Pacific Islands: extinct animals, 168; population history of, 352–53. *See also* Polynesians
Pacific Ocean, 157, 159, 160, 233
packaging, *see* plastics
Pakistan, 310–11, 196, 256, 358, 378, 384; groundwater depletion in, 130
panda, 169
panther, 113
Papago, 362
paper, 47, 119–20, 137, 145, 246, 325

Paraguay, 196, 268, 280–81, 285
Paraná River, 281
Park, Mungo, 334
parrots, 22, 23, 66–67, 166, 171, 185–86, 307
passenger pigeon, 170, 184–85
pasteurization, 18
pastoralists, *see* herders
Pattazhy, Sainudeen, 177
peace walk, *see* Great Peace March for Global Nuclear Disarmament
pearls, 272, 306, 311, 312
peat, 79, 80
Pehuenche, 140
pelicans, 67, 188
Pelto, Pertti, 287
Penan, 28, 265, 267, 268–71
penguins, 22, 153, 166, 190; radio tracking devices on, 213
pepper, 115, 262, 312, 313, 334, 346
per- and polyfluoroalkyl substances, *see* PFAS
Persian Gulf War, 152
Perú, 119, 156, 161, 196, 282, 285, 347, 348, 349, 350, 383, 384
pesticides, *see* biocides
petroleum, 17, 80, 83, 85, 119, 163, 247, 262, 263, 295, 325, 396; and acid rain, 102; and greenhouse effect, 82–95; groundwater pollution from, 134; and oxygen depletion, 58, 79–82; pollution of oceans by, 151–53, 154, 156, 191; raw material for plastics, 216, 218; seismic exploration for, 235
PFAS, 104
Phoenicians, 112, 306, 333, 347
Philippines, dams in, 142
photosynthesis, 40, 52, 58, 62, 79, 109; balance between respiration and, 82–84; described, 82; energy available from, 302–303; impact of increasing carbon dioxide on, 85; impact of ozone depletion on, 191
pig, 22, 115, 170, 260
Pigafetta, Antonio, 344
pigeon, 194; passenger, 170, 184–85; homing, 176; carrier, 176
pirarucu, 173
plague, 294, 308, 326
plankton, 156, 157, 190, 192
plasma, 33–34
plastics, 216–24, 396, 397; additives in, 219–20; air pollution from, 106, 220; foam, and ozone depletion, 96, 97; and food poisoning, 223; history of, 218; packaging, 218, 222–23, 296; polluting the seas, 153–54, 156, 191, 221; toxicity of, 216, 220–24, 396. *See also* microplastics
Plato, 2, 116, 292, 326
Pliny, 117, 345
plow, 15, 16, 17, 188, 255, 256, 295
plutonium, 159, 160
Pokagon, Chief Simon, 184–85, 193–94
polar bear, *see* white bear
polar ice caps, 57, 84, 85, 124
polarized light, 38
Polo, Marco, 116, 307, 322
Polynesians, 118, 346, 352–53, 359–60, 365, 383
population, 265, 299–357; causes of increase, 12, 13, 15–16, 17, 26, 263–64, 290, 292, 299–300, 303, 354, 357; government policies concerning, 300–301; problems of overpopulation, 2, 3, 17, 19, 24, 95, 121, 263–64, 302–304, 327, 386; regulation of, among animals, 24–26; religion and, 372; sex and, 382

population history: America, 341–52; Australia, 353; Central Asia, 319–24; China, 306–309; Congo, 337; Egypt, 313–15; Europe, 303, 326–28; Hawai'i, 353; India, 309–11; Iraq, 313–14, Ireland, 328–29; Israel, 314; Japan, 315–19; Java, 260, 306; Mexico, 348; New Zealand, 353; Siberia, 324–26; Sri Lanka, 311–13; Tahiti, 353; Tasmania, 353
porcupine, 165, 232
Portuguese: in Africa, 331–32, 334; in China, 308; in India, 260, 310; in Japan, 316; in Java, 260, 306; in Sri Lanka, 312
Portuguese man-of-war, 50, 173
posture, vertical, 23–24
poverty, causes of, 263–64, 291, 295–96, 310
power lines, 131, 144, 211–12, 231, 287
prairie, 117, 163, 166, 170, 171, 186–87, 239, 247, 256, 273; dependence on fire, 111, 121; origins of, 14; wildlife
prairie dog, 67, 170
predation, 54; vs. competition, 63–64
prescribed burns, 120–21
primates, 22, 24, 61, 68. *See also* gorilla, lemur, monkey, orangutan, great apes
productivity, earth's, 302–303
prokaryotes, 41
proteins, 39, 40
pulp and paper, 47, 119–20, 137, 145, 246, 325

qi, 38, 210
Quakers, 347, 381
Québec, dams in, 144–45, 146
Queen Charlotte Islands, 47, 48, 246

rabbit, 52, 115
raccoon, 22, 163
race, myth of, 320–21
radar, 174, 180, 183, 204, 212, 214, 394
radiation, *see* electromagnetic radiation; ionizing radiation; non-ionizing radiation; radio waves
radio, 38, 175, 180, 182, 183, 212
radio collars, 213, 288, 394
radio frequency (RF) radiation, *see* radio waves
radio waves, 32, 176, 178, 183, 196, 212–13, 235
radioactive fallout, 102, 105
radioactive wastes, 134, 159, 191, 213–214
radioactivity, 34, 62, 209; from nuclear testing, 213–14; in our homes, 214
railroads, 127, 211, 230, 239, 242, 243, 287, 262, 310, 312–13, 325
rain, 11, 58, 104–105, 121, 123, 125, 278; and global electrical circuit, 35–36, 115; monsoon, 113–14, 126. *See also* acid rain
Rainbow Gathering, 162
Rapa Nui, 118, 368
rattan, 259, 269, 270, 271
recycling: origins of problem, 13; of plastics, 216, 223–24
red squirrel, 52
redwoods, 48, 49, 56, 122–23
refrigeration, and ozone depletion, 2–3, 96–97
reindeer, 14, 68
reindeer herding, 256, 273, 286–89, 324
religion, 266, 322, 348, 372–81; origins of, 13; population and, 372; sex and, 382; warfare and, 372–81

reptiles: evolution of, 61, 70. *See also* crocodiles, dinosaurs, lizards, snakes, turtles
reservoirs, *see* dams
respiration, 83
rhinoceros, 113, 172, 187, 260; woolly, 14, 170
river otter, 125, 150, 165
roads, 78, 144, 230–31, 239, 262, 280, 287, 312–13; effect on climate, 115, 278; impact on Amazon, 107, 284; impact on wildlife, 163–64, 239
robin, 178
robot, 390
rockets, 10, 85, 98–99, 105, 159, 195–98, 382
Rodriguez, Nemesio, 281
Roman Empire, 112, 138, 303, 315, 320, 321, 326, 327, 333
Romanian Communist Party, Resolution on population, 300
Rondônia, 283–84
rubber, 145, 262, 295, 313; and the Amazon, 273, 283, 338; and the Congo, 335–39
rubber tappers, 118, 273
Russia: conquest of Alaska, 325; conquest in Mongolia, 322; conquest of Siberia, 67, 308, 324–26; deforestation in, 117, 118; environmental movement in, 151; historical relations with China, 308; Russo-Japanese War, 318
Ryan, Hazel, 179

Saami, 286–89; *see also* Skolt
saber-toothed tiger, 14, 166
Sahara: ancient forests, 111–13; barrier to migration, 167; groundwater depletion, 129; trade, 333
Sakai, 28, 363
salamanders, 32, 55, 180, 182–83
salt, groundwater pollution by, 133
Samarkand, 306, 323–24
San, 15, 189, 266, 271
sandalwood, 271, 272, 352
sanitation, 18
saponins, 228
Sarawak, *see* Borneo
Sargasso Sea, 155–57, 209
Sargassum weed, 155, 156
satellites, 98–99, 140, 183, 195–98, 212, 394, 395
Saukamappee, 365–67
savannah: burning of, 103–104, 121, 169; civilizations of African, 333–34; origins of, 69, 169
scavengers, 52
Schumann resonances, 208, 210
Schuyler, Philippa, 336
Scythians, 309, 320
sea level, rise of, 85
sea otter, 325
seabirds: driftnets and, 154, 160; ocean pollution and, 153–54; radio-frequency radiation and, 178
seals, 166, 173; Antarctic, 190; communication among, 234, 236; effects of noise pollution on, 234; hunting of, 324, 344; Weddell, 234
Seattle, Chief, 73–76
second law of thermodynamics, 62, 243–44
Seeger, Pete, 162
Semang, 28, 363
separation and individuation, 388–89
Serengeti Plain, 260
Sespe Hot Springs, 130–32, 202, 237
sex, 382–87, 388–89; and vertical posture, 24

shark, 209, 213; basking, 173
sheep, 14, 127, 165, 170, 256, 323, 324
shifting agriculture, 255, 256, 257, 268, 271, 272, 299
Shineberg, Dorothy, 361
Shinto, 380
ships: impact on marine mammals, 238; noise pollution from, 234–35
shipwrecks, 151–53
shrew, 232, 237
Siberia: population history of, 324–26; reindeer hunters and herders, 273, 324; Russian conquest of, 67, 308, 324–26; settlement of, 13
Sikhism, 67, 378–79
silk, 261, 262, 270, 306, 307, 308, 310, 311, 316, 324, 346
Silk Road, 305 *map*, 306, 313, 321
Sinnamary River, 140
Sitka spruce, 46
Sinhalese, 312
Sitting Bull, 275
Skolt, 287–89
skunk, 22, 163, 232
slash-and-burn agriculture, *see* shifting agriculture
slavery, 257, 312, 334–39, 352, 368–71
sloth, giant ground, 14, 166, 172
smallpox, 18, 294, 351, 367
smelting, 17, 272; acid rain from, 102–103; air pollution from, 105; aluminum, 145; iron, and deforestation, 118; water usage in, 145
smog, *see* air pollution
snags, 54–55, 56
snakes, 11, 19, 32, 52, 163, 171, 173, 260, 336, 390; radio tracking devices on, 213
snowbush, 55

snowmobiles, 77, 190, 247, 273, 287–89
soap: natural substitues, 223–29; synthetic substitutes, 225–28
socialism, 291, 293
soil: erosion, 91, 104, 110, 114, 115, 116–17, 121, 125–26; origin of, 63
solar cells, 93–94, 397; chemicals used in, 93–94; groundwater pollution by, 93
solar farms, 93
solar wind, 33, 34
solvents, industrial, *see* industrial solvents
sonar: animal, 23, 236, 237; human, 161, 190, 235
songbirds, endangered, 183–34
Songhay, 333, 334
sound, 233–37. *See also* noise pollution
South America: agriculture, 15, 255; native peoples, 273, 279–86
South Equatorial Current, 340 *map*, 341
Southeast Asia: early agriculture, 14; dams, 140–41; settlement of, 12. *See also* Indonesia
space, pollution of, 195–98. *See also* satellites
Space Shuttle, 195, 196
Spanish: in America, 348–49; in China, 308; in Japan, 316
sparrow, house, 177
species: endangered, 19, 174–93; extinct, 1, 14, 45, 121, 166–72; number of, 192–93, 302–303. *See also* biodiversity
speed, 1, 136, 239, 245
Spence Hot Springs, 132–33
Spice Islands, *see* Maluku
spice trade, 259–61, 306

spiders, 11, 53, 232, 390, 394
sponge, 45
spotted owl, 52
squid, giant, 173
squirrel, 21–22, 52, 54, 67, 163; red, 52
Sri Lanka, 138; groundwater depletion in, 129; population history of, 311–13
Stanislaus River, 147
Stanley, Henry, 332–33, 337
Stanhill, Gerald, 106
Stansbury, Howard, 67
stars, 10, 20, 30–33, 34, 36, 57, 77, 175, 197, 198, 208, 209, 210, 211, 212
steam engine, 17
steamship, 262
Stefanson, Halli, 161
Stellar's sea cow, 173
steppe, *see* prairie
Strabo, 117
strawberry, 50, 166
submarines, 17, 159, 204, 212, 214
Sudan, warfare in, 359
Sudbury, Ontario, 102–103
Suez Canal, 167, 262, 314
sugar, 261, 283, 310
sugar maple, 68
sulfuric acid: in rain, 102
sulfur oxides, 58, 102–103, 105, 106
Sumatra, 28, 259, 261, 267, 271, 383; oilfields, 262
Summers-Smith, James Denis, 177
sun, 11, 33, 34, 35, 36, 40, 63; source of nourishment, 95; ultraviolet from, 95, 96, 101, 191, 210. *See also* photosynthesis
Sunda Islands, 190
supersonic jets, 97
Suriname: dams in, 141–42; deforestation, 119

swan, giant, 170
swidden agriculture, *see* shifting agriculture
synthetic chemicals, *see* chemicals, synthetic
synthetic fibers, 218
Syria, 143, 313, 321, 383; groundwater depletion in, 130

Tahiti, 368; population history of, 353; warfare in, 359–60
Taiwan, 137, 160, 196, 267, 308, 316, 318
Tamerlane, 322
Taoism, 380
Tashkent, 306, 324
Tasman Sea, 158
Tasmania, population history of, 353
taxation, 259, 261, 262, 336,
tea, 261, 308, 310, 313
Teal, John and Mildred, 156
technology: as addiction, 399–400; and agriculture, 295; and extinctions, 168–170; and freedom, 1, 395; history of, 11–20; and population growth, 12, 13, 16, 17, 353–54; and religion, 375–76; and sexuality, 382, 386–90; and slavery, 338; and unemployment, 263–64; and warfare, 17, 351, 352, 354, 363–67; and wilderness, 297–98, 390. *See also* individual technologies: automobiles, computers, etc.
telephone, 1, 18, 38, 107, 237, 242, 397, 388; underwater, 235; *see also* cell phones
television, 1, 18, 38, 212, 388, 395, 396; electromagnetic radiation from, 137, 175
temperature of earth's surface, 57, 84
Tertullian, 303–304, 327

Thailand, 196, 214, 256, 267, 271, 294, 316, 368, 379; groundwater depletion in, 129
Thar desert, 113–14
The Geysers, 132
thistle, 55, 356
Thompson, Francis, 32
thunder, 11, 19, 29–30, 34–36, 40, 182, 198, 209
Tibet, 255, 273, 291, 308, 321, 322; deforestation, 116
tiger, 172, 260, 390; saber-toothed, 14, 166
Timbuktu, 333–34
Timor, 258 *map*, 272
tobacco, 189, 221, 261, 262, 269, 270, 271, 289, 346, 353
toilet, *see* flush toilet
toads, 180, 181, 182
tools, specialization of, 11, 12, 13, 20
Torres Strait, 258 *map*, 272
tortoise, giant, 171, 173
Townsend chipmunk, 52
toxic chemicals, *see* chemicals, synthetic
toxic wastes, 2, 133–36, 206; quantity produced, 135–36, 206; plastics and,
trade, 220; in agricultural economies, 257; in foraging economies, 268, 269–72, 278–79; and population growth, 264, 299; by Saami, 286; spice, 259–61, 306; in subsistence economies, 260, 264; and warfare, 363–64, 365
trade routes, 16, 259, 260, 270, 271, 272, 305 *map*, 306–353, 364, 365, 371,
traders, 291–94
Trans-Amazon Highway, 107
transportation, 16, 18, 166, 238–39, 388. *See also* airplanes; automobiles; motorboats; railroads; roads; ships
trees: oxygen production, 109; size, 173. *See also* forests
tropics, ozone depletion in, 101
truffle, 50, 52, 54
tuberculosis, 281, 353
Tucson, Arizona, 126–28
Tunisia, 196, 382; ancient forests of, 112
Tupac Inca, 345
Turkestan: population history, 319–24. *See also* Silk Road
Turkey: dams in, 143; groundwater depletion in, 130
turkeys, 15, 125
Turkish empire, 322
Turks, 310, 315, 320, 321
turtles, 22, 163; driftnets and, 160; extinct, 141; ocean pollution and, 154; radio collars on, 213
twentieth century disease, 201–205
Tyndall, John, 83
Tyler, Michael, 180–81

Ukraine, 196, 213, 321; deforestation of, 117
ultrasound, 202
ultraviolet, 32, 34, 202, 210, 213; ozone depletion and, 95, 96, 101, 191
Uncertainty Principle, 69
unemployment, 263–64, 295–296, 318
United States: dams in, 138, 139, 147–48; deforestation, 117–18; historical relations with Japan, 317, 319
uranium, 34, 209; mining of, 282–83
Uru-eu-wau-wau, 284–85
Uzbekistan, 323, 324

vaccination, 18
Van Allen radiation belts, 34
Veddahs, 28, 363
veneral diseases, 353
Venezuela, 196, 268, 280, 349, 350; dams in, 148; deforestation, 119, 148
Vietnam, 196, 214, 256, 267, 309; groundwater depletion in, 129
vine maple, 55
viruses, 41–43, 45, 59
voles, 52, 54, 56
Volta River, 138, 142, 148
vultures, 52, 131–33, 166, 170

Wake, David, 179
Wallace, Alfred Russel, 53–54
Wallace, Joseph, 183–84
Wallace, Robert, 303
walrus, 166, 170, 173, 324, 344
warfare, 358–367; in America, 362–63; in Australia, 363; economics and, 296, 386; history of, 363, 293–94, 335; in India, 358–59; in Melanesia, 360; in New Guinea, 361–62; origins of, 13, 16, 17, 26–28, 293, 320, 322, 363–67; religion and, 372–81; sex and, 382; in Sudan, 359; in Tahiti, 359–60; technology of, 17, 351, 352, 354, 363–67; trade and, 363–64, 365; in West Africa, 359; wilderness and, 279–80, 297–98, 390. *See also* guns
water, 124–61; consumption, 145–46; global cycles, 35–36, 40, 58, 129
water pollution, 2, 10, 77, 133–36, 137, 149, 150–59; by biocides, 142, 230; by detergents, 77, 150, 225, 226–27; by plastics, 153–54, 156, 191, 216, 220, 221, 222
water power, 17

water table, 125, 128, 139, 143
wealth: distribution of, 259–60, 296; origins of, 16; source of, 257, 291, 293. *See also* money
weapons: hunting, 11, 12, 13–14; of war, 26–27, 293–94, 296, 317, 318, 354, 364, 365–67. *See also* guns; warfare
weasel, 67, 165
weather: global weather report, 87–91; satellites, 197, 204. *See also* climatic change.
Weddell seal, 234
Wedgewood, Camilla H., 360–61
West Africa, warfare in, 359
western red cedar, 55
wetlands, 125, 188, 285
whales, 23; blue, 173, 190; collissions with ships, 238; communication in, 234–36; fin, 236; humpback, 155, 190, 236; hunting of, 190, 324, 344; impact of noise on, 234; International Whaling Commission, 190; ocean pollution and, 154; radio tracking devices on, 213; right, 238; sperm, 158, 190
wheel, 16
white bear, 19, 166, 172, 247; radio collars on, 213
Whiteway, R.S., 358–59
Wichi, 281–82
Wi-Fi, 212
Wild and Scenic Rivers System, 147
wilderness: Antarctica as global, 189–91; and inner wildness, 297–80, 395; technology and, 297–98, 390
wildlife, 163–64, 165–93. *See also* names of specific animals
Wilson, William, 359–60
wind farms, 92–93, 131
wind power, 17

wireless communication, xii, 137, 175, 203, 204, 214, 242
wolf, 17, 19, 52, 125, 165, 166, 170, 247, 279, 390, 394; dire (giant), 14; radio collars on, 213; *See also* dog
wolverine, 165
wombat, giant, 170
wood as fuel, 17, 117
work day, length of, 15
World War I, 169, 182, 262, 295, 309, 310, 313, 319, 325
World War II, 102, 104, 125, 201, 206, 218, 225, 273, 296; and colonialism, 262–63
Worm, Boris, 160
worms, giant, 173

Xenophon, 292
Xingu National Park, 285
Xinjian, 308. *See also* Turkestan
X-rays, 32, 34, 202, 213

yak, 256, 257
Yangzi River, 307; dams on, 140–41, 145, 146
Yosemite National Park, 181, 237
Yukon, 247, 290, 362–63, 393–94
Yumbri, 267, 271

zebra, 187, 188
Zimbabwe, 148, 196, 331
Zoroastrianism, 378